MICROBIAL GROWTH

*Other Publications of the
Society for General Microbiology**

THE JOURNAL OF GENERAL MICROBIOLOGY
THE JOURNAL OF GENERAL VIROLOGY

SYMPOSIA

* Published by the Cambridge University Press, except for the first Symposium, which was published by Blackwell's Scientific Publications Limited.

MICROBIAL GROWTH

NINETEENTH SYMPOSIUM OF THE
SOCIETY FOR GENERAL MICROBIOLOGY
HELD AT
UNIVERSITY COLLEGE LONDON
APRIL 1969

CAMBRIDGE
Published for the Society for General Microbiology
AT THE UNIVERSITY PRESS
1969

Published by the Syndics of the Cambridge University Press
Bentley House, 200 Euston Road, London N.W.1
American Branch: 32 East 57th Street, New York, N.Y.10022

Library of Congress Catalogue Card Number: 69–19381

Standard Book Number: 521 07509 2

Printed in Great Britain
at the University Printing House, Cambridge
(Brooke Crutchley, University Printer)

CONTRIBUTORS

BARTH, P. T., Department of Genetics, University of Leicester.

BROCK, T. D., Department of Microbiology, Indiana University, Bloomington, Indiana.

CLARKE, PATRICIA H., Department of Biochemistry, University College London.

COLLINS, J., Department of Genetics, University of Leicester.

FORREST, W. W., Division of Nutritional Biochemistry, Commonwealth Scientific and Industrial Research Organization, Adelaide, South Australia.

GOODWIN, B. C., School of Biological Sciences, University of Sussex, Brighton.

HOBSON, P. N., The Rowett Research Institute, Bucksburn, Aberdeen.

LILLY, M. D., Biochemical Engineering Section, Department of Chemical Engineering, University College London.

LLOYD, D., Department of Microbiology, University College, Cardiff.

MANDELSTAM, J., Microbiology Unit, Department of Biochemistry, University of Oxford.

MARR, A. G., Department of Bacteriology, University of California, Davis, California.

MAYNARD SMITH, J., School of Biological Sciences, University of Sussex, Brighton.

NILSON, E. H., Department of Bacteriology, University of California, Davis, California.

NORTHCOTE, D. H., Department of Biochemistry, University of Cambridge.

PAINTER, P. R., Department of Bacteriology, University of California, Davis, California.

PAUL, J., The Beatson Institute for Cancer Research, Glasgow.

PIRT, S. J., Department of Microbiology, Queen Elizabeth College, University of London.

CONTENTS

ERRATA

The correct magnifications of the plates facing page
344 are as follows: Plate 1, figs 1 and 2 × 1100; Plate 2,
fig. 1 × 25,000, figs 2 and 3 × 38,000; Plate 3, fig. 1
× 57,000, fig. 2 × 62,000, fig. 3 × 39,500, fig. 4 × 53,000;
Plate 4, fig. 1 × 59,000, fig. 2 × 41,000, fig. 3 × 42,000;
Plate 5, fig. 1 × 100, fig. 2 × 380, fig. 3 × 135; Plate 6,
fig. 1 × 14,500, fig. 2 × 33,500, fig. 3 × 12,500, fig. 4
× 25,000; Plate 7, fig. 1 × 12,500, fig. 2 × 34,000, fig. 3
× 12,500, fig. 4 × 25,000.

EDITORS' PREFACE

At first sight it might appear surprising that the Society should have waited until this, its 19th Symposium, to consider a subject so fundamental to Microbiology as Growth. But it is probably just because it is all-encompassing and also because for a long time it was not studied quantitatively, that it was neglected in favour of previous topics. The early studies of microbial growth were limited to recording whether or not it occurred, although as Monod* pointed out, the recording of + or 0 in a laboratory book has led to many great discoveries and it may still produce valuable information particularly for the microbial geneticist. A new era in studies of growth began with Monod's monograph.† This study showed clearly, for the first time, that growth of cell populations should and could be formulated in terms of the basic parameters; lag period, specific growth rate, growth yield and maximum cell population. Until this time the importance of growth yield (ratio of amount of organisms produced to substrate consumed) had not been appreciated, and the phenomenon of lag was the most fashionable aspect of studies of microbial growth. Besides showing a marked lack of balance, the earlier studies failed to provide any significant insight into the process of growth and we can now see that this was inevitable because our understanding of the synthesis of nucleic acids and enzymes and, perhaps more important, of their regulation was lacking. Nowadays the lag period attracts little attention and even less thought is applied to lack of growth, but perhaps with today's knowledge a fresh look at these two phenomena could yield valuable information about cell growth.

One may take the attitude that growth of cells and microbes is purely a question of Methodology; 'Study of the growth of bacterial cultures . . .is the basic method of Microbiology' (Monod, 1949). However, we must not forget that the methodology can help to study the mechanism of growth, both of individual cells and of populations. The contributors to this Symposium were invited to consider not only the methodology of growth with its increasing complexity, but also, as far as possible at the molecular level, various aspects of the problem of how cells grow in nature as well as in the laboratory.

Despite the fact that growth is the basic technique of Microbiology, there is probably no other technique whose principles are so often ignored with the result that the experiments are seriously limited if not meaningless. For example, growth of a bacterial culture may be de-

---

* Monod, J. (1949). *Ann. Rev. Microbiol.* 3, 371.

† Monod, J. (1942). *Recherches sur la croissance bactériennes*. Paris: Masson.

scribed as 'better than the control' as measured by turbidity at 16 hr. But this could mean that the lag period was shorter, or that the specific growth rate was higher, or that the decline rate was less than the control. Clearly use of the term 'growth was better' should be severely restricted in quantitative growth studies. Another aspect of growth methodology not sufficiently widely appreciated is the aeration of cultures to obtain aerobic conditions. Many workers seem to assume that oxygen is not growth-limiting without taking the elementary precautions laid down by Monod (1942) or applying the more advanced principles developed since. Monod (1942) discussed the narrow limits of the shake-flask culture technique and much engineering effort has since been devoted to the development of the 'stirred fermenter' which has overcome these limitations and has made possible many of the contributions to this Symposium. The growth studies described here also reveal the impact of other new techniques, notably the continuous culture of micro-organisms and their synchronous division. Both these techniques have been undergoing development for the last few years, but they can now be widely used throughout biological and biochemical laboratories. The chemostat is not merely a means of mass producing bacteria as some microbiologists have supposed. It is a unique, and in some cases indispensable, tool for the investigation of many aspects of the physiology of microbial growth and its regulation.

We have divided the Symposium into two main parts. In the first, growth of microbial populations is considered and in the second part the discussion is concerned with the growth and differentiation of individual cells and their components. Since we can now apply microbiological culture methods to tissue cultures of plant and animal cells, it seemed appropriate to include them in a Symposium devoted primarily to Microbial Growth. It is clear that, at least in some respects, current understanding of the growth of plant and animal cells lags behind that of microbial cells, but we should be able to look forward to cross-fertilization of ideas between microbiologists and cell biologists.

Many of the approaches described in this Symposium are new and tentative, from which we gather that knowledge of the subject is on the threshold of advance rather than at the end of a phase of expansion and consolidation. Challenging problems emerge which are all aspects of the central problem of how living cells propagate and in what direction they will evolve, either naturally or by man's conscious direction.

PAULINE M. MEADOW
S. J. PIRT

University College London
Queen Elizabeth College, London

LIMITATIONS ON GROWTH RATE

J. MAYNARD SMITH

School of Biological Sciences, University of Sussex, Brighton

THE PROBLEM

What determines the maximum rate at which an organism can grow? This is different from the question which is usually asked, i.e. what regulates the rate of growth of an organism? The distinction between regulation and what one might call 'design limitation' can be seen by analogy with machines. Thus at any moment the speed of a motor car is regulated by the accelerator and the brake, yet there is an upper limit which cannot be exceeded regardless of the state of the accelerator. A designer would be aware of the factors which set this upper limit—for example, the capacity of the cylinder, which could not be increased without increasing fuel consumption and so running costs. These factors would not normally depend on the way in which the speed was to be regulated.

In this article, I want to discuss what are the features of a cell which limit the rate at which it can grow. I shall do so from the point of view of one whose original training was as an engineer, and who has subsequently worked mainly in evolutionary biology. I hope that this background has led me to ask slightly different questions from those usually asked by microbiologists.

I shall make one general assumption. This is that, other things being equal, it is selectively advantageous for at least some cells and organisms to grow as rapidly as possible. For single-celled organisms it is clear that natural selection will favour rapid multiplication. Of course, if a genetic change favouring rapid multiplication brings other less desirable changes—for example an inability to adapt to changes in the environment—then it might not be established. The important point for my argument is that if a genetic change increasing growth rate without causing other major alterations is possible in any species it will be established. It follows that the growth rate of any species is unlikely to be limited by one single factor (e.g. the shortage of a particular enzyme) which could readily be altered. Instead, growth rate will be limited either by many different factors, so that to alter one would not help unless all the others were altered, or by some factor (e.g. the amount of some substance which forms a large part of the biomass) which could not readily be altered. This is equivalent to assuming that the cell resembles a factory which has been subjected to critical path analysis and in which curable bottlenecks have been remedied.

It is not immediately obvious that cells forming part of multicellular organisms have been selected for minimum doubling time. Some cell types do not divide at all. Further, if the growth rate of the whole organism were limited by, for example, the availability of food, then the minimum doubling time of cells in a growing organism might be much longer than is required solely for the replication of cells of that degree of complexity. There are, however, contexts in which metazoan cells might be expected to grow as rapidly as possible. Examples are regeneration, wound healing and the replacement of lost blood.

Summarizing, it will be assumed that at least some cells and organisms are growing as rapidly as possible, in the sense that no simple genetic change could enable them to grow more rapidly without sacrificing some other advantage.

DOUBLING TIMES

In growth, we are concerned simply with an increase in the amount of material present, without alteration in the kind of material. The simplest measure of growth rate is the time taken for the population to double in size, which in organisms reproducing by binary fission is, neglecting deaths, the same as the cell cycle time. In *Escherichia coli*, grown in broth at 37° C., the doubling time is approximately 20 min. This compares with a minimum doubling time for mammalian cells at the same temperature of approximately 10 hr.; this limit holds for cells in tissue culture or in the body, and it holds for malignant or non-malignant cells.

Doubling times for protozoa are very variable, but are usually of the same order as for mammalian cells. However, some protozoa can grow much more rapidly. The amoeba stage of the cellular slime moulds *Dictyostelium* and *Polysphondilium* have doubling times at 25° C. of 2·4–3·5 hr. (Bonner, 1967). Whether it is appropriate to reduce these times by a factor of two to correct for the temperature difference depends upon whether the rates of chemical reaction or of diffusion are limiting. But in either case it is clear that the doubling times of the fastest-growing protozoa are intermediate between those of bacteria and mammalian cells.

THE GROWTH OF BACTERIA

Why should it be that a bacterium can replicate in 20 min., whereas it takes the most rapidly dividing mammalian cell 10 hr.? Before tackling this question, one must understand what limits the growth of a bacterium. In the account which follows, I have leant heavily on the data and the ideas of Maaloe and Kjeldgaard (1966).

What must be done in the course of a single cell cycle? The situation is complicated by the fact that in a rapidly growing bacterium the chromosome is partly replicated before the cell divides. The possible relevance of this fact to the limitation of growth rate is discussed below under the heading 'gene replication'. Consider a cell cycle as lasting from the moment when a cell arises by division to the moment when that cell divides. At the start of a cycle there is one copy of each gene (or two copies, if the gene has already replicated), n_1 molecules of a particular species of m-RNA, $n_1 \times n_2$ molecules of the corresponding protein, n_3 ribosomes and n_4 molecules of a particular t-RNA. Then in one replication cycle, each gene must replicate once, and the following must be synthesized: n_1 m-RNA transcripts of the relevant gene; n_2 protein molecules per m-RNA molecule; n_3 ribosomes, and n_4 molecules of the particular t-RNA.

It seems likely that it is the time taken for these macromolecular syntheses which limits growth rate. It is possible to be more precise than this. Consider the different syntheses in turn:

Gene replication

There is no reason to think that gene replication is limiting. It does in fact take longer (40–50 min.) for a single growing point to traverse a chromosome than it does for the cell to divide in optimal conditions. However, it is possible for there to be two or even three growing points per chromosome. If there were no other factor limiting cell-replication rate, the number of growing points could presumably be increased, as it is in higher organisms. Werner (1968) has shown that there are multiple growing points in T_4 phage replicating within *Escherichia coli*, so growing points on the bacterial chromosome could be spaced more closely.

Differences in the numbers of copies of different genes might however cause difficulties if the number of growing points were raised beyond a certain point. Thus if there are n growing points per chromosome in a rapidly growing bacterium, then there are $n + 1$ copies of genes at one end of the chromosome compared to one copy of genes at the other, whereas in the same bacterium growing slowly there would be one copy of each gene. This might cause difficulties in regulating the levels of gene products. The problem is solved in eukaryotes because each chromosome consists of many replicons.

m-RNA synthesis

In normal circumstances the synthesis of a particular species of m-RNA is unlikely to be limiting. In constitutive mutants in which the synthesis of the corresponding m-RNA is maximal, the concentration of the corresponding protein (and therefore presumably of the messenger) is many times higher than normal, and may constitute up to 8% of the total protein. It follows that there is considerable reserve capacity at this stage in normal cells. This is because control of the relative amounts of different proteins to be synthesized operates at this stage, and there must be reserve capacity if control is to operate.

Constitutive mutants grow more slowly than wild-type, but not dramatically so. In fact the proportional increase in doubling time is of the same order of magnitude as the additional protein they are synthesizing gratuitously (i.e. 8%). This is consistent with the view that doubling time depends on the total amount of protein to be synthesized per ribosome, rather than on the kinds of protein to be synthesized.

It is interesting from a design point of view that control should operate at the transcription rather than the translation level. It is clearly much less wasteful of RNA synthesis to control the cell by regulating the amounts of RNA made than by regulating the rates at which different messengers are translated. If it turns out that control operates also at the translation level, it will be worth asking why the advantages of economy have been sacrificed. A possible reason, discussed further below, is the need for rapidity of response. It is a curious fact which may be relevant here that in mouse liver cells some 80% of the RNA synthesized is short-lived, and never leaves the nucleus; this RNA seems to have different base sequences from that in the cytoplasm (Shearer & McCarthy, 1967).

Synthesis of ribosomal and transfer RNA

The rate at which ribosomes travel along a messenger molecule is more or less independent of the medium and hence of the growth rate of the cell. When bacteria are transferred from minimal to enriched medium, their first response is to increase the rate of RNA synthesis. To a first approximation, then, the rate of protein synthesis is determined by the number of ribosomes, each ribosome synthesizing protein at a constant rate.

It follows that the rate of growth is regulated by the rate of synthesis of r-RNA. When growth rate is maximal, there are three possibilities:

(i) The rate of synthesis of r-RNA is maximal and limiting; as a corollary, if the rate of synthesis of r-RNA could be increased, the rate of cell growth and replication would also increase. This seems unlikely to be the case. There are multiple copies of the ribosomal genes— perhaps 10 copies each of the genes for 16S and 23S components. If r-RNA synthesis were limiting, the number of copies would presumably be increased.

(ii) It is more likely that the rate of synthesis of r-RNA could be further increased, but there is no point in doing so because an increase in the number of ribosomes per cell beyond some limiting value would not result in a further increase in the rate of protein synthesis. This possibility is considered further below.

(iii) It may be that when this maximal useful concentration of ribosomes has been reached, and growth rate is maximal, the rate of synthesis of r-RNA has also reached a limit. If so, the rate of protein synthesis and of r-RNA synthesis are jointly limiting.

I have as yet said nothing about whether the synthesis of t-RNA might be limiting. Whether the synthesis of t-RNA or of r-RNA is the more likely to be limiting depends on the number of molecules produced per cycle, the number of nucleotides per molecule, the number of gene copies per genome, and on the number of polymerase molecules which can transcribe a single gene at any one time. Numerical calculations are made below (p. 7). But it seems unlikely on general grounds that t-RNA synthesis would be limiting, since if it were there would be multiple copies of the t-RNA genes as there are of the r-RNA genes.

Protein synthesis

As was mentioned earlier, the rate at which amino acids are added to protein by a single ribosome is constant and does not vary with the medium. Presumably this rate depends on the concentration of the reacting substances, and primarily of t-RNA-amino acid complexes. One is therefore obliged to ask why it is that the concentration of these complexes is not increased, if this would be sufficient to bring about an increase in the rate of protein synthesis. I have already argued that the rate of t-RNA synthesis is probably not limiting, and in broth there should be no shortage of amino acids. The answer presumably is that a further increase in t-RNA-amino acid complexes would not increase the growth rate. This corresponds to my earlier conclusion that a further increase in the concentration of ribosomes would likewise not increase growth rate.

If these conclusions are correct, it must be because the cell is already so congested that there is no room for further reactants, or rather that the addition of further reactants would slow down the reaction. This conclusion seems plausible. The addition of an amino acid to a polypeptide requires an accurately oriented collision between two large molecular complexes, the polysome and a particular one out of forty kinds of t-RNA molecule. It is easy to recognize that this 'queuing problem' might become so severe that an increase in molecular concentrations might actually slow down the arrival of the appropriate amino acid. To give an analogy, the number of cars passing a particular point on a road decreases if the number of cars per yard rises above a certain point.

It follows from this argument that the phenomenon of 'wobble', by decreasing the required number of t-RNA species, increases the maximum rate of synthesis of protein per unit volume by 50%.

Since I am arguing that the queuing problem is the main limitation on growth rate, it is worth pursuing a little further. There are three time intervals to be considered, which I will call the shunting time (T_s), the backing out time (T_B) and the collision time (T_c). Starting from the moment when a codon on the m-RNA is properly exposed in a ribosomal slot, a time T_c will elapse before there is a collision with a properly oriented anticodon on a t-RNA molecule. This anticodon either is or is not complementary to the m-RNA codon. In the former case, there follows a time interval T_s during which the m-RNA is shifted relative to the ribosome, and at the end of which a new codon is exposed. In the latter case, there follows a time interval T_B at the end of which the t-RNA molecule has moved away sufficiently for the codon to be again exposed.

If in any collision the probability that the codon and anticodon are complementary is $1/N$ (e.g. if there are N equally frequent t-RNA species) then the average time to incorporate an amino acid is $T_s + (N-1) T_B + NT_c$. To a first approximation, one would expect T_s to be independent of t-RNA concentration, T_c to decrease with increasing t-RNA concentration (although at high concentrations T_c may not decrease as rapidly as the reciprocal of the concentration) and T_B to increase with increasing t-RNA concentration. Thus my argument rests on the assumption that at sufficiently high t-RNA concentrations, T_B will increase more rapidly than T_c decreases with concentration.

The tentative conclusion then is that the rate of protein synthesis in a given volume is limited by the fact that the reactants are large molecules whose concentrations cannot efficiently be raised above a certain level. If so, it is not immediately obvious why the doubling time should be longer for large cells than for small ones. But before tackling that problem, it is useful to consider estimates of the time taken for the various unit steps in macromolecular synthesis in bacteria. The values, based on Maaloe and Kjeldgaard (1966), are for *E. coli* at 37°C.

(i) *Amino acid incorporation.* An upper limit of 60 msec. is obtained by assuming all ribosomes in the cell are active.

A lower limit of 20 msec. is obtained by adding labelled leucine to a culture previously stopped by leucine starvation. The calculation rests on the assumption that with leucine starvation each ribosome stops immediately before a leucine, and hence that the first amino acid incorporated after adding label is always leucine.

(ii) *Incorporation of bases into RNA.* The data in Table 1 are taken from Mandelstam & McQuillen (1968). At first sight it seems that the synthesis of r-RNA is more likely to be limiting, and this is probably the case. However, a complication arises because more than one polymerase molecule may be transcribing a gene at any one time. If the number transcribing 23S RNA is n, then the average time taken to add a nucleotide is $n/3 \cdot 4$ msec.

Table 1. *Rate of synthesis of RNA molecules*

		Nucleotides per molecule	Molecule per genome	No. of molecular species	No. of genes per molecular species	Nucleotides per millisecond per gene
23S	r-RNA	3400	12,000	1	10	3·4
	t-RNA	80	100,000	40	1	0·17

(iii) *Incorporation of bases into DNA.* Assuming 3×10^6 base pairs per genome, and 50 min. for a single growing point to traverse a chromosome, gives a time of 1 msec. per base pair.

It is striking that the time taken to add an amino acid to a protein is many times longer than to add a base to a nucleic acid. This is presumably a consequence of the relatively low concentrations of t-RNA-amino acid complexes, which is in turn a consequence of their large size.

THE GROWTH OF EUKARYOTES

Eukaryotic cells typically have doubling times many times greater than bacteria, although the cellular slime moulds are an interesting exception. The first obvious difference between bacterial and eukaryotic cells is that the latter are very much larger. But mere increase in size without increase of complexity is obviously irrelevant to growth rate; 1000 bacteria can turn into 2000 as quickly as one bacterium can turn into two. It is the increased complexity that goes with increased size which must in some way be responsible for the difference. There are a number of ways in which eukaryotic cells are more complicated, and which might therefore explain their slower growth. Although these various ways are functionally interdependent, it is convenient to classify them for the purposes of discussion, as follows:

(i) Biochemical complexity, i.e. a larger number of different chemical substances, and in particular of proteins, are synthesized.

(ii) Temporal complexity, i.e. in the course of a complete division cycle the cell may go through a larger number of different states.

(iii) Morphological complexity, i.e. different biochemical processes are carried out in different places. This could influence growth rate in two ways: (a) the distance which substances have to travel, and consequently the time they take to do it, will be greater; and (b) the construction of organelles out of their component molecules may take time.

These possibilities will be discussed in turn. But before doing so, there are some difficulties of interpretation which arise because eukaryotic cells are not only complicated in themselves, but may be parts of complicated multicellular organisms.

The first difficulty is that tissue cells may be capable (perhaps with minor modifications) of multiplying much more rapidly, but do not do so because there is no selective advantage to be gained. Clearly this is the case for most cells in most organisms. Nevertheless, I think that some cells (e.g. erythroblasts) multiply as rapidly as possible, in the sense that no simple genetic change could enable them to do so more rapidly without causing other selectively disadvantageous changes. What I have to say in this section applies only to cells which are in this sense multiplying as rapidly as possible.

A more serious difficulty is as follows. An erythroblast carries within it a genome which in different circumstances is adequate to control the development of a whole organism. It may be that its doubling time would be much less were it not for this fact. The main argument in favour of such a possibility is that slime mould amoebae double in a much shorter time than do tissue cells. This suggests that the factor limiting the growth rate of tissue cells is the presence within them of genetic material which is irrelevant to their own growth, but which is necessary for the differentiation of other cell types. It is difficult to see why the mere presence of additional DNA should cause a major reduction in growth rate. But it may be that an important factor is the greater complexity of the control mechanisms needed in a tissue cell to prevent the realization of other patterns of differentiation.

The possible relevance of the various types of complexity of eukaryotic cells will now be considered in turn.

Biochemical complexity

There is one type of increased biochemical complexity which would, if the preceding argument is sound, slow down the rate of growth. This is an increase in the number of different types of t-RNA-amino acid complexes. This however does not take place.

The most general way of measuring biochemical complexity would be in terms of the number of different protein species synthesized. At present we have no way of knowing this number for mammalian cells. It could be deduced from the DNA content of a cell if we knew how many copies of the genetic message there are, and if we knew how much if any of the DNA does not code for protein. We do not know these things, but our ignorance does not matter in the present context, because there does not seem to be any reason why the need to synthesize a large number of protein species should slow down growth.

What is relevant is the ratio of the total mass of protein which must be produced in one cell generation to the number of ribosomes available for its synthesis. For many types of mammalian cell, including rapidly dividing ones, this is of the order of 15 mg. of protein per mg. of ribosomal RNA (Leslie, 1955), whereas in rapidly growing bacteria the ratio is 2·5 mg. of protein per mg. of r-RNA. This could account for a six-fold difference in replication time.

Temporal complexity

Information on this point is scarce, but it may be highly relevant to the problem. Thus suppose that a cell in the course of a division cycle passes through a number of states. A state would be specified by which genes were active and which repressed, by the species of m-RNA present, and so on. I am not implying that one such state would be succeeded instantaneously by another—indeed it is the impossibility of such instantaneous transitions which is the whole point of this section—but that a series of states may succeed one another, the transitions being gradual. The functional significance of such temporal organization might be, for example, that in order to build a particular organelle it is important to synthesize its component molecules in the right sequence (see Sussman & Sussman and Mandelstam, this Symposium).

If it is accepted that such a temporal sequence exists, then growth rate may be limited by the time taken for one state to succeed another. Supposing for the moment that control operates at the level of transcription and not of translation, this will depend on the rate at which the population of m-RNA molecules can be changed. This in turn will depend on the rate at which new m-RNA molecules can be synthesized, and, more critically, on the rate at which existing molecules can be removed. Thus there is at present no reason to think that a cell can recognize and remove specific types of m-RNA at particular times. Hence a limit is placed on the rate at which existing species can be removed by the requirement that newly synthesized species should survive in adequate numbers. If this picture is essentially correct, it might place a lower limit on the length of a cell cycle which had to pass through a number of states.

However a control system which depends on the random removal of m-RNA molecules does not seem wholly convincing, and it may be that the system is more efficient in bringing about rapid state transitions. There is, in fact, some evidence that in liver cells different species of m-RNA may have different life spans, although this by itself is not a solution to the control problem, since what is required for rapid transitions is that the relative life spans of different messengers should alter at different stages of the cell cycle. Possible mechanisms are: (*a*) an 'ageing' phenomenon whereby a messenger molecule is more likely to be degraded if it has been in existence a long time; (*b*) recognition and removal of specific types of messenger at particular times in the cycle; (*c*) control of which messengers are translated at a ribosome. Of these possible mechanisms, the first would be only a minor improvement. It is not easy to see how the second or third would operate, but their operation would not necessarily depend on the recognition of specific base sequences. Thus it might be that all messenger molecules synthesized during a particular stage of the cell cycle are 'labelled' in some way, the label being used later in the cycle to determine whether the molecules are to be translated or degraded. The phenomenon of host-induced modification suggests how such labelling might be brought about. What would be needed would be some temporal control of the labelling of newly synthesized messenger.

But despite these possibilities, it may well be that an important factor limiting growth rate is the time taken by complex cells to change from one state to another.

Morphological complexity

It may be that the time taken to assemble macromolecules into organelles is relevant to our problem, but too little is known about this process for it to be fruitful to discuss it further. The question which can be considered in more detail is whether the greater distances over which substances must travel are in any way relevant. At first sight this seems unlikely. Thus if x is the root mean square distance travelled by a molecule in solution in time t, then, in three dimensions, $x^2 = 6Dt$, where D is the diffusion constant. In water at $20°$ C., some values of D are:

	D (cm.^{2}/sec.)	Approx. path in 1 sec. (cm.)
Glycine	95×10^{-7}	10^{-2}
Haemoglobin	$6\cdot2 \times 10^{-7}$	2×10^{-3}
TMV	$0\cdot53 \times 10^{-7}$	$0\cdot5 \times 10^{-3}$

Thus even the largest molecules can travel across a cell in a fraction of a second. But before drawing the obvious conclusion that diffusion times are irrelevant, we should distinguish two situations. In the first (Fig. 1 a) some substance is manufactured in one part of the cell and is transported to another region which is large, and which increases in volume as the volume of the cell increases. An example would be the passage of m-RNA from the nucleus to the ribosomes. Any ribosome will do, and in a large cell this 'target area' is also larger. In this situation, diffusion times will be too short to be relevant.

The second situation is shown in Fig. 1 b. A specific substance manufactured in one part of the cell must travel to a specific target, which is the same size whether the cell is large or small. For example, a specific repressor molecule manufactured at a ribosome must travel to the gene which it represses. In this case we are not concerned with the time taken for a molecule to travel a distance x in any direction, but in one specific direction. Clearly the time taken will increase as x^3, and may be appreciable in large cells.

It is not easy to see what the effects of this might be for the rate of cell growth, particularly since we do not know very much about the way in which eukaryotic cells are regulated. The problem arises mainly for molecules involved in controlling specific genes. Its general effect appears to be that as the number of different genes to be controlled increases, and as the size of the cell increases, one of two things will happen. Either the total number of repressor molecules will increase disproportionately; i.e. their total concentration will increase. We shall, if you like, finish up with a motor car whose throttle is bigger than its engine.

Alternatively the time delays in cellular control circuits will increase; if so, it may be necessary to reduce the rates of the various synthetic processes if the whole system is not to run into a divergent oscillation. By analogy, it is unwise to drive a motor car too fast if there is a lot of

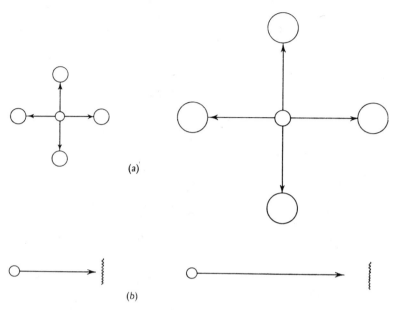

Fig. 1. The effects of scale on diffusion time. (a) It is supposed that substances manufactured in the centre circle must diffuse to the outer circles, and that in the larger cell all linear dimensions are increased proportionately. Thus if, for example, the outer circles represent the region of the cell containing ribosomes, then in a cell of twice the diameter there would be eight times as many ribosomes. It is to this situation that the Table giving paths traversed in one second is relevant. (b) It is supposed that substances synthesized in the circle must travel to a target organ, which may be a gene. In the larger cell the distance to be traversed is increased, but the size of the target is not. In this situation, the length of the path traversed by a molecule in one second is misleading, since most molecules will travel in the wrong direction.

backlash in the steering gear. In the absence of numerical calculation this is mere speculation. The calculations would be worth doing, since they would also be relevant to the suggestion, which has always appeared to me to be somewhat implausible, that in mammals 99 % of the genes are required to regulate the remaining 1 %.

EPILOGUE

I have argued that in bacteria in enriched medium growth rate is limited by the requirements of protein synthesis. Ribosomes and t-RNA-amino acid complexes reach a limiting concentration, such that any further increase would actually slow down the rate of synthesis. I have suggested a number of reasons why eukaryotes may replicate more slowly. These include the fact that the ratio of protein to ribosomal RNA is greater in eukaryotes; the possibility that the cell cycle may involve transition through a number of biochemical states, the rate of transition being limited by the time taken to synthesize new m-RNA and to remove existing m-RNA species; and the possibility that the large size and complexity of eukaryotic cells may introduce longer time delays into their control circuits.

In conclusion, I want to introduce the only new fact to appear in this article; it is a fact which might be held to invalidate much of the preceding argument. If fruit flies (*Drosophila subobscura*) are kept at a low temperature (10° C.), and are then exposed to a lethal temperature (35° C. in wet air) they die after approximately 25 min. But if they have previously been kept at 20° C., they can survive at 35° C. for 50 min. (i.e. twice as long). By keeping flies at 10° C., then transferring them to 20° C., and testing them at 35° C. at various times after the transfer, it is possible to show that this process of acclimatization is completed in 6–8 hr. The awkward fact is that it is completed just as quickly in flies previously fed cycloheximide, at a concentration known to reduce protein synthesis to 10% of its normal level in all tissues.

Here then is a change of state requiring 6–8 hr. for its completion, but apparently not requiring protein synthesis. The nature of the change involved in acclimatization is not known. I mention it only to remind you that macromolecular synthesis and its control may not be the only factor limiting the growth rate of cells.

REFERENCES

BONNER, J. T. (1967). *The Cellular Slime Moulds*. Princeton University Press.

LESLIE, I. (1955). The nucleic acid content of tissues and cells. In *The Nucleic Acids*, vol. II. Ed. E. Chargaff and J. N. Davidson. New York: Academic Press.

MAALOE, O. & KJELDGAARD, N. O. (1966). *Control of Macromolecular Synthesis*. New York: W. A. Benjamin.

MANDELSTAM, J. & McQUILLEN, K. (1968). *Biochemistry of Bacterial Growth*. Oxford: Blackwell.

SHEARER, R. W. & McCARTHY, B. J. (1967). Evidence for ribonucleic acid molecules restricted to the cell nucleus. *Biochemistry*, **6**, 283.

WERNER, R. (1968). Distribution of growing points in DNA for bacteriophage T4. *J. molec. Biol.* **33**, 679.

MICROBIAL GROWTH UNDER EXTREME CONDITIONS

T. D. BROCK

Department of Microbiology, Indiana University, Bloomington, Indiana, U.S.A.

One day I rode to a large salt-lake, or Salina, which is distant fifteen miles from the town. During winter it consists of a shallow lake of brine, which in summer is converted into a field of snow-white salt. . . . The border of the lake is formed of mud: and in this numerous large crystals of gypsum . . . lie embedded. . . . The mud is black, and has a fetid odour . . . the froth which the wind drifted on shore was coloured green, as if by confervae [i.e. blue-green algae]. . . . Parts of the lake . . . appeared of a reddish colour, and this perhaps was owing to some infusorial animalcula. . . . How surprising it is that any creatures should be able to exist in brine, and that they should be crawling among crystals of sulphate of soda and lime! . . . Thus we have a little living world within itself, adapted to these inland lakes of brine. A minute crustaceous animal . . . is said to live in countless numbers in the brine-pans at Lymington; but only in those in which the fluid has attained, from evaporation, considerable strength—namely, about a quarter of a pound of salt in a pint of water. *Well may we affirm, that every part of the world is habitable! Whether lakes of brine, or . . . warm mineral springs—the wide expanse and depths of the ocean—the upper regions of the atmosphere, and even the surface of perpetual snow—all support organic beings.* [Charles Darwin, *Voyage of H.M.S. Beagle*, ch. IV.]

INTRODUCTION

What is an extreme environment? It is not appropriate to define it anthropocentrically, as we should be the first to admit that human life is not everywhere possible. More appropriate is its definition as a condition under which some kinds of organisms can grow, whereas others cannot. If we accept this definition it means that an environmental extreme must be defined taxonomically. Instead of looking at single species, or groups of related species, we must examine the whole assemblage of species, microbial and multicellular, living in various environments. When we do this we find that there are environments with high species diversity and others with low species diversity. In some environments with low species diversity we find that whole taxonomic groups are missing. For instance, in saline and thermal lakes there are no vertebrates and no vascular plants, although they may be rich in micro-organisms, and very high in the numbers of organisms of the species which do live there. In many of the most extreme conditions we find conditions approaching the pure culture, with only a single species present.

Such environments are clearly of enormous interest to the microbiologist, especially one interested in the ecological and evolutionary relationships of organisms. In such biologically simple environments experimental microbial ecology is most easy to carry out, since microbial interactions are minimized or absent. Compare the problems of the soil microbiologist, faced with thousands of species in a mere handful of dirt, with those of the thermal microbiologist, who can often lift from his hot waters grams of essentially monotypic bacterial protoplasm!

Since by definition many organisms cannot grow in the extreme environment, we can ask two sets of questions: (1) how does the environmental extreme affect organisms which cannot tolerate it? (2) How is it possible for organisms which are adapted to overcome the effects of the extreme factor? One point which we must keep in mind, however, is the other environmental factors of the extreme environment, which must be adequate for the support of life. For instance, the bottom of the ocean floor not only has a high hydrostatic pressure but is also dark and is low in organic nutrient concentration, and the organisms which we find there will thus be those which are adapted to all the relevant factors, of which high hydrostatic pressure is only one. For this reason it is desirable to study habitats in which a gradient exists for the factor of interest, from extreme to normal, with all other factors remaining the same. It is for this reason that in our own work we have tried to study the thermal gradients formed by the cooling of water in hot springs effluents (Brock, 1967a).

ENVIRONMENTAL EXTREMES

The most commonly considered environmental extremes are for the factors temperature, redox potential (Eh), pH, salinity, hydrostatic pressure, water activity, nutrient concentration and electromagnetic and ionizing radiations. A brief general review of these factors has been given by Vallentyne (1963). For virtually all of these factors, it is found that organisms with simple structures can grow under more extreme conditions than organisms with more complicated structures; hence the pre-eminence of micro-organisms in extreme environments. There are three ways in which an organism can adapt to an environmental extreme: (1) it can develop a mechanism for excluding the factor from its structure; (2) it can develop a mechanism for detoxifying the factor; (3) it can learn to live with the factor. In some cases, it is the latter situation which obtains, and in fact it sometimes occurs that evolutionary adaptation is such that the organism actually becomes dependent on the factor which for other organisms is lethal. In what follows we will take some of the environmental factors listed above in turn, and consider the natural habitats where the factor is relevant, and the biochemical mechanisms involved in adapting to it. It should be emphasized, however, that an environmental factor might not always affect an organism directly, but might do so indirectly through its effects on other environmental factors. High temperature, for instance, reduces greatly the solubility of O_2, reduces the viscosity of water, and increases its ionization. We must be certain, therefore, that the effects we are measuring are not due to such indirect effects of the environmental variable.

Saline environments

Aqueous habitats exist with total ion concentration varying from essentially zero (double-distilled water) up to saturation. As ion concentration of the water goes up, some ions precipitate before others. The most soluble are the cations Na^+, K^+ and Mg^{2+}, and the anions Cl^-, SO_4^{2-} and HCO_3^- (or CO_3^{2-} depending on pH). It is thus not surprising that it is these ions which appear in the largest amounts in sea water and the saline lakes. The effect of increasing salinity on the species composition of various waters is shown by the data of Table 1. The limiting effect of salt concentration on species composition is well illustrated, but it is clear that sea water is much less restrictive than hypersaline water. The precise ion composition of a saline lake is determined by the nature of the rocks surrounding the lake basin. Some representative examples are given in Table 2. Great Salt Lake, Utah, and many smaller lakes of the U.S. Great Basin, have essentially the ion proportions of sea water in more concentrated form; such lakes are sometimes called thalassohaline. The Dead Sea, Palestine, is quite different from Great Salt Lake in that it is higher in Mg^{2+} than Na^+, and is low in SO_4^{2-}. Lakes high in bicarbonate-carbonate are often called alkali lakes because they are of high pH, values over 10 being reported. One lake in British Columbia is almost pure $NaHCO_3$ (Table 2). Still other lakes are low in chloride and high in sulphate; one lake in British Columbia (Table 2) is composed mostly of magnesium sulphate. It should also be noted that most of these lakes differ greatly in concentrations of some of the minor but biologically important ions, such as F^-, Br^-, Fe^{3+}, Hg^+, etc. For analyses of a wide variety of interesting

2

Table 1. *Occurrence of microbial and plant groups in waters of various salinities*

Group	Fresh water	Brackish water	Sea water	Hypersaline water
Procaryotes				
Cyanophyceae	+++	++	+	+
Bacteria	+++	+++	+++	+
Eucaryotic algae				
Flagellates	+++	++	+++	±
Dinoflagellates	++	+	+++	0
Silicoflagellates	0	+	+++	0
Coccolithophores	±	+	+++	0
Diatoms	+++	+++	+++	±
Conjugales + Desmids	+++	±	0	0
Siphonales	+	+	+++	0
Characeae	+++	+	0	0
Phaeophyceae	±	+	+++	0
Rhodophyceae	±	+	+++	0
Eucaryotic fungi				
Phycomycetes	+++	±	±	0
Myxomycetes	++	0	0	0
Higher plants				
Mosses	+++	±	0	0
Ferns	+++	0	0	0
Dicotyledons	+++	+	0	0
Monocotyledons	+++	+	+	0

Data based on Gessner (1959).
+++, Many species; +, several species; ±, very few species; 0, no species.

Table 2. *Proportions of the major ions in some lakes of high total ion content*

	Great Salt Lake (1)	Dead Sea (2)	Eighty-Three Mile Lake, B.C. (2)	Basque Lake No. 1, B.C. (2)	Hot Lake Wash. (3)
Anions					
CO_3^{2-}	—	—	—	—	0
HCO_3	—	0·24	53·6	0·3	3·1
SO_4^{2-}	20	0·54	Trace	195	243
Cl^-	150	208	0·8	1·7	1·9
Cations					
Ca^{2+}	—	15·8	Trace	—	0·7
Mg^{2+}	8·0	41·9	Trace	42·4	53·6
Na^+	89·9	34·9	20·9	13·6	16·8
K^+	4·7	7·5	—	1·6	1·5
Total salinity	275	315	>75	>258	392

Concentrations in g./l.
References: (1) Adams (1964); (2) Livingstone (1963); (3) Anderson (1958).

lakes throughout the world, the reader should consult Livingstone (1963), Gessner (1959), Cole & Brown (1967) and the references cited by these authors. Discussion of the chemistry of hypersaline sea water lagoons is given by Copeland (1967), and Hedgepeth (1959) discusses the zoology of inland mineral waters.

Biologically, differences in ratios of the various ions have a great influence on the ability of organisms to grow in saline lakes. Cole & Brown (1967) have pointed out that the brine shrimp, *Artemia salina*, occurs predominantly in chloride (e.g. thalassohaline) waters, and is rare in, or absent from, lakes with other predominating ions. The species composition of Great Salt Lake (which is thalassohaline) is much more diverse than is the Dead Sea, which has an ion balance quite different from sea water (Table 3). However, not all the organisms found

Table 3. *Species lists for Great Salt Lake,
Dead Sea and the Ocean*

Species	Ocean	Great Salt Lake	Dead Sea
Total number	> 10,000	~27	< 10
Kinds	Bacteria to mammals	Bacteria 12 Cyanophyceae 2 Chlorophyceae 2 Flagellates—several Protozoa—several Ciliates 7 Amoeba 2 Arthropoda 2 Insects 2	Bacteria *ca.* 8 Cyanophyceae 2 No higher taxa

Data derived from Boyka (1966) and Gessner (1959).

even in the Great Salt Lake are optimally adapted to it. Flowers & Evans (1966) note that *Chlamydomonas* (Chlorophyceae) shows its maximum development at 13–15 % salt, and *Cladospora* only grows up to 3–4 % salt. The animals probably also prefer lower salinities, since *Artemia* develops better in diluted Great Salt Lake water, and the protozoa grow optimally at 2–3 % salt and grow not at all at saturation. The insects found in Great Salt Lake are probably not living in direct contact with the water, since the larvae breath air by means of tracheal gills, and the adults live on the shore. Thus we see that it is probably only the bacteria and blue-green algae which show optimal development at saturation. Although some information is available on brine shrimp ecology in different saline lakes (Cole & Brown, 1967) we know virtually nothing about the microbial ecology of these interesting habitats. How interesting it would be to know what kind of bacteria live in a lake containing mostly $MgSO_4$, such as Basque Lake, British Columbia! What kind of cell membranes and ribosomes would they have?

It must be emphasized, however, that few workers have established that growth of the organisms was taking place at the salinities at which they were found. The mere presence of an organism in a water body does not indicate that it has grown there; it may have drifted in from fresh water sources. This apparently was the source of some of the algae reported by Elazari-Volcani (1940) to be in the Dead Sea (Gessner, 1959). The optimum salt concentration for a *Dunaliella viridis* isolate from a solar salt pan was about 6%, even though it was isolated by enrichment in a medium containing over 20% salt (Johnson *et al.* 1968). Further, even if the organism can be shown to be alive and growing, the salinity measurements must be made precisely at the site and time at which it is found. Saline lakes vary in salinity considerably throughout the year, due to variations in freshwater drainage. Great Salt Lake, for instance, has a considerably lower salinity in winter than in summer, due to entry of freshwater from snow melt in the surrounding mountains. Also, freshwater is less dense than salt water, and hence will float on top. Organisms developing in this freshwater layer might be mistaken for halophiles. This apparently explains some of the biological findings of Anderson (1958) in Hot Lake, Washington. One clear demonstration of *in situ* growth of bacteria in a hypersaline lake is the study of Smith & ZoBell (1937) in Great Salt Lake, by use of a slide-immersion technique, but further studies in natural saline environments are greatly needed. A suitable programme of study of the physiological ecology of halophiles should include the following points: (1) establishment of the salinity of the habitat of the organism; (2) establishment that the organism is actually growing *in situ* at this salinity, at the time sampled; (3) preparation of cultures of the organism and demonstration of the salinity tolerances for growth and survival.

Even if reproduction is taking place, this does not imply that the organism is optimally adapted to that environment, as we have already noted. Only in the case of the halophilic bacteria is the situation reasonably clear; many of these isolates grow optimally in culture in media of about 25% (w/v) NaCl, and show a minimum requirement for growth of 12–20% NaCl (Larsen, 1967). They are thus true halophiles, and even though experiments have apparently never been performed to show that these organisms grow in nature best at high salt concentrations, we would be very surprised indeed if we did not find that to be so. In salt pans, where solar salt is being made, the water at the time its salinity reaches close to saturation often acquires a brilliant pink cast due to the growth of halobacteria, whereas in the more dilute waters, before evaporation is completed, these organisms are not seen. It seems very unlikely that the appearance of these organisms could be due to their mere concentration from diluter waters, and therefore they must be growing. Most halophilic bacteria in culture have apparently been isolated from solar salt, or from food products preserved with solar salt. Thus, it seems reasonable to conclude that the natural habitats of these extreme halophilic bacteria are salt pans and salt lakes.

Molecular mechanism of halophily. In what ways can high salt concentrations be toxic? If the cell is impermeable to salt, then of course it would be under a severe osmotic strain, and would have to be growing at an intracellular level of hydration much lower than seems possible. On the other hand, if the cell is freely permeable to salt, as indeed seems to be the case for extreme halophiles, the cell must be faced with the problem of avoiding the precipitation of its intracellular proteins, which if similar to those of non-halophiles would be expected to precipitate at salt concentrations this high. In principle, the organism could circumvent this difficulty by synthesizing enzymes with high net positive charge, so that Na^+ would be repelled. Actually, the reverse seems to be the case; the enzymes have a high net *negative* charge. A variety of enzymes of halophiles require a high concentration of Na^+ (or other cation) for activity (Larsen, 1967), and it has been shown in two cases (Baxter, 1959; Holmes & Halvorson, 1965a, b) that in low salt the halophilic enzymes undergo denaturation due to repulsion of electrostatic charges. In one case (Holmes & Halvorson, 1965a) it was shown that the denatured enzyme (malic dehydrogenase) could be reactivated by re-addition of salt, although the renaturation process was slow. In the presence of high NaCl concentrations the malate dehydrogenase could not be precipitated by ammonium sulphate, whereas in the absence of NaCl, precipitation (and subsequent purification) was possible. This was true not only for this enzyme, but for the majority of the proteins of the cell, since the protein of crude extracts could be precipitated by ammonium sulphate in the absence of NaCl, but not in its presence. The conclusion seems to be that the proteins of halophiles have a high negative charge, and in this way avoid the precipitating effect of the salt.

A high sodium ion concentration is also required for the stabilization of the cell envelope in halobacteria (Larsen, 1967). The question of whether or not halobacteria have a cell wall has apparently now been answered in the affirmative (Stoeckenius & Rowen, 1967), although neither muramic acid or diaminopimelic acid have been detected (Larsen, 1967). A stepwise reduction of salt concentration causes lysis by release first of cell-wall material and then of cell membrane (Stoeckenius & Rowen, 1967). Lysis does not depend on an enzymic or osmotic process, but probably arises because the structural components of the cell surface have a high negative charge density and repel each other when the salt is reduced. The first effect of reducing the Na^+ concentration is the conversion of the rod-shaped structure to a sphere. One can assume that the halobacteria have made use of the ever present Na^+ ion to replace a structural function carried out in non-halophiles by co-valent linkages. However, in contrast to the halobacteria, the halophilic cocci do not lyse or show any morphological alterations when placed even in distilled water (Larsen, 1967). The ribosomes of halobacteria also require high salt for stability, but curiously K^+ is required instead of Na^+. This is in line with the fact that halobacteria concentrate K^+ from the medium against a steep gradient, to such an extent that the intracellular concentration of this ion approaches a value close to its solubility limit. The requirement of high K^+ concentration for ribosome stability is in line with the observation that the ribosomes of this organism, in contrast to those of *Escherichia coli* and other 'normal' organisms, have a high content of acidic protein, but why the halobacteria use K^+ instead of Na^+ to stabilize ribosome structure is unknown.

The nucleic acids of halobacteria are apparently normal; the guanine-cytosine content of the DNA is 66–68 %. Since DNA is stabilized against unwinding by high salt concentrations, one wonders how halobacteria manage to divide with internal salt concentrations as high as they are. Perhaps much of the salt is not free in the cytoplasm, but is bound to the negatively charged protein molecules.

Although the extreme halophiles are optimally adapted to their environment, their growth rates are rather slow, with the shortest generation times of halobacteria reported to be 7 hr., and for halococci about 15 hr. (Larsen, 1967). These long generation times may be a property of the organism unrelated to salt concentration, or it may be that even under optimal conditions some toxic effects of salt exist. Extreme halophilism is a genetic characteristic of the organism and not a physiological adaptation as shown by the fact that attempts to adapt non-halophilic and marine bacteria to grow at high salt concentrations have failed (Larsen, 1967; Ingram, 1957).

Aside from the halobacteria and halococci, the only other bacterial taxonomic group reported to have a representative which is an extreme halophile is that of the photosynthetic bacteria. Anderson (1958) reported extensive accumulations of green sulphur bacteria in Hot Lake, Washington, at total dissolved solids of about 400 g./l., and Raymond

& Sistrom (1967) cultured a photosynthetic bacterium of uncertain affinity from Summer Lake, Oregon. In the latter case, the organism grew well at 22 % NaCl and still grew, although abnormally, when the NaCl concentration was reduced to 8 %. Thus it is not as sensitive to reduced salt concentration as the non-photosynthetic halophiles.

Johnson *et al.* (1968) studied the effect of salt on a variety of enzymes of the halophilic green alga *Dunaliella viridis*. Even though the organism grew optimally at 0·8–2·0 M NaCl, all of the enzymes tested were inhibited by NaCl concentrations far lower than this, significant inhibition even being reported at 0·17 M. Thus we cannot assume that the findings with the extreme halophilic bacteria also hold for the less extreme halophiles.

The field of marine microbiology is too vast to be considered here. Although many marine micro-organisms have specific requirements for Na +, they could hardly be called halophilic in the sense we are using it here, since they develop optimally at about the salinity of sea water, 3·5 % NaCl, and are usually inhibited by concentrations greater than 10 % NaCl. Although many marine bacteria have specific Na + requirements (MacLeod, 1965), most marine fungi, apparently do not (Jones & Jennings, 1964: although see Siegenthaler, Belsky & Goldstein, 1967). The marine fungi thus seem to be salt-tolerant freshwater forms that have been able to extend their range into the sea.

Acid environments

Many soils and lakes are mildly acidic, but we shall concentrate here on environments more acidic than pH 3–4. The stomach of many animals has a pH of 1–2, and Shifrine & Phaff (1958) have described an acidophilic mildly thermophilic yeast, *Saccharomycopsis guttulata*, which grows in the stomach of the rabbit. Many other yeasts (Battley & Bartlett, 1966) and fungi (Lilly & Barnett, 1951) are able to grow at pH values down to 2, but usually show optimal growth around pH 5·5–6·0. Many lakes and bogs are acidic, due to the presence of humic acids derived from decomposing plant remains; pH values lower than 3 are rare, however. Many animals, plants and eucaryotic algae can live in these acid lakes and bogs (Heilbrunn, 1952; Gessner, 1959), so by our definition they would not be considered extreme environments. Acid conditions develop in lakes formed in strip mine areas, with a pH of about 3 or less. The acidity arises from the oxidation to sulphuric acid of sulphides associated with the coal wastes; this oxidation is either spontaneous or is brought about by the activities of *Thiobacillus* species. The ability of high concentrations of hydrogen ions to limit the diversity of the biota in such lakes is well established (Harp & Campbell, 1967), but it is significant that eucaryotic algae and invertebrates do live in these waters. Tendepedid larvae live in many of these acid lakes, although they may not be able to complete their life cycle at a pH of less than 3·0 (Harp & Campbell, 1967). However, a note of caution is necessary in interpreting the presence of an organism in water of an acid lake as indicative that it has grown at the ambient pH, since Harp & Campbell (1967) showed that the pH of the bottom mud of an acid lake was up to two pH units higher than the water itself. In acid mine drainages in Indiana, Lackey (1938) reported 11 genera of flagellates, 7 rhizopods and 12 ciliates at pH 3·9 or lower. Some members of the genus *Gammarus* were found in the pH range 2·2–3·2 and several insects, including caddis fly larvae, occurred in pH as low as 2·3. Nine species of green algae and protozoa were found at a pH of 1·8. However, no studies were performed to show that the species were optimally adapted to these pH values, or that they were even growing there.

The most acidic natural environments are those associated with thermal springs. Acidity usually develops due to the oxidation of H_2S in the volcanic gas when it rises to the surface of the earth and meets the air. Where ground water is restricted, the sulphuric acid formed is not greatly diluted and low pH values occur. In some acid springs the acidity is due almost exclusively to sulphuric acid, whereas in others hydrochloric acid is also present. Virtually the only buffering capacity in these springs is from the sulphate ion, but many of them remain remarkably constant in pH. The lowest pH value I have found recorded is about 1·0 (Yoneda, 1952) for certain Japanese hot springs. Acidic springs occur in Iceland (Barth, 1950), the Azores (Brock & Brock, 1968), Yellowstone National Park (Allen & Day, 1935), Japan (Uzumasa, 1965), and many other parts of the world (Waring, 1965). Acid conditions also develop in fumarole areas, where the steam rising to the surface condenses. One classic fumarole area is the Bay of Naples, where at Vesuvius, Solfatara and Ischia pH values of 3 or less are common. Low pH leads to a greatly restricted species diversity (Copeland, 1936), often only single species being found.

In the most acid hot spring waters the interesting alga *Cyanidium caldarium* occurs in almost pure culture. This organism, which is not only acidophilic but also thermophilic, has been found in nature over a pH range from 1·0 (Gessner, 1959) to 5·4 (Copeland, 1936), but not higher and its optimum development is between pH 2 and 3. Although living in an extremely restricted habitat, it is world-wide in distribution, being reported from California, Yellowstone Park (Wyoming),

Dominica (West Indies), Java, Sumatra, Japan (Hirose, 1950) as well as Italy (Carmelo Rigano, University of Naples, personal communication). So far as can be told, the organism from all of these locations is the same in life cycle, pH tolerance and nutrition. In recent times, cultural and biochemical studies have been reported by Hirose (1958), Allen (1959), Fukuda (1958) and Ascione, Southwick & Fresco (1966). The organism is unquestionably a eucaryote, but its anomalous pigmentation and other properties make its taxonomic position uncertain. These and other evolutionary considerations about *C. caldarium* are reviewed by Klein & Cronquist (1967).

The work of Ascione *et al.* (1966) would suggest that the optimum pH for growth of *Cyanidium caldarium* is 2; no growth occurs above 5, and growth still occurs well at pH 1, which is about 0·1 N H_2SO_4. Allen (1959) claims to have grown the organism in 1 N H_2SO_4, but her statement that it will grow at pH 7 is probably erroneous, since Ascione *et al.* (1966) and W. D. Doemel & T. D. Brock (unpublished) have observed that cultures adjusted to pH 7 rapidly reduce the pH towards 2. Thus *C. caldarium* is undoubtedly an obligate acidophile, of an extreme sort. As Allen (1959) has pointed out, the internal pH of *C. caldarium* could not be near its environmental pH, as chlorophyll is unstable to acid and would break down. Indeed, she showed that if *C. caldarium* cells are frozen in their culture medium, the chlorophyll was converted to pheophytin upon thawing the cells, probably because the membrane barrier had been broken by freezing. Other important acid-labile molecules, DNA and ATP, would probably also be destroyed by a low internal pH.

The number of acidophilic bacteria is quite limited. Early compilations of pH optima and ranges (Buchanan & Fulmer, 1930; Porter, 1946) list only *Thiobacillus thiooxidans* and related Thiobacilli. Just recently, Uchino & Doi (1967) have reported acidophilic thermophilic spore-forming bacteria from Japanese hot springs. These organisms, easily referable to the genus *Bacillus*, showed pH optima at 3·5–4·0, with a range from 2·3 to 5·0. Optimal growth temperature was 65° C. with a range from 45° to 71° C. Physiologically these strains resembled *Bacillus coagulans*, but they must differ biochemically in many ways from this species. Hopefully, further work will be forthcoming on this interesting group of Bacilli.

The acidophilic *Thiobacillus*, *T. thiooxidans*, was first described by Waksman & Joffe (1922) and has been the object of considerable study for many years. It is characterized by an ability to oxidize elemental sulphur with the production of sulphuric acid, pH values down to 1·0 or less being reported. The negative pH values occasionally reported have been shown to be probably erroneous (Kempner, 1966), but there seems to be no question that *T. thiooxidans* can metabolize at pH values less than 1, and survive values close to 0, although the pH optimum for growth of *T. thiooxidans* is near 3·5 (Vishniac & Santer, 1957). A closely related species, *T. ferroxidans*, has the ability to oxidize iron as well as sulphur, and a third species, *Ferrobacillus ferroxidans*, oxidizes iron but not sulphur. Both these latter two species are also acidophilic autotrophs, of morphology similar to *T. thiooxidans*, and are probably closely related to it.

Despite the long history of studies on *Thiobacillus thiooxidans*, and the interest it created by its ability to generate low pH values, relatively little information is available about the details of its pH relations, or how it manages to tolerate low pH. Despite the poetic phrase of Umbreit (1962), 'it is the biological submarine living in a hostile world by excluding it', we do not really know that *T. thiooxidans* is impermeable to hydrogen ions. It has been well established that *T. thiooxidans* is permeable to organic compounds (Borichewski & Umbreit, 1966), and it seems strange that glucose, amino acids and similar compounds can enter these organisms, but that the tiny hydrogen ion cannot. The measurement of internal pH is a difficult matter (well reviewed by Heilbrunn, 1952), but apparently even the pH optima of internal enzymes of *T. thiooxidans* have not been well defined. Although the pH optimum for iron oxidation by whole cells of *Ferrobacillus ferroxidans* was 3·0, the pH optimum of the cell-free Fe^{2+}-cytochrome *c* reductase showed a broad optimum around pH 7·0 (Din, Suzuki & Lees, 1967). We may thus tentatively conclude that these acidophiles can grow at low pH because of their ability selectively to exclude H^+ ions.

Doetsch, Cooke & Vaituzis (1967) have recently shown that the flagellum of *Thiobacillus thiooxidans* is highly resistant to both high acidity and high temperature, although the flagella of other bacteria are disaggregated at pH values below 3–4. Since the flagellum is an extracellular organelle, and would be continuously exposed to an acid environment, this discovery is not surprising, but does show that *T. thiooxidans* is able to make at least one acid-stable protein.

The most acid-resistant organisms to have been reported are two fungi described by Starkey & Waksman (1943), *Acontium velatum* and an unclassified Dematiacean fungus, which were found growing at very acid pH values in 4% $CuSO_4$ solution. Growth occurred well at pH 0·2–0·7 and slowly at pH 0. The upper pH limit was not defined, as the fungi quickly lowered the pH from 7 to 3, and then grew well. Although resistant to massive Cu^{2+} concentrations, the fungi grew better in the absence of Cu^{2+}.

Copper resistance may be a general property of acidophiles, since *Thiobacillus thiooxidans* and *Ferrobacillus ferrooxidans* are also resistant to high Cu^{2+} concentrations (Booth & Mercer, 1963). Perhaps this resistance is due to the ability of the H^+ ion to compete with the Cu^{2+} ion on the cell surface, thus rendering the cell impervious to Cu^{2+}. Indeed, resistance to heavy metals in acid mine drainages and mineral springs may have been one of the selective pressures towards the evolution of acidophiles. Further work on this interesting group is clearly warranted.

Temperature

Although organisms growing at low temperature are of considerable interest, we can hardly consider low temperature as an extreme environment, since mammals, fishes and a variety of other poikilotherms thrive in the cold waters of the Antarctic coast, along with sea weeds and other eucaryotic plants. On the other hand, high-temperature environments are quite obviously extreme.

The distribution of natural high-temperature environments on earth has been recently reviewed (Brock, 1967a) and references to earlier work can be sought there. Unnatural high-temperature environments occur in many industrial situations; as for instance sugar manufacture (Klaushofer, 1967), power plant effluents (Warinner & Brehmer, 1966; Mihursky, 1967) and perhaps even melted asphalt (Mateles, Baruah & Tannenbaum, 1967). In addition, self-heating of such products as compost and hay (Miehe, 1907; Festenstein, Lacey & Skinner 1965), stored grain (Wallace & Sinha, 1962), sweating tobacco (English, Bell & Berger, 1967) and bird's nests (Apinis & Pugh, 1967) can lead to the selective growth of thermophiles. Solar heating of soils (Apinis, 1965; Schramm, 1966) probably also leads to temperatures high enough for the selective growth of thermophiles. However, the isolation of thermophiles from low-temperature environments (the first thermophile was isolated by Miquel (1879) from the River Seine) does not mean that they are growing at these low temperatures but rather that they may be present there as dormant structures.

Although there may be some doubt that the cytoplasm of an organism in high salt or low pH is subject to the environmental extreme, there is no doubt that micro-organisms are isothermal with their environments. A thermophile cannot be a biological submarine with an air-conditioning system, since there is no way for a heat pump to operate in an isothermal environment. (On the other hand, it is possible for organisms to be warmer than their environments if they contain their own internal heating systems, driven by metabolic energy; witness the whale and the submarine.) Thus suggestions that thermophiles can survive at high temperatures because of a thick layer of insulation can only be taken with a smile.

There are two ways in which we can demonstrate that high-temperature environments really are biological extremes, by examining the success in them of taxonomic groups of various degrees of complexity, and by examining the species diversity of a single taxonomic group at different temperatures. The first approach is shown by the data of Table 4. As higher and higher temperatures are reached, whole taxonomic groups disappear, with the most complex falling away first; at the very

Table 4. *Approximate upper thermal limits for different groups of organisms*

Organisms	Upper limit (°C.)
Animals, including protozoa	45–51
Eucaryotic micro-organisms (certain fungi and the alga *Cyanidium caldarium*)	56–60
Photosynthetic procaryotes (blue-green algae)	73–75
Non-photosynthetic procaryotes (bacteria)	> 90

Data of Brock (1967 *a*).

extremes only bacteria are seen. The second approach is shown by the study of Copeland (1936), who showed that amongst the blue-green algae there was much more species diversity of this single microbial group at lower than at higher temperatures in hot springs of Yellowstone Park. (There is some question about some of Copeland's taxonomic determinations (Brock, 1968*a*), but this does not alter the over-all picture.) The data of Brues (1932) on the species diversity of water beetles at various temperatures gives a similar picture, although the temperature scale is shifted down about 30° C.

Diversity of thermophilic bacteria. Since the organisms growing at the most extreme temperatures are bacteria, attention naturally focuses on them. From a cursory perusal of the literature one might obtain the impression that spore-forming bacilli such as *Bacillus stearothermophilus* are the only thermophilic bacteria of any significance. When enrichments are incubated at 55–60° C. in the usual culture media using inocula obtained from soils, hot springs (Marsh & Larsen, 1953), or food stuffs (Cameron & Esty, 1926), this organism usually predominates. However, it is clear from the older literature (well reviewed by Miehe, 1907) that other kinds of thermophilic bacteria exist. In recent times, thermophilic forms of other species or genera reported have included:

anaerobic cellulose decomposers (Loginova, Golovacheva & Shcher-bakov, 1966); hydrocarbon utilizers (Klug & Markovetz, 1967; Mateles *et al.* 1967); hydrogen-oxidizing bacteria (McGee, Brown & Tischer, 1967); actinomycetes (Manachini, Ferrari & Craveri, 1965; Craveri & Manachini, 1966); and sulphate-reducing bacteria (Campbell & Post-gate, 1965; Schwartz & Schwartz, 1965). Thermophilic photosynthetic bacteria have probably not been sought often, but a thermophilic purple sulphur bacterium exists at 60° C. in Hunter Hot Springs, Oregon (R. Castengolz, personal communication), and has been cultured (W. Sistrom, personal communication). I have observed heavy accumulations of a purple sulphur bacterium at 60° C. in one spring in Yellow-stone Park growing underneath a thick mat of blue-green algae and recently we have found a new group of extreme thermophilic bacteria, which is discussed in detail below.

The absence of a particular bacterial group from thermal environments could mean either that there is some inherent factor of its metabolism or other phenotypic character which makes it impossible for it to evolve thermophilic strains or it could be that the available thermal environments do not provide the other environmental factors needed for their development. Hot springs are low in organic matter (as are most spring waters as they come from the ground) and would obviously limit the kinds of bacteria which could grow. There is some minimal evidence that temperature alone may be a limiting factor for the development or evolution of acidophilic sulphur bacteria. I have made extensive observations for the presence of accumulations of sulphur bacteria along the thermal gradients of H_2S-rich acid hot springs at Yellowstone Park. Above about 60° C. these organisms are always visually absent, but appear in large amounts at temperatures lower than this. Schwartz & Schwartz (1965) were unable to find acidophilic thiobacilli which could grow at temperatures above 50° C., and most strains grew only at lower temperatures, even when the inocula were taken directly from acidic fumarole areas. Conceivably the combined factors of acid and temperature provide a more restrictive environment than temperature alone, since in alkaline hot springs thermophilic bacteria are found at over 90° C., occasionally in massive accumulations (see below).

Other environmental factors may be able to modify to some extent the effects of high temperature on organisms. Thermophilic fungi are able to grow at higher temperatures in the presence than in the absence of Ca^{2+} ions (Craveri, Craveri & Guicciarde, 1967), and high pressure inhibits the thermal inactivation of some proteins and other macro-molecular structures (Johnson, 1957).

A non-sporulating extreme thermophile. In most of the alkaline hot springs which we have studied, mats of filamentous bacteria develop in association with blue-green algae at temperatures from 73 to 75° C. and lower. We showed (Brock, 1968 b) that the bacteria are probably essential for the development of the algae, in that they form a matrix within which the unicellular algae become embedded, thus making it possible for them to hold on in the flowing water of the hot spring. We have cultured a variety of strains of these bacteria, as they are easily enriched by incubating at 70° C. in media containing 0·1 % tryptone and yeast extract in synthetic spring water (Freeze & Brock, 1968). All of the strains we have isolated are yellow pigmented Gram-negative non-motile rods. At temperatures above 70° C., they grow usually as long filaments, and under these conditions resemble the filamentous organisms which we see in nature. The natural as well as the cultured material contains a yellow pigment, probably a carotenoid. Similar organisms have been isolated from hot tap water. It is perhaps relevant that if the same media and inocula are incubated at 55° C. spore-forming bacilli develop instead, suggesting that this new group of bacteria are more extreme thermophiles than the bacilli.

We have studied one strain (YT-1) in some detail. Its temperature optimum for growth is about 70° C., its maximum at about 79° C. and its minimum at around 40–45° C. The generation time at the optimum is about 50 min. Although growth occurs at 79° C., we have not been able to achieve growth at 80° C. with the medium and aeration conditions used. This organism seems to have no growth factor requirements and grows, although slowly, in a medium containing acetate as sole carbon and energy source, and ammonium as nitrogen source. It will also use some sugars and amino acids as carbon and energy sources. It is lysozyme sensitive. As a simple way of determining whether this extreme thermophile had any unusual macromolecular properties, we determined the sensitivity of a number of strains to growth inhibition by various antimicrobial agents. The strains were all sensitive to cycloserine (100 µg./ml.), streptomycin (10 µg./ml.), penicillin (10 µg./ml.), sodium azide (500 µg./ml.), novobiocin (10 µg./ml.), tetracycline (10 µg./ml.), chloramphenicol (10 µg./ml.), and sodium lauryl sulphate (100 µg./ml.). If we accept that the mode of actions of these agents are the same in these extreme thermophiles as they are in 'normal' organisms, we can conclude that our organism has a peptido-glycan cell wall, a protein-synthesizing machinery using 70S ribosomes, an electron transport chain, and a cell membrane. In fact, for Gram-negative bacteria these organisms are surprisingly sensitive to penicillin and novobiocin. The strains tested derived from inocula taken in Yellowstone Park, Calistoga Hot Springs, California, and Bloomington, Indiana. The similarity of their growth responses to the various inhibitors is striking, considering their dispersed geography. These organisms seem not to have DNA of especially unusual composition, since YT-1 had 67% GC and a strain from Calistoga, California, had 65% CG (both less than the mesophilic *Sarcina lutea*).

Temperature limits of Cyanidium caldarium. The temperature limits of this acidophilic organism are of considerable interest, since they may provide some insight into how the two extremes of pH and temperature interact. Field observations of Copeland (1936) suggesting that *Cyanidium caldarium* grows at temperatures up to 80°C. in Yellowstone Park are probably erroneous. Doemel and I made an extensive survey of acid hot springs in Yellowstone Park in the summer of 1967. Although *C. caldarium* was present in most of these springs, we never found it at a temperature higher than 56°C. (Doemel & Brock, 1968). In some springs the thermal gradient created by cooling of the water as it flowed from the source permitted an accurate measurement of the upper temperature limit. Also, Doemel measured the effect of temperature on photosynthetic uptake of $^{14}CO_2$ for specimens collected from various environmental temperatures. All showed an optimum temperature for photosynthesis at 45–50°C. and a maximum at 55–56°C. These data agree well with the cultural studies of Ascione *et al.* (1966), who reported a temperature optimum for growth of 45°C. and an upper temperature near 60°C. The one field record of Copeland (1936) for the presence of *C. caldarium* at 80°C. was from the edge of the Great Sulphur Spring of the Crater Hills in Yellowstone Park. Doemel and I visited this spring and found its thermal and chemical characteristics unchanged since the 1930's (Copeland, 1936; Allen & Day, 1935), but *C. caldarium* was absent. A careful reading of Copeland's paper suggests that perhaps he mistook the greenish yellow deposit of sulphur present at this spring for *C. caldarium*. The lower limit for *C. caldarium* in nature is about 30–34° C. Although it will grow at lower temperatures than this in the laboratory, in nature it apparently cannot compete with a variety of flagellate algae which occupy the lower temperature niche (W. D. Doemel & T. D. Brock, unpublished observations).

Molecular basis of thermophily. It should be emphasized, however, that although high temperature seriously restricts taxonomic diversity, many of the organisms which are found at these extremes are not struggling to survive, but are optimally adapted (Brock, 1967b). Two questions arise: (1) How do thermophiles survive and grow at high temperatures which are rapidly lethal to low-temperature organisms?

(2) Why is it that they are obligate thermophiles, and cannot grow at all or grow only slowly at mesophilic temperatures? These two questions, although perhaps related, are really independent of each other, and need to be considered separately. The molecular basis of thermophily has been reviewed recently by Brock (1967a) and Friedman (1968) and only the salient points will be discussed here. There seems to be no question that the enzymes and protein-synthesizing machinery of thermophiles are more heat resistant than those of mesophiles. As an addition to the previous literature on this topic, I might point out that Freeze in my laboratory has shown that strain YT-1 has an aldolase which is even more thermostable than that of *Bacillus stearothermophilus* (Thompson, Militzer & Georgi, 1958; Thompson & Thompson, 1962). It shows maximal rate of activity at 90°C., and is stable in the absence of substrate in crude extracts at least up to 75°C. In contrast to the *B. stearothermophilus* aldolase, that of YT-1 is inhibited by EDTA, and hence fits into the class II of Rutter (1964) where other microbial aldolases fall. The K_m and pH optimum of the YT-1 aldolase are not especially unusual for a microbial aldolase. Whether the enzyme is intrinsically stable or is stabilized by non-enzymic components remains to be seen.

I have suggested (Brock, 1967a) that it may be the cell membrane of thermophiles which is most likely to be the crucial thermostable structure. Evidence supporting this suggestion has been recently forthcoming. Golovacheva (1967) has shown that *Bacillus stearothermophilus* strains spontaneously form protoplasts which are remarkably stable both to heat and to osmotic shock, and N. E. Welker (personal communication) has shown that protoplasts of his strain of *B. stearothermophilus* made by the action of a cell-wall digesting enzyme are also remarkably stable. (Such stability is also implicit in the studies of Abram (1965).) Although Golovacheva was not certain that her forms had completely lost their cell wall Welker could find no trace of wall components in his spherical bodies. Paul Ray in my laboratory has shown that spheres of YT-1 produced by lysozyme action were stable to dilution into distilled water or to boiling for up to 1 hr. These spheres were rapidly lysed by sodium lauryl sulphate at concentrations which did not cause lysis of the normal rods, thus showing that these spheres are not structures held together by covalent bonds. On the other hand, the protoplasts of mesophilic *Sarcina lutea* are readily lysed by osmotic shock or temperatures of 60°C. It thus seems that the cell membrane of thermophiles is unusually heat stable and is also more stable to osmotic shock. The chemical nature of these cell membranes will be of interest to determine.

Like *Bacillus stearothermophilus*, strain YT-1 forms spheres spontaneously in culture, presumably through the action of an autolytic enzyme. (Spontaneous formation of spheres in mesophiles may also occur, but the spheres, being osmotically sensitive, would probably lyse and thus would not be seen microscopically.) The formation of spheres also occurs readily in the hot spring thermophiles growing in nature. I have seen them often in the thick bacterial-algal mats growing at temperatures of 50–65°C., but they are seen most dramatically in the filamentous bacteria growing at temperatures near boiling in some of the Yellowstone hot springs. Pl. 1, Fig. 1, shows a picture of a spring we call pool A, whose water courses out into the narrow channel at a temperature about 91·5°C. In the effluent channel, large masses of pink filamentous bacteria develop in the temperature range of 80–88°C. and are shown in close-up in Pl. 1, Fig. 2. If microscope slides are placed on the bottom of this channel they become covered within several days with the same kind of filamentous bacteria, and when the slides become crowded with filaments spheres are often seen in large numbers (Pl. 2). Frequently, several spheres can be seen forming along the same filament. The ability of these spheres to survive temperatures near 90°C. testifies to their inherent heat stability.

Based on these observations and on the data presented in Table 4, the following hypothesis seems tenable. Membranes with larger pores, through which larger molecules can pass, must be constructed loosely. Nuclear membranes must have pores large enough for messenger RNA and ribosomes to pass through, and hence should be very loose. Mitochondrial and chloroplast membranes must have holes large enough for ATP to pass through, and perhaps proteins. It thus seems likely that the internal membrane systems of eucaryotes would be fairly heat labile, and this fact might explain why eucaryotes cannot grow at temperatures as high as procaryotes. The destruction of nuclei and mitochondria of pea seedlings by heat has been shown by Famin (1933), and a similar phenomenon in animals was commented upon by Brues (1932). Perhaps Brues's cautious conclusion is worth quoting here: '. . . it does not seem wise at this time to go further than to state that the absence of mitochondria and the generally associated imperfect differentiation of nuclear structure are characteristic of those organisms that have been able to adapt themselves to life at temperatures above 50° or thereabouts'.

Even in procaryotes, differences in membrane structure must exist, to account for the differences in thermostability of membranes of mesophilic and thermophilic species already noted. These differences might arise from a necessity for a more efficient solute uptake in organisms living at lower temperatures. But the key point is that the internal membranes of eucaryotes *must* be loose (and hence thermolabile) if the necessary functions of the organelles they surround are to be maintained, whereas in procaryotes only the plasma membrane is involved, and this could be made more rigid and hence heat stable if necessary to enable the organism to grow at high temperature. Thus I postulate that there is an inherent fundamental limitation in eucaryotic membrane structure which cannot be overcome by further evolutionary changes, if the organism is to remain a eucaryote.

We thus may conclude that a variety of macromolecular structures of thermophiles are more stable than those of mesophiles, and that this thermostability is responsible for the ability of these organisms to grow at high temperatures. Why is it that these organisms are unable to grow at temperatures of 37° C. or lower, at which mesophiles grow rapidly? Although we have no firm data on this point, one hypothesis might be that growth at low temperatures requires the existence of heat labile structures, perhaps because these latter, being more flexible, are able to act more efficiently at low temperatures. Indeed, the problem of why low-temperature organisms form heat labile macromolecules deserves detailed study.

Other environmental extremes

Space does not permit more than a brief mention of some other kinds of extreme environments which have been studied microbiologically. Environments of high hydrostatic pressure in the depths of the oceans yield not only microbes but higher animals, including vertebrates. Thus, by our definition such environments are not extreme, although they may be biologically interesting. Surprisingly, although bacteria capable of growing at high hydrostatic pressure have been isolated it is not certain that true barophilic bacteria exist. In some 'deep-sea samples, just as many and often more bacteria were found that grew at 1 atm. than at higher pressures. Very few pure cultures isolated from enrichment cultures have proved to be obligate barophiles. Growth and reproduction of bacteria at increased pressures . . . are very slow' (ZoBell, 1964). Another review of effects of hydrostatic pressure on micro-organisms has been presented by Morita (1967).

Anaerobic environments including environments high in H_2S concentration are undoubtedly extreme. Of the animals, only certain protozoa, and a few other invertebrates, can live anaerobically (Heilbrunn, 1952) and of the plants probably only the blue-green algae. Many bacteria are of course obligately anaerobic, although the reasons for this are still obscure and deserve investigation by modern methods

(Hewitt, 1957). Highly anaerobic environments are found in lake and estuarine sediments, animal intestinal tracts, the rumen, sewage plants, and a few other areas. Although H_2S is more toxic than cyanide to most aerobes, many micro-organisms tolerate high amounts. Lackey, Lackey & Morgan (1965) have discussed this problem in an interesting article which deserves wider attention. They provide an extensive list of pro-caryotic and eucaryotic algae and protozoa which can tolerate H_2S. A number of species remained alive (i.e. were still actively motile) after 48 hr. in water through which a stream of H_2S gas was passed continuously. Although it is usually assumed that H_2S, like cyanide, affects the cyto-chrome system, many of the organisms Lackey *et al.* used had cyto-chromes, either in photosynthetic or respiratory systems. Sulphur bacteria such as *Thiothrix*, which only live in habitats where free H_2S is present, are strict aerobes, and must have a respiratory system unusually resistant to H_2S.

Another environmental extreme, which has been inadvertently ex-plored by several generations of continuous culturists, is the essential nutrient in limiting quantities. Beginning with Monod (1942) and con-tinuing up until this symposium, the effect of limiting substrate on growth rate has been explored primarily as a laboratory phenomenon but, as has been pointed out by van Uden (1967) among others, substrate-limited growth is a common natural phenomenon in the sea. The reasons why growth rates are slower under substrate limitation are probably complex and are only vaguely perceived. The inability of transport systems to promote accumulation of limiting substrates is probably an important part of the story (Hamilton, Morgan & Strick-land, 1966; van Uden, 1967), but perhaps not all. It seems clear however (Hamilton *et al.* 1966; Jannasch, 1968) that some marine bacteria are able to grow at substrate concentrations considerably lower than the bacteria which usually form colonies on organically rich laboratory media.

The effects of ultraviolet light on algae have been reviewed by Halldal (1967). Although we usually assume that ultraviolet light is not a natural environmental factor, this is not so, as near ultraviolet is present in significant amounts, especially at high altitudes. Findenegg (1966) has shown that algal photosynthesis in high alpine lakes is definitely in-hibited in surface waters by near ultraviolet light from the sun, although not all lakes showed this phenomenon. Adaptation to near ultraviolet by the reduction of total pigment concentration or by the production of a photo-protective pigment which resides at the cell periphery seems possible. The role of carotenoids in the protection of bacteria which live

in habitats where they are exposed to bright visible light is well established (Mathews, 1963; Mathews & Sistrom, 1959). It is interesting to
note that sensitivity to ultraviolet and visible light is more likely to occur
in micro-organisms than in macro-organisms, since the latter have only
a small fraction of their cells exposed to the outside world.

FINAL COMMENTS

The study of the adaptations of micro-organisms which make it possible
for them to live in extreme environments provides us with some of the
clearest insights into the phenomena of microbial evolution. This is
because we can concentrate on the single extreme factor to the virtual
exclusion of all others. Organisms living in such environments reveal for
us the extremes to which evolution can be pushed. We have seen that in
general procaryotic cells are able to adapt more readily to extremes of
temperature and Na^+ concentration than eucaryotes, whereas as far as
extreme H^+ concentration is concerned, three eucaryotes, an alga and
two fungi, are most successful. The limited species diversity in extreme
environments suggests that many organisms are unable to evolve
resistant forms. On the other hand (and I am trying not to reason in a
circular fashion), it is the limited species diversity in a particular environment which tells us that it is extreme. In an illuminating article, Vallentyne (1963) has argued that the limited species diversity in a habitat does
not necessarily indicate that it is extreme. 'One can instructively reverse
the point of view that has been taken here and ask why it is that most
organisms live under "common" conditions. The answer is, of course,
because life as a whole is selectively adapted to growth in common
environments. If the waters of the earth were predominantly acid,
growth at neutral pH values would be regarded as an oddity. Thus, the
fact that most living species conform physiologically and ecologically to
average earth conditions should not be taken to indicate any inherent
environmentally based physicochemical conservatism of living matter.
Adaptation has taken place' (Vallentyne, 1963).

Although I appreciate Vallentyne's argument fully, I do not agree
with it. I believe that even if acid pH values were common, there would
still be a restricted species composition due to the potential effects of the
H^+ ion on the stability of macromolecules and cellular structures. It is
a fact that DNA, ATP, and a variety of other essential cell constituents
are acid labile, and no amount of biological evolution can alter the
inherent properties of such biochemical compounds. Unless an organism
were to evolve with an entirely new ground plan in which these acid

labile compounds could be dispensed with, it seems unlikely that acid environments will ever be other than biological oddities. The same can be said for high-temperature environments. Since thermal environments, in particular, have probably existed as long as the earth has, there has been plenty of time for the evolution of organisms adapted to these environments. The fact that we do not find photosynthetic procaryotes living at temperatures above 73–75°C. suggests that there is some inherent physiocochemical limitation, perhaps of the photosynthetic apparatus, which it is impossible for these organisms to overcome by further evolutionary changes.

We are just at the beginning of an exciting era in the study of evolution. Molecular biology, microbial ecology, and paleo-microbiology are converging to provide us with new insights into the origin, evolution and nature of life itself. Micro-organisms, with their great diversity, will prove the most useful objects of evolutionary study, and those living in extreme environments will be especially relevant.

REFERENCES

ABRAM, D. (1965). Electron microscope observations on intact cells, protoplasts, and the cytoplasmic membrane of *Bacillus stearothermophilus*. *J. Bact.* **89**, 855.

ADAMS, T. C. (1964). Salt migration to the northwest body of Great Salt Lake. *Science*, **143**, 1027.

ALLEN, E. T. & DAY, A. L. (1935). Hot springs of the Yellowstone National Park. *Publs Carnegie Instn*, no. 466.

ALLEN, M. B. (1959). Studies with *Cyanidium caldarium*, an anomalously pigmented chlorophyte. *Arch. Mikrobiol.* **32**, 270.

ANDERSON, G. C. (1958). Some limnological features of a shallow saline meromictic lake. *Limnol. Oceanogr.* **3**, 259.

APINIS, A. E. (1965). A new thermophilous *Coprinus* species from coastal grasslands. *Trans. Br. mycol. Soc.* **48**, 653.

APINIS, A. E. & PUGH, G. J. F. (1967). Thermophilous fungi of birds' nests. *Mycopath. Mycol. appl.* **33**, 1.

ASCIONE, R., SOUTHWICK, W. & FRESCO, J. R. (1966). Laboratory culturing of a thermophilic alga at high temperature. *Science*, **153**, 752.

BARTH, T. F. W. (1950). Volcanic geology, hot springs, and geysers of Iceland. *Publs Carnegie Instn*, no. 587.

BATTLEY, E. H. & BARTLETT, E. J. (1966). A convenient pH-gradient method for the determination of the maximum and minimum pH for microbial growth. *Antonie van v. Leeuwenhoek*, **32**, 245.

BAXTER, R. M. (1959). An interpretation of the effects of salts on the lactic dehydrogenase of *Halobacterium salinarium*. *Can. J. Microbiol.* **5**, 47.

BOOTH, G. H. & MERCER, S. J. (1963). Resistance to copper of some oxidizing and reducing bacteria. *Nature, Lond.* **199**, 622.

BORICHEWSKI, R. M. & UMBREIT, W. W. (1966). Growth of *Thiobacillus thiooxidans* on glucose. *Archs Biochem. Biophys.* **116**, 97.

BOYKA, H. (1966). *Salinity and Aridity*, p. 16. The Hague: Dr W. Junk Publications.

BROCK, T. D. (1967a). Life at high temperatures. *Science*, **158**, 1012.

BROCK, T. D. (1967b). Microorganisms adapted to high temperatures. *Nature, Lond*. **214**, 882.

BROCK, T. D. (1968). Taxonomic confusion concerning certain filamentous blue-green algae. *J. Phycol.* **4**, 178.

BROCK, T. D. Vertical zonation in hot spring algal mats. *Phycologia* (in the Press).

BROCK, T. D. & BROCK, M. L. (1968). The hot springs of the Furnas Valley, Azores. *Int. Revue ges. Hydrobiol. Hydrogr.* **52**, 545.

BRUES, C. T. (1932). Further studies on the fauna of North American hot springs. *Proc. Am. Acad. Arts Sci.* **67**, 185.

BUCHANAN, R. E. & FULMER, E. I. (1930). *Physiology and Biochemistry of Bacteria*, vol. II. Baltimore, Maryland: Williams and Wilkins.

CAMERON, E. S. & ESTY, J. R. (1926). The examination of spoiled canned foods. 2. Classification of flat sour, spoilage organisms from nonacid foods. *J. Infect. Dis.* **39**, 89.

CAMPBELL, L. L. & POSTGATE, J. R. (1965). Classification of the spore-forming sulfate-reducing bacteria. *Bact. Rev.* **29**, 359.

COLE, G. A. & BROWN, R. J. (1967). The chemistry of *Artemia* habitats. *Ecology*, **48**, 858.

COPELAND, B. J. (1967). Environmental characteristics of hypersaline lagoons. *Contr. Marine Science (Texas)*, **4**, 207.

COPELAND, J. J. (1936). Yellowstone thermal myxophyceae. *Ann. N.Y. Acad. Sci.* **36**, 1.

CRAVERI, R., CRAVERI, A. & GUICCIARDI, A. (1967). Ricerche sulle proprietà ed attività di eumiceti termofiili isolati dal terreno. *Annali Microbiol.* **17**, 1.

CRAVERI, R. & MANACHINI, P. L. (1966). Contenuto GC del DNA di *Streptomyces argenteolus* e di *Thermoactinomyces vulgaris* coltivati a differenti temperature. *Annali Microbiol.* **16**, 1.

DIN, G. A., SUZUKI, I. & LEES, H. (1967). Ferrous iron oxidation by *Ferrobacillus ferrooxidans*. *Can. J. Biochem.* **45**, 1523.

DOEMEL, W. D. & BROCK, T. D. (1968). The temperature range of *Cyanidium caldarium* in nature. *Bact. Proc.* p. 45.

DOETSCH, R. N., COOK, T. M. & VAITUZIS, Z. (1967). On the uniqueness of the flagellum of *Thiobacillus thiooxidans*. *Antonie van Leeuwenhoek*, **33**, 196.

ELAZARI-VOLCANI, B. (1940). *Studies on the microflora on the Dead Sea*. Ph.D. Thesis, University of Jerusalem.

ENGLISH, C. F., BELL, E. J. & BERGER, A. J. (1967). Isolation of thermophiles from broadleaf tobacco and effect of pure culture inoculation on cigar and aroma mildness. *Appl. Microbiol.* **15**, 117.

FAMIN, M. A. (1933). Action de la température sur les végétaux. Quatrième partie. *Rev. gen. Bot.* **45**, 574.

FESTENSTEIN, G. N., LACEY, J. & SKINNER, F. A. (1965). Self-heating of hay and grain in Dewar flasks and the development of farmer's lung antigens. *J. gen. Microbiol.* **41**, 389.

FINDENEGG, I. (1966). Die Bedeutung kurz-welliger Strahlung für die planktische Primärproduktion in den Seen. *Verh. Internat. Verein. Limnol.* **16**, 314.

FLOWERS, S. & EVANS, F. R. (1966). The flora and fauna of the Great Salt Lake Region, Utah. In *Salinity and Aridity*, p. 367. Ed. H. Boyka.

FREEZE, H. & BROCK, T. D. (1968). A nonsporulating extreme thermophile. *Bact. Proc.* p. 45.

FRIEDMAN, S. M. (1968). Protein-synthesizing machinery of thermophilic bacteria. *Bact. Rev.* **32**, 27.

FUKUDA, I. (1958). Physiological studies on a thermophilic blue-green alga, *Cyanidium caldarium*. *Bot. Mag., Tokyo*, **71**, 79.

GESSNER, F. (1959). *Hydrobotanik*, vol. II. Berlin: Deutscher Verlag der Wissenschaften.

GOLOVACHEVA, R. S. (1967). Spontaneous generation of spherical forms in obligate thermophilic bacilli. *Microbiology (U.S.S.R.)*, **36**, 560.

HALLDAL, P. (1967). Ultraviolet action spectra in algology. A review. *Photochem. Photobiol.* **6**, 455.

HAMILTON, R. D., MORGAN, K. M. & STRICKLAND, J. D. H. (1966). The glucose uptake kinetics of some marine bacteria. *Can. J. Microbiol.* **12**, 995.

HARP, G. L. & CAMPBELL, R. S. (1967). The distribution of *Tendipes plumosus* (Linné) in mineral acid water. *Limnol. Oceanogr.* **12**, 260.

HEDGEPETH, J. W. (1959). Some preliminary considerations of the biology of inland mineral waters. *Arch. Oceanogr. Limnol. Roma* (supp.), **11**, 111.

HEILBRUNN, L. V. (1952). *An Outline of General Physiology*, 3rd ed. Philadelphia: Saunders.

HEWITT, L. F. (1957). Influence of hydrogen-ion concentration and oxidation-reduction conditions on bacterial behaviour. *Symp. Soc. gen. Microbiol.* **7**, 42.

HIROSE, H. (1950). Studies on a thermal alga, *Cyanidium caldarium*. *Bot. Mag., Tokyo*, **63**, 107.

HOLMES, P. K. & HALVORSON, H. O. (1965*a*). Purification of a salt-requiring enzyme from an obligately halophilic bacterium. *J. Bact.* **90**, 312.

HOLMES, P. K. & HALVORSON, H. O. (1965*b*). Properties of a purified halophilic malic dehydrogenase. *J. Bact.* **90**, 316.

INGRAM, M. (1957). Micro-organisms resisting high concentrations of sugars or salts. *Symp. Soc. gen. Microbiol.* **7**, 90.

JANNASCH, H. W. (1968). Growth characteristics of heterotrophic bacteria in sea-water. *J. Bact.* **95**, 722.

JOHNSON, F. H. (1957). The action of pressure and temperature. *Symp. Soc. gen. Microbiol.* **7**, 134.

JOHNSON, M. K., JOHNSON, E. J., MacELROY, R. D., SPEER, H. L. & BRUFF, B. S. (1968). Effect of salts on the halophilic alga *Dunaliella viridis*. *J. Bact.* **95**, 1461.

JONES, E. B. G. & JENNINGS, D. H. (1964). The effect of salinity on the growth of marine fungi in comparison with non-marine species. *Trans. Br. mycol. Soc.* **47**, 619.

KEMPNER, E. S. (1966). Acid production by *Thiobacillus thiooxidans. J. Bact.* **92**, 1842.

KLAUSHOFER, H. (1967). Die moderne mikrobiologische Kontrolle in der Zuckerfabrikation. *Zentbl. Bakt. ParasitKde* (I. Abt., Suppl. 2), 185.

KLEIN, R. M. & CRONQUIST, A. (1967). A consideration of the evolutionary and taxonomic significance of some biochemical, micromorphological and physiological characters in the thallophytes. *Q. Rev. Biol.* **42**, 105.

KLUG, M. J. & MARKOVETZ, A. J. (1967). Thermophilic bacterium isolated on n-tetradecane. *Nature, Lond.* **215**, 1082.

LACKEY, J. B. (1938). The flora and fauna of surface waters polluted by acid mine drainage. *Publ. Hlth Rep., Wash.* **53**, 1499.

LACKEY, J. B., LACKEY, E. W. & MORGAN, G. B. (1965). Taxonomy and ecology of the sulfur bacteria. *Bull. Florida Engineering and Industrial Experiment Station, Gainesville, Florida*, Ser. no. 119.

LARSEN, H. (1967). Biochemical aspects of extreme halophilism. *Adv. Microbial physiol.* **1**, 97.

LILLY, V. G. & BARNETT, H. L. (1951). *Physiology of the Fungi*. New York: McGraw-Hill.

LIVINGSTONE, D. A. (1963). Chemical composition of rivers and lakes. *U.S. Geol. Survey Prof. Paper*, no. 440–6.

LOGINOVA, L. G., GOLOVACHEVA, R. S. & SHCHERBAKOV, M. A. (1966). *Microbiology*, **35**, 675.

MACLEOD, R. A. (1965). The question of the existence of specific marine bacteria. *Bact. Rev.* **29**, 9.

MANACHINI, P. L., FERRARI, A. & CRAVERI, R. (1965). Forme termofile di Actinoplanaceae. Isolamento e caratteristiche di *Streptosporangium album* var. *thermophilum. Annali Microbiol.* **15**, 129.

MARSH, C. L. & LARSEN, D. H. (1953). Characterization of some thermophilic bacteria from the hot springs of Yellowstone National Park. *J. Bact.* **65**, 193.

MATELES, R. I., BARUAH, J. N. & TANNENBAUM, S. R. (1967). Growth of a thermophilic bacterium on hydrocarbons: a new source of single-cell protein. *Science*, **157**, 1322.

MATHEWS, M. M. (1963). Studies on the localization, function, and formation of the carotenoid pigments of a strain of *Mycobacterium marinum. Photochem. Photobiol.* **2**, 1.

MATHEWS, M. M. & SISTROM, W. R. (1959). Function of carotenoid pigments in non-photosynthetic bacteria. *Nature, Lond.* **184**, 1892.

MCGEE, J. M., BROWN, L. R. & TISCHER, R. G. (1967). A high-temperature, hydrogen-oxidizing bacterium—*Hydrogenomonas thermophilus*, n.sp. *Nature, Lond.* **214**, 715.

MIEHE, H. (1907). *Die Selbsterhitzung des Heus.* Jena: Gustav Fischer.

MIHURSKY, J. A. (1967). On possible constructive uses of thermal additions to estuaries. *BioScience*, **17**, 698.

MIQUEL, P. (1879). *Bulletin de la Statistique municipale de la ville de Paris.*

MONOD, J. (1942). *Recherches sur la croissance des cultures bactériennes.* Paris: Librairie Scientifique, Hermann.

MORITA, R. Y. (1967). Effects of hydrostatic pressure on marine microorganisms. *Oceanogr. Mar. Biol. Ann. Rev.* **5**, 187.

PORTER, J. R. (1946). *Bacterial Chemistry and Physiology.* New York: John Wiley.

RAYMOND, J. C. & SISTROM, W. R. (1967). The isolation and preliminary characterization of a halophilic photosynthetic bacterium. *Arch. Microbiol.* **59**, 255.

RUTTER, W. T. (1964). Evolution of aldolase. *Fedn Proc.* **23**, 1248.

SCHRAMM, J. R. (1966). Plant colonization studies on black wastes from anthracite mining in Pennsylvania. *Trans. Am. phil. Soc.* **56**, part 1.

SCHWARTZ, A. & SCHWARTZ, W. (1965). Über das Vorkommen von Mikroorganismen in Solfataren und heissen Quellen. *Z. allg. Microbiol.* **5**, 395.

SHIFRINE, M. & PHAFF, H. J. (1958). On the isolation, ecology and taxonomy of *Saccharomycopsis guttulata. Antonie van Leeuwenhoek*, **24**, 193.

SIEGENTHALER, P. A., BELSKY, M. M. & GOLDSTEIN, S. (1967). Phosphate uptake in an obligately marine fungus: a specific requirement for sodium. *Science*, **155**, 93.

SMITH, W. W. & ZOBELL, C. E. (1937). Direct microscopic evidence of an autochthonous bacterial flora in Great Salt Lake. *Ecology*, **18**, 453.

STARKEY, R. L. & WAKSMAN, S. A. (1943). Fungi tolerant to extreme acidity and high concentrations of copper sulfate. *J. Bact.* **45**, 509.

STOECKENIUS, W. & ROWEN, R. (1967). A morphological study of *Halobacterium halobium* and its lysis in media of low salt concentration. *J. cell Biol.* **34**, 365.

THOMPSON, P. J. & THOMPSON, T. L. (1962). Some characteristics of a purified heat-stable aldolase. *J. Bact.* **84**, 694.

THOMPSON, T. L., MILITZER, W. E. & GEORGI, C. E. (1958). Partial denaturation of a bacterial aldolase without loss of activity. *J. Bact.* **76**, 337.

PLATE 1

Fig. 1

Fig. 2

(*Facing page* 40)

PLATE 2

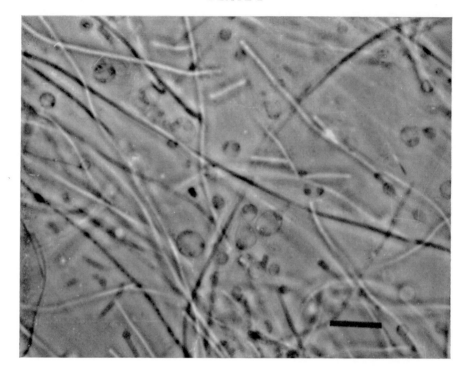

Fig. 1

UCHINO, F. & DOI, S. (1967). Acido-thermophilic bacteria from thermal waters. *Agric. Biol. Chem.* **31**, 817.

UMBREIT, W. W. (1962). The comparative physiology of autotrophic bacteria. *Bact. Rev.* **26**, 145.

UZUMASA, Y. (1965). *Chemical Investigations of Hot Springs in Japan.* Tokyo: Tsukiji Shokan.

VALLENTYNE, J. R. (1963). Environmental biophysics and microbial ubiquity. *Ann. N.Y. Acad. Sci.* **108**, 342.

VAN UDEN, N. (1967). Transport-limited growth in the chemostat and its competitive inhibition; a theoretical treatment. *Arch. Mikrobiol.* **58**, 145.

VISHNIAC, W. & SANTER, M. (1957). The thiobacilli. *Bact. Rev.* **21**, 195.

WAKSMAN, S. A. & JOFFE, J. S. (1922). Microorganisms concerned in the oxidation of sulfur in the soil. II. *Thiobacillus thiooxidans*, a new sulfur-oxidizing organism isolated from the soil. *J. Bact.* **7**, 239.

WALLACE, H. A. H. & SINHA, R. N. (1962). Fungi associated with hot spots in farm stored grain. *Can. J. Plant Sci.* **42**, 130.

WARING, G. A. (1965). Thermal springs of the United States and other countries of the world—a summary. *U.S. Geol. Survey Prof. Paper*, no. 492.

WARINNER, J. E. & BREHMER, M. L. (1966). The effects of thermal effluents on marine organisms. *Air and Water Pollut. Int. J.* **10**, 277.

YONEDA, Y. (1952). A general consideration of the thermal Cyanophyceae of Japan. *Mem. College Agric. Kyoto Univ. Fisheries Ser.* no. **62**.

ZOBELL, C. E. (1964). Hydrostatic pressure as a factor affecting the activities of marine microbes. *Recent Researches in the Fields of Hydrosphere, Atmosphere and Nuclear Geochemistry (Tokyo)*, p. 83.

EXPLANATION OF PLATES

PLATE 1

Fig. 1. Photograph of pool A, a small hot spring in the White Creek valley of Yellowstone National Park. The temperature of the source is 91·5 °C, and the temperature in the outflow channel near the stake varies from 80–88 °C.

Fig. 2. Photograph of the effluent channel of pool A near the source (see Fig. 1), showing the large filamentous masses of bacteria growing attached in the rapidly flowing water. The colour of the bacterial masses is pink.

PLATE 2

Fig. 1. Phase photomicrograph of bacteria which had grown attached to a microscope slide immersed in the effluent channel of pool A for 48 hr. Note the large spherical bodies. The length of the marker is 10 μ.

GROWTH OF MIXED CULTURES AND THEIR BIOLOGICAL CONTROL

P. N. HOBSON

The Rowett Research Institute, Bucksburn, Aberdeen, Scotland

INTRODUCTION

All populations of organisms, whether macro or micro, have a limit to their development. In a closed system there must obviously be a limit to the population, and this may be set by the food supply, but in open systems where food supplies are adequate there seem to be other factors which limit the population, and the limiting factor is often not the space available. Once at this limit the population then stabilizes and such a population of one species will often exclude foreign organisms. In populations formed initially of mixed species either one species will increase and finally exclude the others, or there will develop a stable, mixed population which may contain all, or only a proportion of, the original species present. The study of the factors affecting the macro populations has long been the work of the ecologist, and this work is becoming increasingly important with the increasing world population and decreasing (in some cases) populations of animals and plants.

Study of the factors affecting micro populations may provide information of importance not only to those studying the causes and cure of diseases but also to the industrial microbiologist who wishes to use a stable culture. While much attention has been paid to the control of micro-organisms by external factors, usually with the aim of complete elimination of the organisms, not so much attention has been paid to the effects of one micro-organism on another, although it is known that these may be important. In the present paper some aspects of the control of one micro-organism by another are discussed. An exhaustive review of the literature would be long and so references are given only to selected examples. More workers than those cited here have contributed to the literature examples of microbial control and there may be omissions, but it is hoped that the paper will give at least an indication of the many ways in which one organism can control the growth of another.

In the laboratory the microbiologist generally deals with a 'pure culture' of micro-organisms, but it has often been pointed out that a pure culture is an anachronism; it is almost always a laboratory-made artifact from which we hope to deduce properties of the micro-organisms in their original habitat. For how often do micro-organisms occur naturally in 'pure culture'? Experience tells us that probably only in extreme environments (see Brock, this Symposium) do single species occur. We may isolate only one species of organism from a habitat, but this may be due to limitations set by our isolation procedures; either we use media which will only support the growth of a limited number of organisms, and this may be due to our ignorance of the growth requirements of the other organisms present as well as to design, or we ignore organisms that are present only in small numbers. However, even supposing our media could support the growth of all the species present in an environment we may still be selecting variants of the species; those which will grow best under our artificial conditions, and on this basis our natural pure culture may be a mixture of these and other varieties of the same species. In the few cases where we may have a naturally occurring pure culture at any one time this will be most likely only a transient state, as most, if not all, natural cultures are subject to contamination at all times and may often have started off as mixed cultures. However, although there are a vast number of types of micro-organisms ready to settle in any natural environment, any environment usually contains only a few of these types, the others are excluded. The environmental conditions such as presence or absence of air, and utilizable energy and nitrogen sources, obviously exclude many organisms, but even so the numbers of types of organisms theoretically capable of existing in a particular environment are almost always larger than the numbers actually present, and one of the reasons for this is that the organisms in a mixture control each other's growth.

The control of growth of one micro-organism by another is a subject which has had comparatively little laboratory investigation up till now, although it may come into routine procedures such as testing for antibiotic production, so much of this paper will be based on investigation of naturally occurring mixed cultures. Although we shall be concerned mainly with the control of one organism by another, the environment will have to be considered, as the properties of the environment modify the interactions of the micro-organisms.

OPEN AND CLOSED SYSTEMS

There are two types of natural environment to consider, the 'closed' and the 'open'. (I am using throughout the word 'natural' very broadly to include anything other than laboratory cultures.) By 'closed' environment I mean one in which there is only a limited time of growth for the organisms and in which the microbial culture will eventually die off. In a closed environment there will eventually come a time when all nutrients are used up and this will terminate the microbial growth, but many natural cultures are terminated long before this happens and it is these cases which I wish to discuss. An infected animal body thus becomes a closed environment as the culture of organisms may kill off the body and so terminate the growth of the culture, although individual organisms may transfer to some other body. So we have the first of our control systems in which the organisms destroy their habitat through their own metabolism and thus control themselves even though they may not have utilized all the nutrients available to them. If the animal body has been destroyed by toxin production the organisms then have indirectly limited their growth by reducing the temperature and aerobiosis of their habitat on the death of the host. Another simple case which comes to mind is the old one of the ignition of hay stacks. Here the heat of metabolism of the organisms is such that they destroy their environment by its combustion. So we have examples of how the environment must be considered in any discussion of self-regulation of micro-organisms. The environment, especially in a closed system, has only a limited ability to absorb or dissipate the products of microbial metabolism. This may eventually result in death of the entire culture, but it may also result in a succession of predominant microbial types before the final breakdown. This 'final breakdown' is often, of course, only final so far as one succession of organisms is concerned. The death of the host animal which terminates the growth of the pathogenic organisms converts the body into a habitat suitable for another succession of putrefactive organisms.

In an 'open' environment, on the other hand, we have a culture of organisms in which the time available for growth is not limited by the environmental conditions, and many natural cultures are of this type. They are types of continuous culture. They may correspond to the laboratory-type continuous stirred fermentor in that we may have, as in the rumen, a system in which the culture medium flows into the fermentation chamber, where mixing takes place and the microbial cells and end products flow out. This may have feed-back as in an activated sludge plant. Or, the culture may, as in the intestines or a flowing stream, correspond more to a tubular fermentor. Or, again, we may have fixed organisms with a medium flowing over them. However, no natural continuous culture corresponds exactly to the simple laboratory type, they are all more complex in one way or another.

There is yet another type of natural 'unlimited-time' mixed culture and that is the one in which two or more organisms have developed in such a way as to form a whole integrated structure. There are a number of examples of this such as lichens and plant nodules and in some cases the combination of organisms has become so interdependent that there may be doubts as to whether they are two or one organism. Cases like that of the bacteroid body in the protozoon *Chrithidia oncopelti* come to mind. However, in this paper I do not wish to consider this latter type of symbiosis, but only populations of free-living organisms which inhabit the same environment.

CONTROL OF MIXED POPULATIONS

In both the 'open' and 'closed' environments the controlling factors will be similar. On a broad basis this control may be 'positive' or 'negative'. In positive control one organism makes the conditions suitable for another by regulating the environment to a suitable Eh, pH, temperature or level of nutrients, or by providing growth factors, energy sources, nitrogen, or combinations of these. In negative control one organism may so alter the environment that conditions become unsuitable for growth of the other. This type of control may operate so as to exclude contamination of a natural culture and thus stabilize it, or it may produce large oscillations, or a complete breakdown, in the natural culture. In both cases the controlling factors involved are similar, but in negative control the environment is more directly involved as it is usually a change in nutrient intake into a continuous culture and the limited ability of the environment to dissipate metabolic products that leads to a breakdown of the culture. These latter factors are some of the difficulties involved in investigating natural continuous cultures. The conditions are seldom as stable as in laboratory cultures and although the change in environmental conditions may be obvious (e.g. overfeeding of an animal or storm loading of a sewage plant) some changes in microbial population may be due to unnoticed changes in environment and not primarily to interactions of the organisms. Furthermore, although a natural system may have no obvious feedback of organisms there may be more contaminating organisms than one suspects and continued inoculation may be what is keeping one species of the population growing, although the effects of the rest of the population may tend to suppress its growth. As will be mentioned later, few natural continuous cultures attain a steady state and there may be short periods of growth and decline of organisms all within an apparently long turnover time. Sometimes very detailed investigation of the culture is needed to show this.

Although there are known ways in which intermicrobial control is exerted there are many examples where the control mechanism is unknown. An example of this is the exclusion of *Escherichia coli* from the rumen. Theoretically *E. coli* should grow well in the rumen, and it must be continuously inoculated into the rumen, but the highest counts of this organism in the rumen fluid are only a few thousand per ml. compared with total bacterial counts around 10^{10}/ml. Hollowell & Wolin (1965) made a study of factors affecting growth of *E. coli* in the rumen using batch and continuous culture techniques. Growth was not inhibited by the microbially produced factors Eh, pH, CO_2, volatile fatty acids and bacteriophage nor by the temperature; but an unknown inhibitor for *E. coli* growth was found in the fluid. The presence of this inhibitor made it difficult to show whether lack of suitable nutrients was responsible for lack of growth of *E. coli* in the rumen, but it was thought that this was unlikely. However, this shows that microbial control mechanisms are not easy to establish and may well be complex.

Control by Eh

In any mixed microbial culture one might suggest that the major control system to operate is that which leads to the culture becoming aerobic or anaerobic and in this the environment must be considered. If the environment is such that diffusion of oxygen is unimpeded, say in a turbulent stream, then growth of aerobic organisms will not lower the oxygen tension sufficiently and anaerobic organisms will be excluded. However, if entry of air is limited then we shall obtain finally either a completely anaerobic population or a population with mixed aerobic and anaerobic metabolism. This control of Eh is, of course, used in the laboratory to obtain anaerobic conditions, but a few natural examples will be given. In silage-making, aerobic organisms on the original plants greatly outnumber facultative and anaerobic bacteria. When the herbage is piled for silage, conditions rapidly become anaerobic and growth of aerobic bacteria is suppressed. The anaerobic conditions could be brought about by aerobic bacteria, but, according to Whittenbury, McDonald & Bryan-Jones (1967), they result

from respiration of the herbage and metabolism of aerobic bacteria contributes little to the process. Another example of microbial lowering of Eh
is found in emulsion oils for metal cutting, although here again a number
of factors are involved and the promotion of growth of sulphate-reducing
bacteria by aerobic bacteria may involve effects other than reduction of
Eh (Isenberg & Bennett, 1959).

Probably one of the most extensively studied mixed microbial
populations is that of the rumen and this system will be mentioned a
number of times in this Chapter. The principal rumen bacteria are
strict anaerobes which are unable to reduce a medium to an Eh suitable
for their growth (Hobson & Summers, 1967). However the rumen
culture remains at a low Eh and the oxygen partial pressure in the gas
phase is low, in spite of the fact that air must enter the rumen with the
animals' feed, or, under experimental conditions, in spite of bubbling
air through the rumen (Broberg, 1957) or exposing rumen contents *in
vitro* to air. Oxygen is removed by microbial metabolism, almost
certainly that of the facultative anaerobes such as *Streptococcus bovis*,
and possibly of anaerobes such as *Veillonella alcalescens* which can take
up oxygen under experimental conditions (Hobson & Summers, 1967;
Rogosa, 1964) and which are present in numbers around 10^5/ml. Oxygen
may even be removed by aerobic bacteria which are present in a few
hundreds or thousands/ml. as compared with the strict anaerobes which
are present in numbers around 10^9/ml. These bacteria then form a stable
mixed population in the continuous culture system of the rumen.

The control of Eh may work the other way and Schaede (1962)
described associations of colourless and green bacteria in which the
latter provided oxygen for the former. A similar relationship exists in
sewage lagoons where oxygen is supplied to aerobic organisms both
by diffusion of air and growth of algae.

Control by temperature

Temperature of a microbial habitat is usually primarily regulated by
the environment, but unless the heat of metabolism of the micro-
organisms is dissipated, conditions can become warmer and a succession
of more thermophilic organisms be obtained (see Forrest and Brock,
this Symposium). In the rumen system, heat of fermentation helps the
animal's regulatory mechanism to maintain the rumen microbes at
their optimum temperature. The protozoa especially are rather temperature sensitive, and a fall in rumen temperature of a few degrees
caused by ingestion of cold drinking water can cause a diminution in
activity of the micro-organisms.

Control by pH

Micro-organisms can grow only over certain pH ranges, and for some organisms the pH range may be quite limited (see Brock, this Symposium). During growth micro-organisms may produce acids, which can lower the pH of their 'medium', or bases, such as ammonia, which can increase the pH, and these pH changes will affect their growth. Here again the environment plays a part as the magnitude of the pH changes produced by microbial growth will depend on the ability of the environment to counteract these changes, either by the buffering power or rate of through-flow of medium or the absorption of acids or bases by the 'container'. There are many examples of pH changes controlling the growth of micro-organisms. In a continuous culture of a single bacterial type the cell numbers may be made to vary cyclically at one flow rate because of pH changes caused by the fermentation of the cells (Finn & Wilson, 1954; Hobson, 1965). The lowering of pH causes a fall in metabolic rate of the bacteria allowing washout both of the bacteria and of the acid fermentation products. The pH then increases until rapid metabolism of the cells again takes place. If a similar pH change is brought about by one organism in a mixture, then the growth of another may be affected, thus probably accounting, in part, for cyclic variations in viable rumen bacterial numbers (Hobson, 1965).

Although cyclic variations in microbial numbers in a continuous-type culture can be obtained with a steady feed of nutrients, fermentation-induced pH changes, which lead to gross changes in dominant types of organisms in a mixture, are generally brought about by changes in the concentration or rate of flow of nutrients into the system. In the rumen system, decreases in the average pH are caused *in vivo* by changes in the diet of the ruminant towards a higher starch content. This provides more fermentable sugar for the micro-organisms. The concentration of fermentation acids then overcomes the buffering action of the ruminant's saliva and the absorptive capacity of the rumen wall. Work in our own laboratories has shown that at the low rumen pH values of high-starch diets, growth of many of the normal rumen bacteria is inhibited and a population in which *Bacteroides* spp. and other relatively aciduric bacteria become prominent is obtained. Slyter, Bryant & Wolin (1966) found that, in a continuous-flow artificial rumen system, when the 'feed' nutrients were kept of constant composition but the pH of the system was lowered by alterations in the pH of the inflowing buffer solution, the flora changed to one in which *Bacteroides* spp., lacto-

4

bacilli and unidentified Gram-positive rods were predominant. At higher pH values these were present only as minor constituents of the population. This suggests that pH alone is a major controlling factor, but as Jowitt (1968) pointed out, altering the pH in complex buffer systems may alter a number of ionic concentrations and these may affect microbial growth.

In vivo a very low pH in the rumen can result in streptococci and then lactobacilli becoming the dominant flora. Krogh (1959) investigated changes in the rumen flora of animals fed increasing quantities of carbohydrates and his results from increased feeding of sucrose are typical. As the sucrose feed increased the pH remained about 6·8 and the flora was normal until a certain amount of sucrose was fed when over about 2–3 days the streptococcal count rose from $2·5 \times 10^6$/ml. to $3·8 \times 10^7$/ml. and then fell to 1×10^2/ml. and the rumen pH fell to 5·3 and then 4·0. As the rumen pH fell from 5·3 to 4·0 and streptococci decreased the lactobacillus count rose from *ca.* 0 to 7×10^7/ml. Finally as the pH remained at 3·9–4·0 the lactobacillus numbers began to fall and yeasts increased from *ca.* 0 to $1·7 \times 10^4$/ml. At this stage the animal died. In our own laboratories (S. O. Mann unpublished) we found that in the rumen of a heifer overfed on barley the viable counts of lactobacilli increased from $1·21 \times 10^4$/ml. to $1·96 \times 10^9$/ml. over 24 hr. as the rumen pH dropped from 5·7 to 4·5. The numbers of lactobacilli increased from controlled growth at about 0·001 % of the population to about 60 % of the viable population. At the same time rumen lactic acid concentration rose from the usual rumen level of almost undetectable to 951 mg./100 ml. In these experiments no increase in streptococcal numbers was observed, but any increase in the first few hours after feeding could have been missed as samples were not taken at this time. In this and in Krogh's experiment, growth of streptococci (which also produce lactic acid) could have been controlled by lactate concentration as well as, or instead of, rumen pH.

The 'steady-state' numbers of different organisms in a mixture may be affected by the changes in pH of the culture. For instance, Purser & Moir (1959) found that the number of protozoa in the sheep rumen was related to the minimum pH of the rumen itself controlled by bacterial fermentation. Similarly, in our own laboratories we have found that protozoa cannot be established in the rumens of animals fed *ad lib.* on barley rations where there is rapid bacterial fermentation and the pH is for comparatively long periods as low as 5 (Eadie, Hobson & Mann, 1967). When the starchy ration is fed at a more controlled level the pH remains generally higher, because the rate of bacterial fermentation is reduced, and protozoal populations can be established.

The effects of lactobacilli in lowering the pH can be seen in a number of systems. Lactic acid production by lactobacilli in the intestines may be important in controlling the growth of *Escherichia coli* and production of diarrhoea in young mammals. In the batch culture conditions of silage making, acid conditions are needed in order to control the growth of putrefactive bacteria and the more rapid the fall in pH the better the silage in that clostridial growth is inhibited. Whittenbury, McDonald & Bryan-Jones (1967) found that, under laboratory conditions, growth of an inoculum of *Streptococcus faecalis* rapidly lowered the pH of grass to a level which favoured the growth of lactobacilli and these latter then further lowered the pH to about 4. Lactobacilli alone apparently did not grow fast enough to compete with other bacteria for the sugars available, and the resultant fall in pH was not so great or so rapid as with the mixed streptococcus-lactobacillus flora. Further fermentation of silage lactic acid to butyric acid may raise the pH, producing favourable growth conditions for the putrefactive clostridia.

This brings us from 'negative' to 'positive' control when one organism produces conditions suitable for the growth of another. In the normal rumen lactic acid is produced in the primary fermentation of carbohydrates and is converted in a secondary fermentation to acetic and propionic acids by bacteria such as *Veillonella alcalescens*, *Selenomonas ruminantium* and *Peptostreptococcus elsdenii*. When lactic acid production becomes uncontrolled it is presumably being produced faster than it can be metabolized by the number of these bacteria present and their growth must be limited by factors other than lactic acid concentration. But this may not be the only reason for the accumulation of lactic acid. It can be produced in both D- and L-forms, but some of the lactic acid-fermenting bacteria are specific for a particular isomer. A change in type of lactobacillus may then cause a change to the production of an isomer of lactic acid which cannot be metabolized by the lactate-fermenters and will thus accumulate and lower the pH.

Control of microbial growth by pH is a complex process but is presumably bound up with the cells' ability to assimilate nutrients, and with the activity of autolytic enzymes and of fermentation systems which provide energy not only for growth but for cell maintenance.

Control by organic acids

Control of micro-organisms by acid production may not be a simple matter of the pH. The organic acids produced by fermentation, lactic, formic, acetic and propionic, have all been shown to have growth-inhibitory effects separate from pH changes.

The control of growth of bacteria by the presence of lactobacilli has been noted for a number of environments including the rumen and silage. But Kao & Frazier (1966) and Hattori, Misawa, Igarashi & Sugiya (1965), for instance, showed that lactic acid-forming bacteria could inhibit the growth of *Staphylococcus aureus* and *Vibrio comma* and Sabine (1963) showed that *Lactobacillus acidophilus* had an 'antibiotic effect' on *Escherichia coli*. Tramer (1966) followed up this work and showed that the 'antibiotic effect' was due to lactic acid which was more inhibitory than hydrochloric acid at the same pH (3·6). Furthermore, the lactic acid was more effective in inhibiting growth of *E. coli* in broth at pH 3·6 than at higher pH values. Hentges (1967 .a, b) showed that the growth of *Shigella flexneri* was inhibited by formic and acetic acids produced by *Klebsiella* and this inhibition of growth and finally death of the bacteria was greatest under anaerobic conditions. Again lowering the pH had a potentiating effect. Inhibition was greater at pH 6·0 than at 6·5 or 7·0 but the pH had to be as low as 5·5 to cause inhibition of growth in the absence of the organic acids. Ichikawa & Kitamoto (1967) found a similar pH dependence for the inhibition of *E. coli* by propionate. They also found propionate to be more inhibitory than acetate or butyrate and suggested that the inhibition was due to 'the antagonistic properties of the former (propionate) to the β-alanine and pyruvate metabolism'.

Meynell & Subbaiah (1963) showed that growth of *Salmonella typhimurium* was inhibited by the normal flora of the mouse gut. In further work Meynell (1963) showed that the fermentation products acetic, propionic and butyric acids in concentrations similar to those found in the gut inhibited *in vitro* growth of *Salmonella*, particularly under anaerobic conditions. He also pointed out that passage of digesta through the mouse gut is quite rapid, and this underlines the fact that, whatever the mechanism involved, even if only a slight decrease in growth rate of one organism caused by another can be demonstrated in batch cultures *in vitro*, this may be sufficient entirely to wash out the organism from a natural continuous culture such as the gut.

There is thus evidence that inhibition of growth of bacteria by organic acids is Eh- and pH-dependent. These acids are all products of bacterial fermentations, and product inhibition of growth, especially in continuous cultures, is well established in laboratory experiments. Is the type of inhibition of one bacterium by another cited above then a reflection of product inhibition? If this is so then inhibition by fermentation acids should be maximum in anaerobic growth. This possibility is difficult to substantiate at the moment as not enough is known about fermentation products of many of the bacteria shown to be inhibited, under the conditions pertaining at the time. Jerusalimsky (1967) showed that propionate added to the culture medium of *Propionibacterium shermanii* in increasing concentration progressively decreased the growth rate and the rate of formation of propionate, the principal fermentation product, but did not greatly affect the rate of acetate production. Thus the fermentation product inhibited its own formation. The potentiating effects of low pH may be partly a reflection of the degree of dissociation of the acids since undissociated molecules may be able to penetrate the cell to a greater extent than dissociated ones. However, other factors could be involved, in particular the fermentation products themselves may be changed by the pH. In *E. coli* Paege & Gibbs (1961) showed that at low pH the fermentation product was essentially lactic acid which inhibits its growth (Tramer, 1966). Similarly, Gunsalus & Niven (1942) showed that lactic acid formed a greater part of the fermentation products of *Streptococcus liquefaciens* at low pH (5) than at higher pH values. If this type of pH effect were general in systems where an organism is inhibited by a lactic-acid bacterium then product inhibition would always be greater at low pH. Essentially the only fermentation pathway of the organism would be inhibited and the low pH would also increase the amount of lactic acid produced by the lactic-acid bacterium. However, the examples discussed show that amongst bacteria producing lactic acid some are affected to a greater extent than others by the acid. This could be brought about in at least three ways: the controlling bacterium could be less permeable to lactic acid; or more able to excrete lactic acid against a concentration gradient; or it could be better able to move its fermentation pathways into other products. The reduction of pyruvate to lactate does not of itself produce ATP needed for growth, but it is a method of removing pyruvate and reduced nucleotides, the end-products of the initial stages of sugar fermentation during which ATP is generated. Inhibition of growth by lactate is, therefore, possibly caused by inhibition of lactate formation which leads to an accumulation of pyruvate and reduced nucleotides and thus cessation of the ATP-yielding reactions. If the pyruvate and reduced nucleotides can be dissimilated to other products, then fermentation and growth of the bacteria can continue. As in the case of lactic acid accumulation in the rumen, fermentation-product inhibition of microbial growth in a stable mixed culture can be overcome by a secondary fermentation. Inhibition by formic, acetic and propionic acids could be controlled by secondary fermentation to methane, in anaerobic sewage fermentation for example, or by uptake of acetic acid by butyric acid-forming bacteria and so on.

Control by antibiotics

Inhibition of growth of some organisms caused by antibiotic substances produced by another organism is well known, and these substances may be 'broad-spectrum' or of limited activity like the colicins and lactobacillins. This type of control mechanism is almost always 'negative', the effect being to suppress contamination of the antibiotic-producing culture rather than to lead to a stable mixed population. However, this type of culture might be stabilized by a third organism producing an enzyme like penicillinase which destroys the antibiotic. In many cases the antibiotic substance has been identified and studied. However, there are cases (e.g. Troller & Frazier, 1963; Di Giacinto & Frazier, 1966) where only partially identified antibiotic substances have been implicated. These authors showed that growth of some common food contaminants, including coliforms, might repress growth of the food-poisoning staphylococci. However, they found that competition for nutrients in batch cultures was also a factor in suppressing growth of staphylococci.

Control by nutrients

Negative control. At first sight competition for a limiting nutrient might be thought to be the most likely way in which control of one organism may be exerted on another. However, although this may occur to some extent in natural 'closed' cultures, such as the silage system mentioned earlier, in many cases it seems possible that factors other than nutrient availability may decide which organisms out of an initial population take the lead in growth. Once this growth is established then these organisms will remain dominant and consume a larger proportion of available nutrients even if the rate of uptake of nutrients per cell is similar in the different types present. The factors deciding whether an inoculum cell will grow or not are a complication in investigating intermicrobial control in continuous cultures. Powell (1958) pointed out that there is a finite probability that an organism contaminating a continuous culture will be washed out before it has a chance to divide. Thus it is possible that the lack of growth of an organism in a continuous-type natural culture is due to the fact that the inoculum organisms are unable to adjust themselves to the culture conditions before they are washed out, and not to any control mechanism exerted by the stable components of the system. However, in natural 'open' systems I would suggest that little, if any, control of the stable population is exerted by competition for essential nutrients. Powell (1958) deduced from the simple theories of continuous culture that in a culture of two organisms interacting only in competition for a growth-limiting nutrient, the one which grows at the lowest substrate concentration at the culture dilution rate will completely displace the other. The substrate may be a large molecule or some trace element or other minor constituent of the medium. For instance Tempest, Dicks & Meers (1967) showed that *Aerobacter aerogenes* rapidly outgrew *Bacillus subtilis* in magnesium-limited continuous culture at one dilution rate, owing to the greater ability of *A. aerogenes* to take up magnesium at low concentrations of these ions. In further experiments using *B. subtilis* and *Torula utilis* a differential effect of amino acids in the medium on Mg^{2+} uptake was found. These experiments again emphasize the difficulties of comparing laboratory results with natural mixed cultures where the precise composition of the 'medium' may be unknown. However, competition for nutrients may serve to control contamination of a natural culture or it may lead to its complete breakdown. It could not be the decisive factor in determining the population of the stable culture as it seems unlikely that the dilution rate imposed on a natural continuous culture would be the one at which the populations of organisms could, in theory, all exist in competition for nutrients.

The large lactate-fermenting coccus LC (*Peptostreptococcus elsdenii*) is found in large numbers (10^9/ml.) together with lactobacilli and streptococci in the rumens of young animals feeding on starchy concentrates, but almost disappears along with lactobacilli in the rumens of older animals fed on a less starchy diet (Hobson, Mann & Oxford, 1958). It was thought that the disappearance of LC might be due to the lower production of lactic acid in the rumens of older animals and the inability of LC to compete with other lactate-fermenting bacteria. However, experiments (Hobson & Mann, 1961) showed that lack of lactate was not the reason for the disappearance of LC and even when large inocula were given into the rumen of a sheep together with lactic acid the LC failed to multiply and form a stable population above the numbers (10^2–10^4/ml.) in which it naturally existed. Here substrate competition was apparently not a controlling factor. Most natural continuous cultures (including man-made ones such as sewage systems) are very complex with respect to the medium so that there is a large number of nutrients available to the organisms, and most of the organisms can take advantage of this. For instance one of the main functions of the rumen bacteria (under natural feeding conditions) is the hydrolysis and fermentation of cellulose. We can isolate a number of types of cellulolytic bacteria from any rumen and one might suppose that these would be competing for cellulose. However, the problem is more complex because the cellulolytic bacteria can all ferment a number of sugars. Since the substrate they are digesting (plants) is not pure cellulose, different cellulolytic types may at any one time be fermenting different plant polysaccharides. Diauxic growth may occur in which the same organism(s) uses different sugars at different times as has been reported in an *in vitro* aerobic sludge system (Bhatla & Gaudy, 1965), and some organisms may even be utilizing two sugars at the same time. In some batch culture experiments Hobson & Fina (unpublished) showed that *Selenomonas ruminantium* could use both starch and lactate during growth, although with glucose and lactate glucose was utilized preferentially. Catabolite repression may be involved here, but Mateles, Chian & Silver (1967) showed that this phenomenon was not important in laboratory continuous cultures of *Escherichia coli* and *Pseudomonas fluorescens* growing at low growth rates on mixed substrates (see Clarke & Lilly, this Symposium). In many natural cultures the over-all growth rate is very low; in the rumen it is about one generation in 18 hr. However, even if two types of rumen cellulolytic bacteria are fermenting the same cellulose as energy source it may still not be the growth-limiting substance. The types of cellulolytic bacteria vary in their requirements for other nutrients including nitrogen sources, branched-chain fatty acids, sulphur sources, vitamins and other growth factors. Supply of these substances could be controlling the growth rate and cell concentration of the different bacteria, and so result in a stable mixed population.

However, we must note that a natural mixed continuous culture system is rarely, if ever, stable in the sense of a single component laboratory culture. There are day-to-day variations in numbers of particular species, or in, for instance, serological type of the same species (Hobson *et al.* 1958). In the rumen system, although the over-all growth rate of the organisms must be the same to maintain the population, studies show that total numbers of organisms vary during the day, and Warner (1962) and others have shown that the growth curves of individual organisms do not coincide. This is to be expected if the growth of different organisms is controlled by different factors, some of which are supplied by other organisms.

Positive control. An organism may not only produce substances which control growth by being toxic to other organisms, but it may produce substances which stimulate or are necessary for the growth of others. We have already noted the way in which a secondary fermentation could remove acids which would otherwise inhibit growth, but the primary rate of formation of acids, or other fermentation products, could also act as a control mechanism for the growth of the secondary bacteria. *Veillonella alcalescens* was mentioned as an example, in the rumen, of an organism which can ferment only the fermentation acids of another organism. In the rumen system the methane bacteria (*Methanobacterium ruminantium*) grow on carbon dioxide and hydrogen, and the hydrogen is a sugar-fermentation product of some of the other bacteria present. Hydrogen is rapidly utilized and it could be that rate of hydrogen production limits the growth of methane bacteria. Gaffney (1965) recently reported that in mixed cultures from sewage sludge a small proportion of CO_2 in the atmosphere enhanced biological oxidation of the sewage organic matter. The effects of CO_2 on growth of organisms are discussed by Wimpenny (this Symposium). The branched-chain fatty acids and most of the other possible controlling factors for growth of ruminal cellulolytic bacteria mentioned in the previous paragraph are products of the metabolism of other organisms. Similarly, *iso*butyrate produced by an oral diphtheroid organism could allow the growth of an oral strain of *Treponema microdentium* (Hardy & Munro, 1966). Few of the lactobacilli isolated from the rumens of cattle on high-starch feeds were amylolytic (S. O. Mann, unpublished), so they must depend for utilizable maltose on the amylolytic action of other bacteria, such as the *Bacteroides* spp. which occur in large numbers in these rumens. In *in vitro* cultures non-cellulolytic rumen *Borrelia* sp. migrated through agar and grew around colonies of cellulolytic bacteria where they were utilizing products of cellulolysis (Bryant, 1952). In general in any natural culture, organisms which cannot hydrolyse polymeric carbohydrates or proteins will have to depend on substances released by the exoenzymes of other organisms for their substrates. Some rumen bacteria require B vitamins and these are produced by other rumen bacteria, and this kind of interdependence is not limited to the rumen. For instance, Lochhead & Thexton (1952) showed that a group of soil bacteria depended for maximum growth on vitamin B_{12} produced by other soil bacteria. A more complex relationship exists between *Lactobacillus leichmannii* which requires B_{12} and produces citrovorum factor and *Leuconostoc citrovorum* which requires the factor and produces B_{12} (Doctor & Couch, 1954). Nevin

(1960) showed that growth of *Borrelia vincenti* was stimulated by acetyl phosphate produced by an oral diphtheroid, and this could be a factor controlling the numbers of *B. vincenti* in oral infections.

Other substances produced by one organism may protect another organism from the action of inhibitory factors. For instance, Bennett & Bauerle (1960) showed that H_2S produced by *Desulphovibrio desulphuricans* decreased the sensitivity of an associated pseudomonad to mercurials by precipitating the mercury. Truby & Bennett (1966) also showed that a number of Gram-negative bacteria were able to protect *Staphylococcus aureus* against the action of trichlorophenol as the lipid of the Gram-negative bacterial cells could absorb the phenol. Although these are not microbially produced inhibitors such a mechanism might be important in some inter-microbial associations. Long-chain fatty acids have both stimulatory and inhibitory effects on growth, and an example of both was found by Pollock (1948) where *Haemophilus pertussis* produced autotoxic long-chain fatty acids. Growth of the haemophilus was stimulated by the presence of a contaminant bacterium which utilized these fatty acids for growth. In most mixed cultures, because of the different growth cycles of the organisms, there will be not only growth but death and lysis of cells occurring continuously. Lysis of cells may release compounds, which can control the growth of other cells in that they can be used as metabolites, or which, with extracellular enzymes from cells, may interact not only with nutrients and dead cell components but with other enzymes. The digestion of extracellular amylases, cellulases, etc., by proteinases of other organisms may serve to limit the sugars available to the amylolytic or cellulolytic bacteria.

Control by protozoa

In systems containing protozoa, such as the rumen or activated sludge, the protozoa ingest large numbers of bacteria. In the rumen system Eadie & Hobson (1962) and Kurihara, Eadie, Hobson & Mann (1968) showed that ingestion of bacteria by protozoa was a factor controlling the numbers of bacteria; thus the 'steady-state' bacterial population is lower in the presence of protozoa. The latter authors showed how the bacterial numbers fell as the protozoal numbers increased and related the volume of the protozoa to the volume of the bacteria 'displaced' by the protozoa. Coleman (1964) had earlier shown that *Entodinium caudatum* could ingest almost its own volume of bacteria, using them as a source of amino acids. The change in bacterial numbers seemed to be mainly a result of indiscriminate ingestion by the protozoa. In continuous cultures *in vitro* Curds (1967) showed that ingestion of bacteria (*Klebsiella aerogenes*) by protozoa (*Tetrahymena pyriformis*) caused cyclic changes in the numbers of bacteria and protozoa, but this could have been caused by a purely physical mechanism. The protozoa must have a maximum rate at which they can swim through, and ingest, the medium and the number of bacteria which they can ingest will then depend on the concentration of bacteria in the medium. If the protozoa ingest the bacteria faster than the latter grow then they will reduce the numbers of bacteria to a level at which the protozoal growth rate will be affected. This will then allow regrowth of the bacteria and the cycle will start again. The results of Kurihara *et al.* (1968) coupled with those of Coleman (1964) on the rate of ingestion of bacteria by *Entodinium caudatum* suggested that at their most active the rumen protozoa were ingesting bacteria faster than the bacteria were dividing, but this active phase did not extend over the whole growth cycle of the protozoa. However, since the numbers of rumen bacteria even in the presence of large populations of protozoa were about 10^{10}/ml. it seems unlikely that the previously mentioned physical factors were controlling growth of the protozoa; availability of carbohydrate substrates or other factors must have been involved as well.

Physical factors similar to those involved in protozoal-bacterial relationships must also influence the relationship between the 'prey' bacteria and the 'predatory', motile *Bdellovibrio*.

Mixed continuous cultures in the laboratory

In the foregoing review a number of examples of laboratory cultures designed to elucidate mechanisms of microbial control have been mentioned. These were mostly batch cultures, but the natural mixed cultures of greatest interest are continuous ones. While the behaviour of a single type of organism in continuous culture can often be predicted reasonably accurately from batch cultures, prediction from pure batch culture of the behaviour of an organism in mixed continuous culture may not always be possible. For instance, Parker (1966) found the minimum generation time of *Streptococcus salivarius* was greatly reduced when it was in culture with *Veillonella alcalescens* and *Staphylococcus aureus*. The *Veillonella* generation time was also reduced from that in pure continuous culture. The factors causing this were possibly complex and were not elucidated in these experiments, but variation in generation times was also caused by variations in medium. However, this system does not correspond to the majority of natural mixed cultures as the mixed culture was fed as a second stage from three pure continuous cultures of the bacteria, and the mixed culture was run at high dilution rates. Hetling, Washington & Rao (1965) reported higher yields of organisms in mixed culture than pure culture in a simulated aerobic sludge system.

Although some experimental work has been done on mixed continuous cultures of bacteria using defined populations, in general this work has been limited to the determination of whether or not various populations grew. These observations are useful in showing that intermicrobial control can exist in a laboratory culture as well as in a natural culture, but much more defined conditions, and many analyses, will be needed to explain the control mechanisms actually involved. On the other hand, in investigations of aerobic sewage systems laboratory models have been used in which measurements of various characteristics of the systems have been made, but the microbial population has been undefined. These systems may serve as models for obtaining optimum sludge treatment, but cannot be used as models for elucidating microbial control mechanisms. For study of microbial control mechanisms in continuous culture the culture needs to be run for a considerable length of time, and at a low dilution rate, to simulate many natural conditions. Although rapid overgrowth of one organism by another may occur under certain conditions, in systems such as the rumen we have found that some weeks may be necessary for a new 'steady-state' to be established after inoculation of a new organism into the 'culture' (see, for instance, Kurihara et al. 1968). Although this process might be speeded up by increasing the dilution rate of the simulated system *in vitro*, the results obtained in this way will not necessarily apply to natural culture. We (Henderson, Hobson & Summers, 1969, and previous papers cited there) have found that extracellular enzyme production in pure cultures varies with growth rate, and if one of these enzymes (say the amylase of *Bacteroides amylophilus*) were providing substrate for growth of another organism then changes in dilution rate might alter the whole relationship between the organisms. Similarly, an uncoupled fermentation or polysaccharide formation by one organism at low growth rates may decrease amounts of fermentable sugar available to other organisms. Very low growth rates have also been shown to result in death of a proportion of the bacteria in a continuous culture (Tempest, Herbert & Phipps, 1967). This leads to a growth rate of the remaining bacteria more rapid than the dilution rate (shown theoretically by Powell, 1965). It could also lead to lysis of the dead cells and liberation of growth factors for other organisms, as mentioned previously. The change in fermentation products which can take place with change of culture pH has already been mentioned, but fermentation products may also change with growth rate and this may influence inhibitory or stimulatory actions of these products on other organisms.

It may be doubted whether a natural mixed multicomponent culture attains a true steady state and it may be found impossible to run a mixed culture *in vitro*, even on defined substrates, in which all organisms are in a steady state as usually understood for single component cultures. Mathematical analysis would then become complex. It may be possible to obtain some functions describing approximately the relationships between over-all cell yield, degradation of substrate, dilution rate, etc., of *in vitro* models of natural systems. These can be used to predict the optimum dilution rate, substrate concentration and so on of the natural system (Gaudy & Gaudy (1966) reviewed such work on sewage systems), but in order to study control mechanisms in the microbial populations more complex analysis is needed. Mathematical models have been made for pure culture stirred single and multistage cultures with and without feed-back, and for tubular fermentors; for substrate-limited growth and end-product-limited growth. The mathematical analysis of a mixed population may be possible only on the basis of treating each organism individually. For instance, an organism fermenting sugar to lactate and one growing on the lactate produced could each be treated as carbon-limited growth, the sugar concentration being the input to the culture and the lactate concentration that produced by the first bacterium growing alone under the particular conditions. This kind of analysis obviously involves a very large amount of experimental work and means that the mixed culture must be built up from single components, but it seems a more fruitful way than attempting to reproduce a natural mixed culture by inoculation of the whole into an *in vitro* system and analysing this. In the case of the rumen system no one has yet found it possible to reproduce completely *in vitro* the *in vivo* system by using inocula of whole rumen contents as the start of a continuous culture and the same statement probably applies to all other attempts to reproduce natural systems.

REFERENCES

BENNETT, E. O. & BAUERLE, R. G. (1960). The sensitivities of mixed populations of bacteria to inhibitors. *Aust. J. exp. biol. med. Sci.* **13**, 142.

BHATLA, M. N. & GAUDY, A. F. (1965). Sequential substrate removal from a dilute system by heterogeneous microbial populations. *Appl. Microbiol.* **13**, 345.

BROBERG, G. (1957). Investigations into the effect of oxygen on the redox potential and quantitative *in vitro* determinations of the capacity of rumen contents to consume oxygen. *Nord. Vet. Med.* **9**, 942.

BRYANT, M. P. (1952). The isolation and characteristics of a spirochete from the bovine rumen. *J. Bact.* **64**, 325.

COLEMAN, G. S. (1964). The metabolism of *Escherichia coli* and other bacteria by *Entodinium caudatum*. *J. gen. Microbiol.* **37**, 209.

CURDS, C. R. (1967). Continuous culture—a method for the determination of food consumption by ciliated protozoa. *Proc. Symp. Methods of Study of Soil Ecology*. Paris: U.N.E.S.C.O.

DI GIACINTO, J. V. & FRAZIER, W. C. (1966). Effect of coliform and *Proteus* bacteria on growth of *Staphylococcus aureus*. *Appl. Microbiol.* **14**, 124.

DOCTOR, B. M. & COUCH, J. R. (1954). An unusual example of symbiosis in bacteria. *Archs Biochem. Biophys.* **51**, 530.

EADIE, J. M. & HOBSON, P. N. (1962). Effect of the presence or absence of rumen ciliate protozoa on the total rumen bacterial count in lambs. *Nature, Lond.* **193**, 503.

EADIE, J. M., HOBSON, P. N. & MANN, S. O. (1967). A note on some comparisons between the rumen content of barley-fed steers and that of young calves also fed on a high concentrate ration. *Anim. Prod.* **9**, 247.

FINN, R. K. & WILSON, R. E. (1954). Population dynamics of a continuous propagator for micro-organisms. *J. agric. Fd Chem.* **2**, 66.

GAFFNEY, P. E. (1965). Carbon dioxide effects on glucose catabolism by mixed microbial cultures. *Appl. Microbiol.* **13**, 507.

GAUDY, A. F. & GAUDY, E. T. (1966). Microbiology of waste waters. *Ann. Rev. Microbiol.* **20**, 319.

GUNSALUS, I. C. & NIVEN, C. F. (1942). The effect of pH on the lactic acid fermentation. *J. biol. Chem.* **145**, 131.

HARDY, P. H. & MUNRO, C. O. (1966). Nutritional requirements of anaerobic spirochaetes. 1. Demonstration of *iso*butyrate and bicarbonate for a strain of *Treponema microdentium*. *J. Bact.* **91**, 27.

HATTORI, Z., MISAWA, H., IGARASHI, I. & SUGIYA, Y. (1965). Effects of lactic acid-forming bacteria on *Vibrio comma* inoculated into intestinal segments of rabbits. *J. Bact.* **90**, 541.

HENDERSON, C., HOBSON, P. N. & SUMMERS, R. (1969). The production of amylase, protease and lipolytic enzymes by two species of anaerobic rumen bacteria. *Folia Microbiol.* (in the Press).

HENTGES, D. J. (1967 *a*). Inhibition of *Shigella flexneri* by the normal intestinal flora. 1. Mechanisms of inhibition by *Klebsiella*. *J. Bact.* **93**, 1369.

HENTGES, D. J. (1967 *b*). Influence of pH on the inhibitory activity of formic and acetic acids for *Shigella*. *J. Bact.* **93**, 2029.

HETLING, L. J., WASHINGTON, D. R. & RAO, S. S. (1965). Kinetics of the steady-state bacterial culture. II. Variations in synthesis. *Purdue Engng Extension Ser.* **49**, 687.

HOBSON, P. N. (1965). Continuous culture of some anaerobic and facultatively anaerobic rumen bacteria. *J. gen. Microbiol.* **38**, 167.

HOBSON, P. N. & MANN, S. O. (1961). Experiments relating to the survival of bacteria introduced into the sheep rumen. *J. gen. Microbiol.* **24**, i.

HOBSON, P. N., MANN, S. O. & OXFORD, A. E. (1958). Some studies of the occurrence and properties of a large Gram-negative coccus from the rumen. *J. gen. Microbiol.* **19**, 462.

HOBSON, P. N. & SUMMERS, R. (1967). The continuous culture of anaerobic bacteria. *J. gen. Microbiol.* **47**, 53.

HOLLOWELL, C. A. & WOLIN, M. J. (1965). Basis for the exclusion of *E. coli* from the rumen ecosystem. *Appl. Microbiol.* **13**, 918.

ICHIKAWA, Y. & KITAMOTO, Y. (1967). The inhibitory properties of propionate and related substances to the growth of *Escherichia coli. J. agric. Chem. Soc. Japan*, **41**, 171.

ISENBERG, D. L. & BENNETT, E. O. (1959). Nature of the relationship between aerobes and sulphate-reducing bacteria. *Appl. Microbiol.* **7**, 121.

JERUSALIMSKY, N. D. (1967). Bottlenecks in metabolism as growth rate controlling factors. In *Microbial Physiology and Continuous Culture*. London: H.M.S.O.

JOWITT, R. (1968). Yields in anaerobic cultures. *Chem. Ind.* p. 235.

KAO, C. T. & FRAZIER, W. C. (1966). Effect of lactic acid bacteria on growth of *Staphylococcus aereus. Appl. Microbiol.* **14**, 251.

KROGH, N. (1959). Studies on alterations in the rumen fluid of sheep, especially concerning the microbial composition, when readily available carbohydrates are added to the food. I. Sucrose. *Acta Vet. Scand.* **1**, 74.

KURIHARA, Y., EADIE, J. M., HOBSON, P. N. & MANN, S. O. (1968). Relationship between bacteria and ciliate protozoa in the sheep rumen. *J. gen. Microbiol.* **51**, 267.

LOCHHEAD, H. & THEXTON, R. H. (1952). Qualitative studies of soil microorganisms and bacteria requiring vitamin B_{12} as growth factors. *J. Bact.* **63**, 219.

MATELES, R. I., CHIAN, S. K. & SILVER, R. (1967). Continuous culture on mixed substrates. In *Microbial Physiology and Continuous Culture*. London: H.M.S.O.

MEYNELL, G. C. (1963). The role of Eh and volatile fatty acids in the normal gut. *Br. J. Expt. Path.* **44**, 209.

MEYNELL, G. C. & SUBBAIAH, T. V. (1963). Kinetics of infection by *Salmonella typhi-murium* in normal and streptomycin-treated mice studied with abortive transductants. *Br. J. Expt. Path.* **44**, 197.

NEVIN, J. (1960). Interaction between *Borrelia vincenti* and an oral diphtheroid. *J. Bact.* **80**, 783.

PAEGE, L. M. & GIBBS, M. (1961). Anaerobic dissimilation of glucose ^{14}C by *Escherichia coli. J. Bact.* **81**, 107.

PARKER, R. B. (1966). Continuous-culture system for ecological studies of micro-organisms. *Biotech. Bioeng.* **8**, 473.

POLLOCK, M. R. (1948). A case of bacterial symbiosis based on the combined growth stimulating and growth inhibiting properties of long chain fatty acids. *J. gen. Microbiol.* **2**, xxiii.

POWELL, E. O. (1958). Criteria for the growth of contaminants and mutants in continuous culture. *J. gen. Microbiol.* **18**, 259.

POWELL, E. O. (1965). Theory of the chemostat. *Lab. Pract.* **14**, 1145.

PURSER, D. B. & MOIR, R. J. (1959). The effect of pH on the ciliate population of the rumen *in vivo. Aust. J. Agric. Res.* **10**, 555.

ROGOSA, M. (1964). The genus *Veillonella*. 1. General cultural, ecological and bio-chemical considerations. *J. Bact.* **87**, 162.

SABINE, D. B. (1963). An antibiotic effect of *Lactobacillus acidophilus. Nature, Lond.* **199**, 811.

SCHAEDE, R. (1962). *Die pflanzlichen symbiosen*, 3rd ed. Fischer: Stuttgart. Quoted by Orenski, P. N., in *Symbiosis*, vol. I. New York: Academic Press.

SLYTER, L. L., BRYANT, M. P. & WOLIN, M. J. (1966). Effect of pH on population and fermentation in a continuously cultured rumen ecosystem. *Appl. Microbiol.* **14**, 573.

TEMPEST, D. W., DICKS, J. W. & MEERS, J. L. (1967). Magnesium-limited growth of *Bacillus subtilis*, in pure and mixed cultures, in a chemostat. *J. gen. Microbiol.* **49**, 139.

TEMPEST, D. W., HERBERT, D. & PHIPPS, P. J. (1967). Studies on the growth of *Aerobacter aerogenes* at low dilution rates in a chemostat. In *Microbial Physiology and Continuous Culture*. London: H.M.S.O.

TRAMER, J. (1966). Inhibitory effects of *Lactobacillus acidophilus*. *Nature, Lond.* **211**, 204.

TROLLER, J. A. & FRAZIER, W. C. (1963). Repression of *Staphylococcus aereus* by food bacteria. II. Causes of inhibition. *Appl. Microbiol.* **11**, 163.

TRUBY, C. P. & BENNETT, E. O. (1966). Role of lipid in the protection of *Staphylococcus aureus* against trichlorophenol in mixed culture. *Appl. Microbiol.* **14**, 769.

WARNER, A. C. I. (1962). Some factors influencing the rumen microbial population. *J. gen. Microbiol.* **28**, 129.

WHITTENBURY, R., McDONALD, P. & BRYAN-JONES, D. G. (1967). A short review of some biochemical and microbiological aspects of ensilage. *J. Sci. Fd Agric.* **18**, 441.

ENERGETIC ASPECTS OF MICROBIAL GROWTH

W. W. FORREST

Division of Nutritional Biochemistry,
Commonwealth Scientific and Industrial Research Organization,
Kintore Avenue, Adelaide, South Australia

One of the striking features of bacterial growth is the very high metabolic activity of bacterial cells. This manifests itself both in a very rapid rate of cell division and in a high rate of catabolism. Associated with these processes, there is a very noticeable evolution of heat.

When such rates of heat production for different organisms are compared, large differences become apparent. Prat (1963) gives a comparative table showing that for *Escherichia coli* during exponential growth the rate of heat production is of the order of 1000 calories per hour per gram of cells, for *Drosophila* 100 calories per gram per hour and for man 10 calories per gram per hour.

Living organisms do the work required of them by utilizing the free energy of chemical reactions, and the heat produced represents that fraction of the total free-energy change which is unavailable to the organisms for the performance of work. Organisms in general do not seem to be very efficient as converters of free energy (Wilkie, 1960) and it has been suggested that the more rapid rate of bacterial metabolism and high rate of heat production indicate a lowered efficiency compared with the slower processes of higher organisms.

Numerous attempts have been made to assess the efficiency of growth of micro-organisms in quantitative terms, but the results have been rather inconclusive. Earlier attempts were based on thermodynamic considerations but, with more recent advances in biochemical knowledge, the biochemical approach has complemented the thermodynamic assessments.

MOLAR GROWTH YIELDS

Monod (1942) studied aerobic growth of *Bacillus subtilis*, *Escherichia coli* and *Salmonella typhimurium* in minimal medium with a number of carbohydrates as carbon and energy source. He showed that the amount of growth with limited energy source was proportional to the amount of added carbohydrate. Monod expressed the yields so obtained as a ratio of weight of cells produced to weight of added energy source.

5

In an extension of this work, Bauchop & Elsden (1960) studied the simpler situation of several organisms growing anaerobically in complex media with limited energy source. They showed that on complex media essentially all the substrate added was used as energy source so that cellular carbon was derived from preformed monomers in the complex media. Thus the free energy required for synthesis, under these conditions, must be required chiefly for polymerization of the preformed monomers to cellular macromolecules. Other energy-requiring processes would be at a minimum, so that the maximum yield of cells would be expected. Bauchop & Elsden expressed their results as molar yield coefficients Y_s, the grams dry weight of cells produced per mole of substrate degraded.

In the fermentations studied, they were able to calculate the amount of adenosine triphosphate (ATP) produced by known metabolic pathways, so that they could also compare the yields of different organisms on the basis of a yield coefficient Y_{ATP}, defined as the grams dry weight of cells produced per (calculated) mole of ATP generated from catabolism. This procedure gave a rational basis for comparison in biochemical terms of the efficiency of different organisms and fermentation pathways on the basis of the energy available for biosynthesis. Their results indicated that several different organisms gave a common value for Y_{ATP} of about 10 g. dry weight of cells produced per mole of ATP; since their original publication there has been ample confirmation of their findings for fermentation reactions.

Results obtained with a wide variety of substrates with a number of organisms with different metabolic pathways have now been reported. Table 1 lists molar growth yields for fermentations with limited energy source under conditions where the amount of incorporation of energy source into cellular materials has been determined and the ATP yields from the fermentations are known. The mean value of Y_{ATP} is $10 \cdot 3 \pm 0 \cdot 3$ g. per mole for 38 determinations. The differences in Y_{ATP} shown in the table for different organisms are real and greater than the experimental error of the determination, but in general there is strong evidence that micro-organisms can use the energy available from catabolism to produce cellular material with about the same maximum efficiency of conversion, that is, in fermentation at least, the degree of coupling between anabolism and catabolism is the same for widely differing organisms and metabolic pathways. Two determinations are involved in calculation of Y_{ATP}, a measurement of dry weight of organisms produced and an estimation of the ATP produced during the fermentation. The good agreement between the values of Y_{ATP} obtained with different

Table 1. *Growth yields from fermentations*

Organism	Substrate	ATP yield moles ATP/mole substrate	Y substrate g.dry wt./ mole substrate	Y_{ATP} g.dry wt./ mole ATP	References
Streptococcus faecalis	Glucose	2·0–3·0	20·0–32·0	10·8 ± 0·2(6)	Bauchop & Elsden (1960), Beck & Shugart (1966), Forrest & Walker (1965), Sokatch & Gunsalus (1957)
	Gluconate	1·8	17·6–20	10·4(2)	Goddard & Sokatch (1964), Sokatch & Gunsalus (1957)
	2-Keto gluconate	2·3	19·5	8·5	Goddard & Sokatch (1964)
	Ribose	1·67	21·0	12·6	Bauchop & Elsden (1960)
	Arginine	1·0	10·2	10·2	Bauchop & Elsden (1960)
	Pyruvate	1·0	10·4	10·4	Forrest (1965)
Streptococcus lactis	Glucose	2·0	19·5	9·8	Boivinet (1964)
Lactobacillus plantarum	Glucose	2·0	18·8	9·4	Oxenburgh & Snoswell (1965)
Saccharomyces cerevisiae	Glucose	2·0	18·8–21·0	10·0 ± 0·2(4)	Bauchop & Elsden (1960), Battley (1960), Bulder (1963)
S. rosei	Glucose	2·0	22·0–24·6	11·6(2)	Bulder (1963), (1966)
Zymomonas mobilis	Glucose	1·0	8·0–9·3	8·5 ± 0·2(5)	Bauchop & Elsden (1960), Belaich & Senez (1965), Dawes et al. (1966), Forrest (1967), Senez & Belaich (1965)
Aerobacter aerogenes*	Glucose	3·0	26·1–29·5	10·7(2)	Boivinet (1964), Hadjipetrou et al. (1964)
	Fructose	3·0	26·7	10·7	Hadjipetrou et al. (1964)
	Mannitol	2·5	21·8	10·8	Hadjipetrou & Stouthamer (1965)
	Gluconate	2·5	21·4	11·0	Hadjipetrou & Stouthamer, cited by Stouthamer (1968)
A. cloacae	Glucose	1·5–2·2	17·7–27·1	11·9 ± 0·5(3)	Hernandez & Johnson (1967a)
Escherichia coli*	Glucose	3·0	25·8	11·2	Stouthamer (1968)
Ruminococcus flavefaciens*	Glucose	2·75	29·1	10·6	Hopgood & Walker (1967)
Actinomyces israelii	Glucose	2·0	24·7	12·3	Buchanan & Pine (1967)
Bifidobacterum bifidum	Glucose	2·5–3·0	37·4	13·1	de Vries & Stouthamer, cited by Stouthamer (1968)

* Corrected for incomplete fermentation.

organisms means therefore that the definition of fermentation pathways in these organisms is also accurate.

This value of Y_{ATP} is considerably less than would be expected if all the ATP used in cellular growth was coupled entirely to biochemical syntheses. Gunsalus & Shuster (1961) have calculated the requirements of ATP for the synthesis of microbial cell material from simple precursors, assuming that all the processes of synthesis can be described biochemically and that physical processes of active transport and structural ordering require no energy in the form of ATP hydrolysis. Their calculation for the production of cell material of the normal composition determined from analytical data gives an estimate of Y_{ATP} of 33 g. per mole. Obviously most of the ATP is not used in ways which can be biochemically defined.

Growth yields of aerobic organisms

Attempts have been made to determine aerobic growth yields in the same way as for fermentations, but the data for yields in aerobic growth do not exhibit the same regularities as the yields for fermentations. Reproducible results can be obtained in terms of grams dry weight of organisms produced from grams of substrate catabolized (Monod, 1942), but comparisons on the basis of Y_{ATP} are very unsatisfactory. In fermentations where ATP is produced by phosphorylation at the substrate level, an accurate assessment of ATP production is possible, but the efficiency of ATP production from oxidative phosphorylation in bacteria is not well defined, so that comparisons on the basis of the P/O ratio, the calculation of Y_{ATP} from oxygen consumption during growth, are not meaningful. If the converse correlation is attempted of assuming Y_{ATP} as 10 and estimating P/O ratios, values ranging from 0·5 to 3 have been reported (Chen, 1964; Hernandez & Johnson, 1967b; Stouthamer, 1968) for a wide range of organisms. In *Zymomonas mobilis* growth on glucose may occur anaerobically or aerobically with respiration, but the growth yield is the same in either case (Belaich & Senez, 1965). This seems to indicate that no energy is derived from oxidative phosphorylation in this organism, a P/O ratio of zero.

The observations suggest that the efficiency of oxidative phosphorylation in micro-organisms actually is low. The alternative explanation that the efficiency of generation of ATP is high, but the coupling between anabolism and catabolism is less effective in organisms with aerobic metabolism seems unlikely, since *Escherichia coli* gives in glucose fermentation a value for Y_{ATP} of 11 (Table 1) but even on the basis of Y_{ATP} of 10 a calculated P/O ratio of only 2 in glucose oxidation (Stouthamer, 1968).

A recent proposal has been made that instead of attempting to correlate oxygen uptake with dry weight of cells produced, as is done at present with only limited success, an approach giving results which would be more directly comparable would be the correlation with 'available electrons' in the substrate or with the enthalpy change during aerobic growth (Mayberry, Prochazka & Payne, 1967).

Not enough suitable calorimetric data are yet available to check the proposal, but a better reference than oxygen uptake is clearly desirable.

Anomalous growth yields

While the value of Y_{ATP} is well established for anaerobic growth there are several causes of serious deviation from this value. Even the determination of Y_{ATP} may not be straightforward. Consider the anaerobic growth of *Streptococcus faecalis* on aliquots of a complex medium to which have been added different amounts of energy source (glucose) (Fig. 1). In the absence of added energy source there is considerable growth on the other medium constituents. At low concentrations of glucose, the degradation products of glucose are lactate and volatile

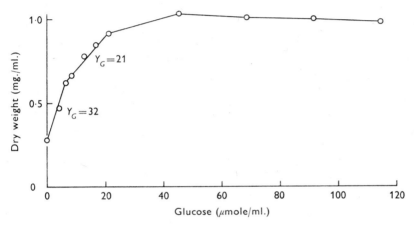

Fig. 1. Growth yields of *Streptococcus faecalis* on complex medium with growth limited by energy source. Different amounts of glucose were added to aliquots of complex medium. (From Forrest & Walker, 1965*a*.)

acids and the net ATP yield is 3 moles per mole of glucose fermented, two from the Embden–Meyerhof pathway and a further one from the dismutation of the pyruvate produced (Forrest & Walker, 1965*a*; Beck & Shugart, 1966); here Y_{G} is 32. At higher concentrations of glucose, the fermentation becomes homolactic and Y_{G} is 21. At still higher concentrations of glucose, reserve materials for endogenous metabolism are accumulated but there is very little further increase in dry weight of organisms. A similar change in growth yield with increasing glucose concentration has been reported with *Aerobacter aerogenes* (Hadjipetrou, Gerrits, Teulings & Stouthamer, 1964), and the general problem of determination of growth yields has been reviewed by Stouthamer (1968).

High growth yields

Very high yields have been reported with rumen organisms (Hungate, 1963; Hobson, 1965; Hobson & Summers, 1967). Hungate (1963) grew *Ruminococcus albus* on cellobiose. This organism accumulated large quantities of storage polysaccharide and gave an apparent molar growth yield of 90·1. Hungate corrected for storage polysaccharide from an analysis of cell nitrogen to obtain a corrected yield $Y_{cellobiose}$ of 74·9. Since the organism was grown in a minimal medium some cellobiose was used as carbon source, for which correction was also necessary. He also proposed a fermentation pathway including splitting of cellobiose by phosphorolytic cleavage giving 9 moles of ATP per mole of cellobiose fermented. However, because of the use of cellobiose both for storage materials and cellular carbon, only enough cellobiose was actually degraded to give by this proposed pathway 7·8 moles of ATP per mole of cellobiose added to the medium. This gives a normal value for Y_{ATP} of about 9·6. The proposed ATP yield in this fermentation seems very high when compared with the data of Table 1; for comparison, the data of Hopgood & Walker (1967) for *Ruminococcus flavefaciens* grown on glucose give, when similarly corrected for storage polysaccharide, a value of Y_G of 29·1 (Table 1), and a calculated ATP yield of 2·75 moles per mole of glucose fermented.

Buchanan & Pine (1967) found that *Actinomyces israelii* carried out a homolactic fermentation for anaerobic growth on glucose giving a normal growth yield in the absence of CO_2 (Table 1). However, in the presence of substrate levels of CO_2, CO_2 fixation occurred with a large increase in growth yield. The fermentation pattern was changed to give products closely similar to those found in *Ruminococcus flavefaciens* (Hopgood & Walker, 1967) with the same calculated ATP yield as determined for *R. flavefaciens*, but the large increase in growth yield suggests that energy additional to that calculated was obtained from CO_2 fixation.

Selenomonas ruminantium grown on glucose in batch culture gave a molar growth yield of about 17, consistent with Y_{ATP} of about 9, but in continuous culture the same organism gave maximum growth yields about 65 (Hobson, 1965). More volatile fatty acid was found under these conditions, but it is difficult to see how an increase in ATP yield could occur to the extent necessary to explain the increase in molar growth yield. Pirt (1965) has analysed Hobson's data for the growth of *S. ruminantium* in continuous culture from the point of view of determination of energy of maintenance. The results indicate a 'true growth yield' so high as to be immeasurable, suggesting that the organism has a very unusual behaviour in continuous culture.

Similar high yields were obtained by Hobson & Summers (1967). The growth yield of lipolytic bacterium 5s on fructose in continuous culture had a maximum value of 60, and *Bacteroides amylophilus* on maltose a maximum of 130. These authors suggest Y_{ATP} values of about 20 for *B. amylophilus* or *S. ruminantium*.

From the calculations of Gunsalus & Schuster (1961) of the amount of cellular material which could be synthesized per mole of available ATP it would seem quite possible to obtain higher yields than Y_{ATP} of 10, but the reason for the high yields mentioned above remains obscure.

It is always possible on complex media for catabolism of other constituents of the medium added primarily as monomers for cellular synthesis to occur concurrently with that of the energy source so that additional ATP is produced over that calculated. Blank determinations of growth yield on the medium without energy source may not reveal such behaviour. For example, *Streptococcus faecalis* can use arginine as energy source for growth only in the presence of glucose (Bauchop & Elsden, 1960).

Conversely, the carbon or energy source can be assimilated either as storage polysaccharides or as normal cellular constituents. The apparent growth yields obtained in rumen organisms from synthesis of storage polysaccharides can be very high. We have found, both in pure culture and whole rumen contents, that added cellobiose and other carbohydrates can be polymerized through glucose-6-phosphate to storage polysaccharide with the expenditure of one ATP per glucose residue (Forrest, 1968; D. J. Walker & W. W. Forrest, unpublished). The process of polymerization seems to be stoichiometric which would give an apparent Y_{ATP} for this simple polymerization of 180. The polymerization is accompanied by little entropy change and it is a reasonable speculation that reserve storage in general where little structural organization is involved may give high efficiencies of use of ATP. For this reason analyses of the cells produced under conditions of high growth yield is of considerable importance. It is perhaps significant that the reported cases of high growth yields occur concurrently with unusual carbon assimilation, either as storage polysaccharides or by CO_2 fixation.

There is some evidence (Senez, 1962) that some of the energy available from catabolism may be coupled to anabolic processes other than through ATP. Where the carbon source for growth is at a higher level of oxidation than the cellular level it must be reduced and energy is required to do this. In the growth of *Desulphovibrio desulphuricans* with pyruvate as carbon and energy source, the carbon is assimilated as pyruvate, oxidation level CHO and reduced to the cellular level, CH_2O. The free energy required to carry out this reduction is 7·7 kcal. which represents nearly all the energy available from ATP; degradation of pyruvate produces one ATP per mole yet the molar growth yield of the organism is 9·6, quite a normal value, so that additional energy is available to carry out the reduction. Senez suggests that the energy for the reduction is provided by reduced electron carriers.

Low growth yields

Low growth yields are much more commonly encountered than high ones. In the situation where the specific growth rate of the organisms is lower than the maximum rate of which the organisms are capable on the medium, the growth yield is commonly reduced in proportion to the lowering of the growth rate. Such a lowered growth yield is often observed in continuous culture systems operating at low dilution rates. Analysis of the data obtained in such systems (Pirt, 1965) shows that the observed rate of catabolism of energy source by unit mass of cells has two components, a small component independent of the specific growth rate and a much larger one proportional to the growth rate.

It is now well documented that organisms require a supply of energy which can be coupled through ATP to maintain their normal functions (Forrest & Walker, 1963; Strange, Wade & Dark, 1963; Forrest & Walker, 1965b; McGrew & Mallette, 1965). It has been proposed that the larger component of degradation of substrate during growth supplies the energy required for biosynthesis, and the smaller the constant requirement of the cells for energy of maintenance (Pirt, 1965). This type of behaviour seems common though the amount of energy diverted to maintenance apparently differs greatly with different organisms. Pirt has analysed continuous culture data for *Aerobacter aerogenes. A. cloacae*, lipolytic bacterium 5s, and *Selenomonas ruminantium* (which gave anomalous results). Similar analysis for growth of *Azotobacter vinelandii* (Aiba, Nagai, Nishizawa & Onedera, 1967) indicate that this organism behaves similarly to those studied by Pirt, but has a very high maintenance requirement. The analysis gives an assessment of the 'true growth yield', the maximum yield possible at infinite growth rate. This situation is practically unattainable, but the values obtained in batch cultures with adequate nutrition (see Table 1) should closely approach this limiting value. Belaich & Senez (1965) grew *Zymomonas mobilis* with glucose as energy source on different media. The molar growth yields so obtained were 8 on complex medium, 5 in synthetic and 4 in minimal medium, and the specific growth rates in batch culture on the three media were in a ratio corresponding to that of the growth yields.

The other cause of lowered growth yield is energetic uncoupling between anabolic and catabolic processes (Senez, 1962). Under normal conditions of growth the efficiency of coupling appears to be at a maximum, but quite small changes in such variables as the composition of the medium or even the temperature of growth may drastically affect the degree of coupling. The extreme form of this uncoupling occurs in washed suspensions of organisms where large quantities of added energy source may be catabolized without any detectable increase in the mass of cells. Direct measurements of the intracellular ATP pool in washed suspensions of *Streptococcus faecalis* (Forrest, 1965) show a very large increase (up to tenfold) in the level of the pool during such uncoupled catabolism of glucose or arginine. Uncoupling in this case occurs after the stage of ATP generation by catabolism, as a result of failure to make use of the energy available as ATP. In growth uncoupling may occur in this way by failure to use ATP which has been produced. For example, when *S. faecalis* was grown in continuous culture under conditions of tryptophan limitation (Rosenberger & Elsden, 1960) this limitation did not affect the rate of catabolism and generation of ATP, but the growth yield was depressed through inability of the organisms to obtain the necessary materials for synthesis.

Alternatively, growth may be limited by failure to have sufficient ATP for synthesis to occur. Senez & Belaich (1965) grew *Escherichia coli* anaerobically on phosphate-limited medium in batch culture and found that exponential growth took place with Y_G 28·9 till phosphate limitation became apparent, then after limitation, though the rate of catabolism of energy source was not reduced, linear growth occurred with Y_G 10·9. The effect of this phosphate limitation was immediately and completely reversible. This appears to be a case where uncoupling occurred through the failure of the organisms under conditions of phosphate limitation to have a sufficient intracellular ATP pool to allow full coupling. Free energy could then not be transferred fast enough for exponential growth to take place. Damoglou & Dawes (1967) found that *E. coli* grown under conditions of phosphate limitation had an ATP pool level very much lower than that normally found during growth in *E. coli*. The finding of immediate complete reversibility suggests a change in the equilibrium situation between inorganic phosphate and ATP.

However, the change in the rate of synthesis of new cellular material during uncoupled growth does not have a direct effect on the rate of catabolism of energy source. Catabolism appears to go on at the maximum rate of which the organisms are capable in the circumstances, irrespective of whether or not the energy so produced can be used for biosynthesis.

As well as uncoupling by chemical limitations, change in temperature of growth may affect the degree of coupling (Senez, 1962; Forrest, 1967).

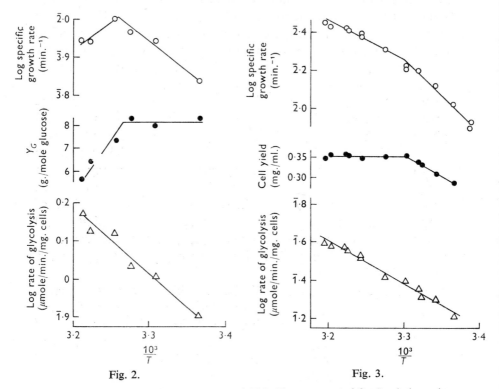

Fig. 2.　　　　　　　　　　　　　　　Fig. 3.

Fig. 2. Energies of activation and growth yield in *Zymomonas mobilis*. Symbols used are: ○, log specific growth rate; △, log rate of catabolism of glucose during growth; ● growth yield. (From Forrest, 1967.)

Fig. 3. Energies of activation and growth yield in *Streptococcus faecalis*. Symbols used are: ○, log specific growth rate; △, log rate of catabolism of glucose during growth; ●, growth yield. (From Forrest, 1967.)

Figures 2 and 3 show the effects of temperature on rate of growth, growth yield and degradation of energy source during growth of *Streptococcus faecalis* and *Zymomonas mobilis*. The rates of catabolism of glucose in both organisms have temperature coefficients, expressed as energies of activation, which over a wide range of temperature behave only in the normal way to be expected for simple chemical reactions; the energies of activation for both organisms grown on closely similar media are about 11 kcal. per mole. At temperatures near the respective accepted temperature optima for the organisms, the energies of activation for growth are the same as those for glycolysis, that is, the coupling between anabolism and catabolism is not affected by temperature and both organisms have a constant growth yield. At temperatures remote from the optima the growth yields fall and the energies of activation for growth become very markedly different from those of glycolysis, which are unaffected by the fall in the growth rate and yield. Coupling thus becomes less effective at temperatures remote from the temperature optima. Measurements of the intracellular ATP pool indicate a different energetic balance above and below the 'critical temperature' at which the energy of activation changes. Z. mobilis exhibits high-temperature uncoupling, and S. faecalis uncoupling at low temperatures, but presumably both forms of uncoupling could be observed in each organism if observations were carried out over a wide enough temperature range.

THE INTRACELLULAR ATP POOL

Measurements of molar growth yields and Y_{ATP} are based on the calculated total production of ATP by catabolic reactions. In batch culture a determination of growth yield consists then of an equilibrium measurement after growth has ceased or a steady-state determination in continuous culture. Kinetic measurements during growth in batch culture (Forrest & Walker, 1964) indicate that the same proportionality between rate of ATP production and rate of cell synthesis obtains throughout growth.

However, all these determinations are based on calculations of ATP yields by known pathways, and it would seem that direct measurements of the intracellular level of ATP of the organisms (the ATP pool) would demonstrate regularities related to the rate of input to the pool from catabolism of energy sources. Similarly, ATP is withdrawn from the pool to provide free energy to drive the processes of biosynthesis so that the level of the pool should also be related to the rate of biosynthesis. From the preceding discussion of growth yields in aerobic organisms, it would appear particularly desirable to obtain such a direct estimation of ATP production in aerobic metabolism, because of the unsatisfactory nature of the calculated ATP production figures.

Measurements of the pool level have been reported with several organisms—*Escherichia coli* (Franzen & Binkley; 1961; Damoglou & Dawes, 1967; Cole, Wimpenny & Hughes, 1967), *Aerobacter aerogenes* (Strange *et al.* 1963), *Streptococcus faecalis* (Forrest, 1965; Forrest &

Walker, 1965*b*), *Thiobacillus* sp. (Kelly & Syrett, 1966) and yeast (Polakis & Bartley, 1966). During aerobic growth a maximum level of ATP may be reached of somewhat under 20 μmoles per g. dry wt. in *E. coli*. This corresponds to a turnover time of the pool of only a few seconds. Measurements of the rate of decay of the ATP in the pool in the absence of significant amounts of biosynthesis show a first-order decrease in pool level with a half-time of about 20 min. in *S. faecalis*

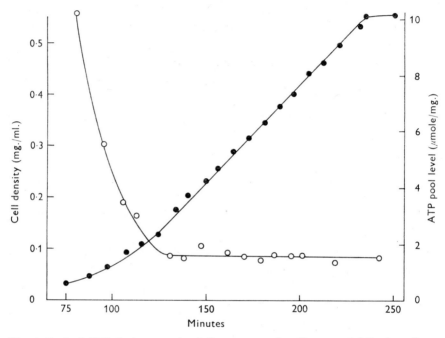

Fig. 4. Pool of ATP during growth of *Streptococcus faecalis* on semi-defined medium. Growth was limited by the energy source (sodium pyruvate). Symbols used are: ●, cell density; ○, pool level. (From Forrest, 1965.)

(Forrest, 1965), so that such decay processes are not significant in the balance of the pool during growth. The level of the pool during growth may be anywhere between the maximum level and a minimum level of around 1 μmole per g.; this minimum level in *S. faecalis* corresponds closely to that observed to be necessary to supply energy of maintenance (Forrest & Walker, 1965*b*). It has been suggested that a minimum level of ATP is necessary to sustain growth and certainly, if the pool level is above the minimum, exponential growth can continue, but if it falls to the minimum level exponential growth can no longer be sustained though linear growth at a lower rate may be observed. Figure 4 shows the behaviour of the ATP pool during growth of *S. faecalis* on a semi-defined medium with pyruvate as energy source. While the pool is above the minimum level a short period of exponential growth takes place, then linear growth begins and continues till all the energy source has been catabolized. At this point exogenous catabolism and growth both cease so that both the rates of input to, and withdrawal from, the pool are reduced to the level of maintenance reactions.

This situation of growth limited by the rate of supply of ATP to the pool appear to correspond to full energetic coupling, and the molar growth yield in the experiment shown in Fig. 4 was 10·4, very close to the mean value determined from Table 1. Similar experiments with glucose as energy source on the same medium gave higher pool levels with exponential growth continuing over the whole course of growth. Glucose gives 2 moles of ATP per mole of glucose fermented, compared with only 1 mole from pyruvate, so that the higher ATP yield was reflected in a higher pool level, and growth was not energy limited.

In *Streptococcus faecalis* (Forrest, 1965) and yeast (Polakis & Bartley, 1966) a fall in the level of the pool corresponding to consistent under-production of ATP during growth in batch culture was always observed, but in *Escherichia coli* four distinct phenomena have been reported (Cole *et al.* 1967) depending on growth conditions. These are: (*a*) ATP production in balance with growth; (*b*) ATP consistently overproduced; (*c*) ATP consistently underproduced; (*d*) cyclic oscillations occur in ATP production.

However, the amount of energy tied up in the ATP in the pool corresponds only to a fraction of a percentage of the total energy coupled through the pool during growth so that such under- or over-production apparently does not represent a significant imbalance between anabolism and catabolism. Exhaustive experiments with *Streptococcus faecalis* (Forrest & Walker, 1964) under conditions where consistent under-production of ATP was observed showed that growth was balanced to within the limits of experimental measurement.

These large differences in the kinetic behaviour of the ATP pool during growth do not support hypotheses which have been advanced for feedback control through the ATP pool of metabolic activity during growth. Polakis & Bartley (1966), from their studies on the ATP pool in yeast, concluded that control of catabolism was exercised by several co-factors which varied in such a way as to produce the maximum possible rate of catabolism, that is the maximum rate of production of ATP, on the lines proposed by the network theorem of Dean & Hinshelwood (1966); the evidence for energetic uncoupling during growth likewise supports the view that ATP is produced regardless of whether or not it can be used.

Regularities in the kinetic behaviour of the pool can be demonstrated and the absolute levels of the pool are usually quite closely reproducible (Franzen & Binkley, 1961; Forrest, 1965). However, several secondary effects can also affect the level of the pool—oxygen tension in the aerobic organisms, extracellular ionic concentration in the Gram-negative organisms, and extracellular phosphate concentration probably in all organisms. Thus quantitative determinations to give a direct measure of ATP yield seem impractical, though qualitative comparisons in similar situations are quite practical; the pool level shows considerable differences between the fermentation of pyruvate producing one ATP per mole, the fermentation of glucose producing 2 or 3 ATP, and oxidative metabolism producing a rather indefinite larger ATP yield.

THERMODYNAMIC EFFICIENCY OF GROWTH

Free energy efficiency

The assessment of efficiency of growth may be made on a thermodynamic basis by determination of the free energy efficiency.

Free energy efficiency %

$$= \frac{\text{Free energy of growth} \times 100}{\text{Free energy produced by metabolism of energy source}}$$

$$= \frac{100\,\Delta F_b}{\Delta F_c}, \tag{1}$$

where the free energy of growth ΔF_b is the quantity of free energy usefully employed for biosynthesis of cellular material during growth.

In most cases ΔF_b cannot be assessed, but in the case of autotrophic bacteria obtaining their energy from simple inorganic compounds and their carbon requirements for growth from reduction of CO_2 a reasonably accurate definition is possible. Baas-Becking & Parks (1927) assumed that the free energy of oxidation of glucose to CO_2 and water was the reverse of the free energy change required to reduce CO_2 and polymerize the reduced products to bacterial cell materials at the level of oxidation of CH_2O. (This assumption is based on the very close correspondence between the heat of combustion of glucose and that of bacterial cells.) They were then able to calculate the free energy efficiency of growth of several organisms from equation (1). From such calculations they obtained values of the free energy efficiency ranging from 30 % for *Hydrogenomonas* down to about 5 % for nitrifying bacteria. Lees (1954), from the more recent studies on nitrifying bacteria, has proposed that these low efficiencies are due to the choice of experimental conditions and that young cultures may reach maximum efficiencies of up to 40 %. As autotrophs grow under more energetically demanding conditions than heterotrophs, the efficiency of heterotrophs should be at least as high.

Table 2. *Free energy changes from degradation of glucose*

(After Senez, 1962.)

Pathway	ΔF kcal./ mole	ATP		Free energy directed into ATP (%)
		Net gain per mole	ΔF kcal.	
Homolactic fermentation (glucose → 2 lactate⁻+2H⁺) via Embden-Meyerhof	−50	2	25	50
Heterolactic fermentation (glucose → ethanol+CO_2+lactate⁻+H⁺) via pentose monophosphates	−56	1	12·5	22
Alcoholic fermentation (glucose → 2 ethanol+$2CO_2$)	−62	2	25	40
Complete oxidation of glucose (liver and muscle) (glucose+$6O_2$ → $6CO_2$+$6H_2O$)	−688	38	475	69
Oxidative phosphorylation (liver and muscle) (NADH+H⁺+$0·5O_2$ → NAD⁺+H_2O)	−52	3	37·5	72

A rather different approach to the problem has been taken by Senez (1962) who calculated the amount of free energy which could be transferred from catabolic to anabolic processes by coupling through ATP (Table 2). The efficiency of ATP formation varies greatly from one pathway to another, but the amount of free energy transferred does not exceed 50 % for the most favourable pathway of glucose fermentation. The maximum attainable efficiency in oxidative metabolism is much higher, but the figures cited refer to mammalian tissues; in most bacteria the P/O ratio for oxidative phosphorylation seems lower than 3.

Though the calculated efficiency in the alcoholic fermentation is only four-fifths that of the homolactic fermentation, the ATP yields (moles per mole of glucose) are the same, and the averages of the yield coefficient Y_{ATP} listed in Table 1 for the organisms catabolizing glucose by the homolactic fermentation (*Streptococcus faecalis*, *S. lactis* and *Lactobacillus plantarum*) are the same as those of the yeasts (*Saccharomyces cerevisiae* and *S. rosei*) catabolizing glucose by the alcoholic fermentation. The calculated difference in thermodynamic efficiency is due to the different balance of fermentation products, but the quantized nature of the production of ATP does not allow the 'more efficient' homolactic fermenters to make use of this additional free energy available to them to produce more ATP.

Microcalorimetric measurements

Numerous attempts have been made to determine directly the amount of energy incorporated into cellular material. In any process there will be a difference in energy between the initial and final states of the system so that during the process energy will be liberated or absorbed in the form of heat. Microcalorimetry is then a completely general method for studying energy changes during microbial metabolism (Forrest, 1968) though the quantity measured is not the free energy change ΔF but the enthalpy change ΔH. The experimentally observed heat production represents a summation of all the processes occurring so that the enthalpy change from endergonic reactions occurring during biosynthesis of cellular material ΔH_b will subtract from the enthalpy change of the exergonic reactions of catabolism (Prigogine, 1961).

Experimentally determined values of heat production during growth of cells of heterotrophs with limited energy source on a nutritionally adequate medium agree to within experimental error (to about $\pm 2\%$) with the value calculated from tables of thermodynamic data of the enthalpy change for the degradation of energy source to the products of catabolism ΔH_c, that is, the enthalpy of growth ΔH_b is too small to be measured. The enthalpy of growth is the difference between the heats of formation of the materials assimilated in the form in which they are assimilated and the heat of formation of the cellular material. When the carbon source is, for example, glucose, which is assimilated to give cellular material at the same level of oxidation (CH_2O), then the difference between these heats of formation ΔH_b would be expected to be small.

This close correspondence between observed heat production and calculated enthalpy change from catabolism has been reported for anaerobic growth on glucose of *Streptococcus lactis* (Boivinet, 1964), *Zymomonas mobilis* and *Escherichia coli* (Belaich & Senez, 1967; Senez & Belaich, 1965) and for aerobic growth of *Aerobacter aerogenes* on succinate (Grangetto, 1963).

If, however, as in the case of autotrophs, the carbon source is CO_2 which must be reduced on assimilation, then the difference in heats of formation between carbon source and cells becomes significant. Meyerhof (1924) found that with nitrifying bacteria, the experimentally measured heat production was 5% less than the calculated value of

ΔH_c for catabolism of energy source. This result, however, seems rather unreliable because of Lees's (1954) criticism of the experimental conditions. Theoretical calculations by Morowitz (1960) give estimates of the enthalpy of growth in *Escherichia coli*. This, according to his estimate, is dependent on the composition of the medium, but even on minimal medium has a maximum value of 125 calories per gram of cells, only about 3 % of the enthalpy of catabolism.

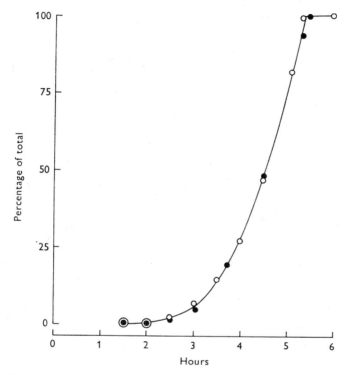

Fig. 5. Heat production by a growing culture of *Streptococcus faecalis*, with growth limited by the energy source (glucose). Glucose fermented 1·8 m-moles; final bacterial dry weight 27 mg.; total heat produced 49 calories. Symbols used are: ○, heat; ●, dry weight. (From Forrest, Walker & Hopgood, 1961.)

The calorimetric data show also that exponential growth in batch culture is generally balanced. The heat produced may be taken as an accurate index of catabolism and the bacterial dry weight as the index of the synthetic reactions which have occurred. Both accurately follow the same exponential relationship (Fig. 5). This relationship has been established with *Streptococcus faecalis* (Forrest, Walker & Hopgood, 1961), *Streptococcus lactis* (Boivinet, 1964) and *Zymomonas mobilis* (Belaich, Senez & Murgier, 1968).

6

More detailed studies with *Streptococcus faecalis* (Forrest & Walker, 1964) have shown that during exponential growth the rate of synthesis of new cellular material, degradation of energy source (glucose) appearance of acidic products of catabolism and heat production are all accurately described by the same exponential function. Using unit mass of cells as the basis for comparison of metabolic activity, we find that the rate of all these processes is much greater during exponential growth than during the other phases of the growth cycle and is constant over the exponential phase of growth. This corroborates with a great deal of other evidence; indeed, the network theorem of Dean & Hinshelwood (1966) is based on the proposition that during growth control mechanisms operate to produce the maximum possible growth rate.

Entropy production

From the point of view of irreversible thermodynamics these constant rates of metabolic activity are the criteria for a steady state (Forrest & Walker, 1964). It must be emphasized that in batch culture this is a steady state of thermodynamic fluxes in an open system consisting of unit mass of organisms and is applicable in this limited sense only. Quantities in the whole mass of the culture are varying widely during the growth process, but in thermodynamic terms the medium is merely the environment of the defined mass of cells. This steady state is the most stable configuration of the system (Denbigh, 1951) and corresponds to the situation of a maximum rate of outflow of entropy from the cells during exponential growth, so that the cells have the capacity to decrease their internal entropy through the outflow.

If we use the rough approximation (Prigogine, 1961; Morowitz, 1955) that the rate of entropy production is measurable by calorimetry, then the entropy of growth for *Streptococcus faecalis* can be very approximately determined from the data of Fig. 5 to be −7 cal. per degree per g. cells, a large quantity compared with the apparently insignificant enthalpy of growth. This is in good agreement with an estimate made by Linchitz (1953) of −10 cal. per degree per g. for the entropy of growth of the autotroph *Bacterium pycnoticus*, for which Baas-Becking & Parks (1927) estimated the free energy efficiency to be 25 %. Linchitz also calculated the informational entropy of each cell and obtained a very high value indicating that the cells have a high degree of organization. The estimates made by Morowitz (1955, 1960) lead to similar high values for the entropy of *Escherichia coli*. It must be pointed out that these estimates are all very approximate, but lead to what seems to be a reasonable value of the thermodynamic entropy of growth.

These figures now enable us to make another estimate of thermo-dynamic efficiency for the growth of *Streptococcus faecalis*. From Table 2 the free energy change ΔF_c for the degradation of one mole of glucose via the Embden–Meyerhof pathway to lactic acid is -50 kcal. Assuming that, in the complex medium used in the experiment of Fig. 5, ΔH_b is close to zero (Morowitz, 1960) which is corroborated by the experimental data for several other organisms, then the free energy of growth ΔF_b is

$$\Delta F_b = \Delta H_b - T\Delta S_b,$$

so that $\qquad\qquad \Delta F_b \approx -T\Delta S_b.$

From the data of Fig. 5, then $-T\Delta S_b = 32$ kcal. at $37°$, and the free energy efficiency

$$\Delta F_b / \Delta F_c \times 100 \approx 60\%.$$

This figure is not to be taken too seriously, but it carries the strong implication that the major portion of the free energy which is available for growth is used to provide a maximum rate of outflow of entropy from the cells, that is, the energy is apparently used mainly for purposes of structural organization, and the very rapid growth rate of micro-organisms requires a large outflow of entropy to allow the organisms to maintain their organization during growth.

CONCLUSIONS

We are able only in the simplest cases of anaerobic metabolism of micro-organisms to make generalized quantitative predictions about the energetic efficiency of the growth process, and even here anomalous results have been observed. The present status is that the growth yield coefficient Y_{ATP} seems to be a limiting value to which anaerobic organisms tend to conform, but that one is quite likely to observe a growth yield deviating from this limiting value. In most fermentations, it seems possible to account for these deviations, but in oxidative metabolism the position is further complicated by the incomplete knowledge of the efficiency of oxidative phosphorylation in micro-organisms. The co-efficient appears to be a biological constant relating anabolic and cata-bolic processes when full energetic coupling occurs, but our present lack of detailed knowledge restricts its potential usefulness as a means of quantitative prediction in energetic studies.

The thermodynamic determinations of energetic efficiency which have been proposed are of more restricted value, though they do indicate that free energy is used mainly in processes where little energy is incorporated into the organisms. That is, there is a high free energy cost involved in synthesis but most of the energy comes out again as heat, a form in which it cannot be put to further use. This situation occurs, for example, in protein synthesis (Wilkie, 1960) where large decreases in entropy occur. The theoretical calculations available for entropy change during microbial growth confirm that the very high degree of organization necessary in the cell would require energetic behaviour of this type. However, the other possible major process requiring free energy is active transport, which can also lead to a large outflow of entropy from the cells. Active transport processes in micro-organisms have been comparatively little studied (Zarlengo & Schultz, 1966) so that it is premature to attempt a more detailed analysis of this aspect of the energetic behaviour of the organisms.

Accordingly, we may conclude that the major factor in the energetic behaviour of the growing microbial cell is the necessity for the cell to maintain a very large rate of outflow of entropy, but that the relative importance of the contributions of biochemical synthesis, structural ordering and active transport cannot at present be accurately assessed.

REFERENCES

AIBA, S., NAGAI, S., NISHIZAWA, Y. & ONODERA, M. (1967). Chemostatic culture of *Azotobacter vinelandii. J. gen. appl. Microbiol. Tokyo*, **13**, 73.

BAAS-BECKING, L. G. M. & PARKS, G. S. (1927). Energy relations in the metabolism of autotrophic bacteria. *Physiol. Rev.* **7**, 85.

BATTLEY, E. H. (1960). Growth-reaction equations for *Saccharomyces cerevisiae. Physiol. Plant.* **13**, 192.

BAUCHOP, T. & ELSDEN, S. R. (1960). The growth of micro-organisms in relation to their energy supply. *J. gen. Microbiol.* **23**, 457.

BECK, R. W. & SHUGART, L. R. (1966). Molar growth yields in *Streptococcus faecalis. J. Bact.* **92**, 802.

BELAICH, J-P. & SENEZ, J. C. (1965). Growth yields of *Zymomonas mobilis. J. Bact.* **89**, 1195.

BELAICH, J-P. & SENEZ, J. C. (1967). Application de la méthode microcalorimétrique à l'étude de l'énergétique de la croissance bactérienne. *Colloques int. Cent. natn. Rech. scient.* **156**, 381.

BELAICH, J-P., SENEZ, J. C. & MURGIER, M. (1968). Microcalorimetric study of glucose permeation in microbial cells. *J. Bact.* **95**, 1750.

BOIVINET, P. (1964). *Utilisation du microcalorimètre en microbiologie.* Thesis, Université d'Aix-Marseille.

BUCHANAN, B. B. & PINE, L. (1967). Glucose breakdown and cell yields of *Actinomyces naeslundii. J. gen. Microbiol.* **46**, 225.

BULDER, C. J. E. A. (1963). *On respiratory deficiency in yeast.* Thesis, Technical University, Delft.

BULDER, C. J. E. A. (1966). Utilization of fermentation energy in petite negative yeasts. *Arch. Mikrobiol.* **53**, 189.

CHEN, S. L. (1964). Energy requirements for microbial growth. *Nature, Lond.* **202**, 1135.

COLE, H. A., WIMPENNY, J. W. T. & HUGHES, D. E. (1967). The ATP pool in *Escherichia coli. Biochim. biophys. Acta*, **143**, 445.

DAWES, E. A., RIBBONS, D. W. & REES, D. A. (1966). Sucrose utilization by *Zymomonas mobilis. Biochem. J.* **98**, 804.

DAMOGLOU, A. P. & DAWES, E. A. (1967). The lipid content and phosphate requirements of glucose- and acetate-grown *Escherichia coli. Biochem. J.* **102**, 37P.

DEAN, A. C. R. & HINSHELWOOD, C. N. (1966). *Growth, Function and Regulation in Bacterial Cells*. Oxford: University Press.

DENBIGH, K. G. (1951). *The Thermodynamics of the Steady State*. London: Methuen.

FORREST, W. W. (1965). The ATP pool in *Streptococcus faecalis. J. Bact.* **90**, 1013.

FORREST, W. W. (1967). Activation energies and uncoupled growth. *J. Bact.* **94**, 1459.

FORREST, W. W. (1968). Microcalorimetry. In *Microbiological Methods*. Ed. D. W. Ribbons and J. R. Norris. London: Academic Press.

FORREST, W. W. & WALKER, D. J. (1963). Calorimetric measurements of energy of maintenance in *Streptococcus faecalis. Biochem. biophys. Res. Commun.* **13**, 217.

FORREST, W. W. & WALKER, D. J. (1964). Change in entropy during bacterial metabolism. *Nature, Lond.* **201**, 49.

FORREST, W. W. & WALKER, D. J. (1965a). Endogenous reserves in *Streptococcus faecalis. J. Bact.* **89**, 1448.

FORREST, W. W. & WALKER, D. J. (1965b). Control of glycolysis in washed suspensions of *Streptococcus faecalis. Nature, Lond.* **207**, 46.

FORREST, W. W., WALKER, D. J. & HOPGOOD, M. F. (1961). Enthalpy changes associated with the lactic fermentation of glucose. *J. Bact.* **82**, 648.

FRANZEN, J. S. & BINKLEY, S. B. (1961). The acid-soluble nucleotides in *Escherichia coli. J. biol. Chem.* **236**, 515.

GODDARD, J. L. & SOKATCH, J. R. (1964). 2-Ketogluconate fermentation by *Streptococcus faecalis. J. Bact.* **87**, 844.

GRANGETTO, A. (1963). *Thermogenèse de croissance aerobie d'Aerobacter aerogenes*. Thesis, Université d'Aix-Marseille.

GUNSALUS, I. C. & SHUSTER, C. W. (1961). Energy-yielding metabolism in bacteria. In *The Bacteria*, vol. II, p. 1. Ed. I. C. Gunsalus and R. Y. Stanier. New York: Academic Press.

HADJIPETROU, L. P. & STOUTHAMER, A. H. (1965). Energy production during nitrate respiration by *Aerobacter aerogenes. J. gen. Microbiol.* **38**, 29.

HADJIPETROU, L. P., GERRITS, J. P., TEULINGS, F. A. G. & STOUTHAMER, A. H. (1964). Relation between energy production and growth of *Aerobacter aerogenes. J. gen. Microbiol.* **36**, 139.

HERNANDEZ, E. & JOHNSON, M. J. (1967a). Anaerobic growth yields of *Aerobacter cloacae* and *Escherichia coli. J. Bact.* **94**, 991.

HERNANDEZ, E. & JOHNSON, M. J. (1967b). Energy supply and cell yield in aerobically grown micro-organisms. *J. Bact.* **94**, 996.

HOBSON, P. N. (1965). Continuous culture of rumen bacteria. *J. gen. Microbiol.* **38**, 167.

HOBSON, P. N. & SUMMERS, R. (1967). The continuous culture of anaerobic bacteria. *J. gen. Microbiol.* **47**, 53.

HOPGOOD, M. F. & WALKER, D. J. (1967). Succinic acid production by rumen bacteria. I. Isolation and metabolism of *Ruminococcus flavefaciens. Aust. J. biol. Sci.* **20**, 165.

HUNGATE, R. E. (1963). Polysaccharide storage and growth efficiency in *Ruminococcus albus. J. Bact.* **86**, 848.

KELLY, D. P. & SYRETT, P. J. (1966). Energy coupling during sulphur compound oxidation by *Thiobacillus. J. gen. Microbiol.* **43**, 109.

LEES, H. (1954). The biochemistry of the nitrifying bacteria. *Symp. Soc. gen. Microbiol.* **4**, 84.

LINCHITZ, H. (1953). The information content of a bacterial cell. In *Information Theory in Biology*. Ed. H. Quastler. Urbana: University of Illinois Press.

McGREW, S. B. & MALLETTE, M. F. (1965). Maintenance of *Escherichia coli* and the assimilation of glucose. *Nature, Lond.* **208**, 1096.

MAYBERRY, W. R., PROCHAZKA, G. J. & PAYNE, W. J. (1967). Growth yields of bacteria on selected organic compounds. *Appl. Microbiol.* **15**, 1332.

MEYERHOF, O. (1924). *Chemical Dynamics of Life Phenomena*, p. 92. Philadelphia: Lipincott.

MONOD, J. (1942). *Recherches sur la croissance des cultures bactériennes.* Paris: Herman.

MOROWITZ, H. J. (1955). Some order-disorder considerations in living systems. *Bull. math. Biophys.* **17**, 81.

MOROWITZ, H. J. (1960). Application of the second law of thermodynamics to cellular systems. *Biochim. biophys. Acta*, **40**, 340.

OXENBURGH, M. S. & SNOSWELL, A. M. (1965). Molar growth yields for the evaluation of energy-producing pathways in *Lactobacillus plantarum. J. Bact.* **89**, 913.

PIRT, S. J. (1965). Maintenance energy of bacteria in growing cultures. *Proc. Roy. Soc.* B, **163**, 224.

POLAKIS, E. S. & BARTLEY, W. (1966). Changes in intracellular concentrations of adenosine phosphates and nicotinamide nucleotides during aerobic growth of yeast. *Biochem. J.* **99**, 521.

PRAT, H. (1963). In *Recent Progress in Microcalorimetry.* Ed. E. Calvet and H. Prat. Oxford: Pergamon Press.

PRIGOGINE, I. (1961). *Thermodynamics of Irreversible Processes.* Second edition. New York: Interscience.

ROSENBERGER, F. & ELSDEN, S. R. (1960). The yields of *Streptococcus faecalis* grown in continuous culture. *J. gen. Microbiol.* **22**, 726.

SENEZ, J. C. (1962). Some considerations on the energetics of bacterial growth. *Bact. Rev.* **26**, 95.

SENEZ, J. C. & BELAICH, J-P. (1965). Mécanismes de régulation des activités cellulaires chez les micro-organismes. *Colloques int. Cent. natn. Rech. scient.* **124**, 357.

SOKATCH, J. T. & GUNSALUS, I. C. (1957). Aldonic acid metabolism. I. Pathway of carbon in an inducible gluconate fermentation by *Streptococcus faecalis. J. Bact.* **73**, 452.

STRANGE, R. E., WADE, H. E. & DARK, F. A. (1963). Effect of starvation on adenosine triphosphate concentration in *Aerobacter aerogenes. Nature, Lond.*, **199**, 55.

STOUTHAMER, A. H. (1968). Molar growth yields. In *Microbiological Methods.* Ed. D. W. Ribbons and J. R. Norris. London: Academic Press.

WILKIE, D. R. (1960). Thermodynamics and the interpretation of biological heat measurements. *Progr. Biophys. mol. Biol.* **10**, 259.

ZARLENGO, M. H. & SCHULTZ, S. G. (1966). Cation transport and metabolism in *Streptococcus faecalis. Biochim. biophys. Acta*, **126**, 308.

QUANTITATIVE RELATIONSHIPS BETWEEN INORGANIC CATIONS AND ANIONIC POLYMERS IN GROWING BACTERIA

D. W. TEMPEST

Microbiological Research Establishment, Porton Down, Salisbury, Wiltshire

INTRODUCTION

Each of the various inorganic cations commonly found in growing bacteria can function as an activator of some enzyme-catalysed reaction (Dixon & Webb, 1964). However, spectrochemical analyses of these elements in vegetative bacteria and spores indicate that potassium, magnesium and, possibly, calcium are present in much higher concentrations than one would expect for them to be functioning in the

Table 1. *Inorganic cation content of some vegetative bacteria and bacterial spores*

g./100 g. dried bacteria ...	K^+	Mg^{2+}	Ca^{2+}	Fe^{3+}	Al^{3+}	Cu^{2+}	Mn^{2+}	Zn^{2+}
Bacillus subtilis	4·9	0·3	0·01	—	—	—	—	—
*Bacillus cereus**	4·6	1·1	0·03	0·02	0·01	0·01	0·01	0·02
B. megaterium	4·0	0·2	—	—	—	—	—	—
B. megaterium†	2·1	1·1	1·1	0·01	0·01	0·01	0·01	—
B. subtilis spores†	0·9	0·5	1·6	0·01	0·01	0·01	0·01	—
B. cereus spores†	0·2	0·3	1·9	0·02	0·04	0·02	0·01	—
B. megaterium spores†	0·6	0·5	1·0	0·01	0·01	0·02	0·01	—
B. macerans†	0·5	0·4	0·2	0·02	0·02	0·01	0·01	—
*Escherichia coli**	1·2	0·6	0·01	0·02	0·01	0·01	0·01	0·01
Aerobacter aerogenes	1·6	0·3	0·02	—	—	—	—	—
Rhizobium trifolii‡	—	0·2	0·2	—	—	—	—	—

* Data of Rouf (1964).   † Data of Curran, Brunstetter & Myers (1943).
‡ Data of Humphrey & Vincent (1962).

growing cell *solely* as enzyme activators. Thus, from the data assembled in Table 1, and making assumptions for the intracellular water content, it is clear that the vegetative organisms listed contained up to 150 mM magnesium and up to 300 mM potassium. The reasons for these high intracellular cation concentrations are not immediately obvious; potassium can, to some extent, be replaced by rubidium (Eddy & Hinshelwood, 1950), but in the absence of rubidium probably all bacteria require potassium for growth (Lester, 1958). On the other hand, magnesium cannot be replaced by any other cation (Webb, 1949, 1966,

1968; Tempest, Hunter & Sykes, 1965). It should be pointed out here that sodium also is present in substantial amounts in non-growing bacteria (Schultz, Epstein & Solomon, 1963) and in halophiles (Christian & Waltho, 1962); however, it is almost absent from bacteria actively growing in media of moderate salinity (Zarlengo & Schultz, 1966) and therefore will not be considered further.

Since magnesium and potassium are present in bacteria in concentrations that make them major cellular components it is easy to arrange chemostat conditions in which each cation is the growth-limiting component of the medium. And when growth of an organism is limited by the availability of such a nutrient the organism may be expected to contain the minimum concentration of that element necessary for growth at the imposed rate in the prescribed environment. Any change in this minimum requirement would be indicated by a change in the growth yield (that is, g. organisms synthesized/g. growth-limiting nutrient consumed) and since at growth rates less than the maximum value ($\mu_{max.}$) nearly all the growth-limiting nutrient present in the culture may be expected to be contained within the organisms, a change in yield would be evident as a change in the steady-state concentration of organisms in the culture. Thus, by varying judiciously the nature of the growth environment (e.g. temperature, pH value, osmolarity, etc.) and growth rate, and by studying the effects of these changes on both the growth yield and the physiology of bacteria in the culture, insight may be obtained into the functional basis of the cellular requirement for each growth-limiting nutrient. This article mainly summarizes the results of such studies on the bacterial requirement for magnesium and potassium; it is concluded that the concentrations of these cations in growing bacteria are related largely to the amounts of certain anionic polymers present in the growing organisms.

MAGNESIUM

Probably no bacteria are able to grow in environments that are completely devoid of magnesium (Young, Begg & Pentz, 1944; Webb, 1949, 1966, 1968) and with a Mg^{2+}-limited chemostat culture of *Aerobacter aerogenes* (growing at a fixed dilution rate, temperature and pH value) it was found that the steady-state bacterial concentration was directly proportional to the culture magnesium content (Tempest *et al.* 1965); thus, a plot of bacterial concentration *versus* magnesium concentration was linear and extrapolated to the origin (Fig. 1). The fact that the extrapolated plot passed through the origin indicated not only that other cations present in the medium could not functionally replace magnesium, but also that almost all of the magnesium present in this culture was contained in the organisms (see Fig. 1).

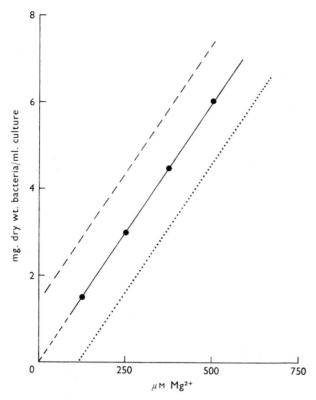

Fig. 1. Plot of bacterial concentration versus culture magnesium concentration for a Mg^{2+}-limited chemostat culture of *Aerobacter aerogenes* (dilution rate of 0·4 hr.$^{-1}$; 37°C; pH 6·5). The upper (broken) line represents the situation that would have prevailed if other substances present in the medium had functionally replaced some magnesium in the organisms. The lower (dotted) line represents the situation that would have prevailed if the extracellular magnesium concentration had been a significantly large fraction of the total culture magnesium concentration. Data from Tempest *et al.* (1965).

With Mg^{2+}-limited chemostat cultures of *Aerobacter aerogenes* (and of *Pseudomonas putida*) the culture bacterial concentration varied with the dilution rate,* and at a rate of about 0·1 hr.$^{-1}$ the yield value was approximately double that at a dilution rate of about 0·8 hr.$^{-1}$ (Tempest *et al.* 1965; Sykes & Tempest, 1965). Similar changes in yield, with dilution rate, were observed with Mg^{2+}-limited cultures of *Bacillus subtilis* var. *niger* (Tempest, Dicks & Meers, 1967) and *Torula*

* Ratio, medium flow rate : culture volume = specific growth rate in steady state.

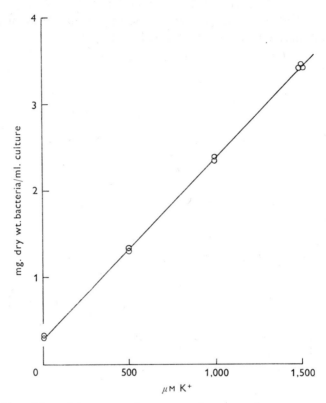

Fig. 2. Plot of bacterial concentration versus culture potassium concentration in a K+-limited chemostat culture of *Aerobacter aerogenes* (dilution rate of 0·4 hr.$^{-1}$; 35°C.; pH 6·5). Failure of the plot to pass through the origin indicates the presence of substance(s) in the culture that functionally replaced some of the potassium (see Fig. 1). Data from Tempest *et al.* (1966).

Table 2. *Approximate yield values for Mg^{2+}-limited chemostat cultures of* Torula utilis, Pseudomonas fluorescens, Bacillus subtilis *and* Aerobacter aerogenes

The yeast and *Ps. fluorescens* cultures were grown at 30°C., the *B. subtilis* culture at 35°C. and the *A. aerogenes* culture at 37°C. In each culture the pH was controlled automatically, at 5·5, 7·2, 7·0 and 6·5, respectively.

Dilution rate	g. organisms formed/g. Mg^{2+} consumed			
(hr.$^{-1}$)	*T. utilis*	*Ps. fluorescens*	*B. subtilis*	*A. aerogenes*
0·1	910	610	720	830
0·2	805	550	570	590
0·4	625	470	450	460
0·6	—	350	390	380
0·8	—	—	—	340

utilis (Table 2). Thus, in all cases the minimum intracellular magnesium requirement varied substantially with growth rate, being greatest at the maximum growth rate. This suggested a quantitative relationship between magnesium and some growth-rate-dependent cellular process, although the yield change could have resulted from a variable synthesis of some storage-type polymer (e.g. glycogen, poly-β-hydroxybutyric acid, etc.) that added to the bacterial dry weight but not to the basic magnesium requirement. However, analysis of Mg^{2+}-limited *A. aerogenes* and *B. subtilis* showed that, at all growth rates, over 80% of the bacterial dry weight could be accounted for as protein, RNA and DNA; clearly the variation in cellular magnesium content with growth rate must have resulted from processes other than accumulation of storage-type compounds.

The only other obvious growth-rate-linked change in the composition of Mg^{2+}-limited bacteria was in their RNA content which varied in a manner that was strikingly similar to the variation in cellular magnesium content. Thus, although with Mg^{2+}-limited cultures of *Aerobacter aerogenes*, *Bacillus subtilis* or *Pseudomonas putida* the cell-bound RNA and magnesium contents varied considerably with growth rate, their molar ratios (RNA-nucleotide:magnesium) remained almost constant (Tempest *et al.* 1965, 1967; Sykes & Tempest, 1965). This relationship suggested an inter-dependence between RNA and magnesium in the growing organisms.

Changes in the RNA content of organisms with growth rate are largely the result of changes in the cellular ribosome content (Wade & Morgan, 1957; Ecker & Schaechter, 1963; Kjeldgaard & Kurland, 1963). The above findings imply that the variation in bacterial magnesium requirement also was due largely to changes in the cellular ribosome content, and this is reasonable since magnesium probably is an integral part of the ribosome structure (Tissieres & Watson, 1958; Rodgers, 1964; Goldberg, 1966) and affects ribosome stability both *in vivo* and *in vitro* (McCarthy, 1962; Suzuki & Hayashi, 1964; Cohen & Ennis, 1967; Kennell, 1967). But in order to establish unequivocally a direct relationship between bacterial magnesium and ribosome contents it would be necessary to show that under all conditions where the bacterial ribosome content was caused to vary, the magnesium content varied similarly. For example, irrespective of the nature of the growth limitation the bacterial magnesium content should vary with growth rate in a parallel manner to the changes in RNA content. With chemostat cultures of *Aerobacter aerogenes* and *Bacillus subtilis*, limited in their growth by the availability of K^+, NH_4^+, PO_4^{3-} or the carbon source (glycerol and glucose, respectively), the bacterial magnesium contents were indeed found to increase progressively with growth rate in a manner that followed closely changes in the cellular RNA content (Table 3).

Table 3. *Magnesium and RNA contents of variously limited chemostat cultures of* Aerobacter aerogenes *(35° C, pH 6·5)*

Organisms were grown in 0·5 l. Porton-type chemostats (Herbert, Phipps & Tempest, 1965) in simple salts media (Tempest, 1965) with glycerol as the carbon source. Magnesium was determined by atomic absorption spectrophotometry and RNA by the orcinol test (Militzer, 1946) on $HClO_4$ extracts (Burton, 1956).

Dilution rate (hr.$^{-1}$)	Magnesium limitation (g./100 g. dried bacteria)					RNA limitation (g./100 g. dried bacteria)				
	Mg^{2+}	C	K^+	NH_4^+	PO_4^{3-}	Mg^{2+}	C	K^+	NH_4^+	PO_4^{3-}
0·1	0·11	0·13	0·12	0·18	0·10	8·4	9·4	7·3	7·5	5·7
0·2	0·17	0·18	0·16	0·22	0·13	11·0	10·5	11·7	10·0	8·0
0·4	0·22	0·20	0·20	0·26	0·18	15·1	14·1	15·0	13·6	12·5
0·8	0·27	0·30	0·24	0·30	0·31	16·4	18·3	17·6	18·3	20·0

Again, if as suggested by Schaechter, Maaløe & Kjeldgaard (1958) and by Ecker & Schaechter (1963), the concentration of ribosomes in growing organisms is directly related to the rate of protein synthesis, then if the latter was fixed (as in a chemostat culture) yet the activity of the ribosomes made to vary (e.g. by varying the incubation temperature) the ribosome content of the organisms must then change in order to maintain the over-all rate of protein synthesis, and hence the growth rate, at the value fixed by the dilution rate. Thus, by varying the temperature of a culture growing in a chemostat, one could provide a means of varying the cellular ribosome content without altering the growth rate. This experiment was carried out (both with Mg^{2+}-limited and glycerol-limited cultures of *Aerobacter aerogenes*) and the changes in bacterial ribosome, RNA and magnesium contents determined (Tempest & Hunter, 1965). The results are summarized in Table 4. Clearly, altering the incubation temperature from 35° to 25°C effected a large increase (about 30%) in the bacterial ribosome and RNA contents (see also Harder & Veldkamp, 1967); the magnesium contents again varied similarly so that the RNA:magnesium ratios remained almost constant.

Thus it is reasonable to conclude that the large requirement of bacteria for magnesium results principally from a specific, stoichiometric requirement of their ribosomes for this cation (see also Edelman, Ts'o & Vinograd, 1960; Goldberg, 1966).

Table 4. *Effect of temperature on the RNA, magnesium and ribosome contents of* Aerobacter aerogenes, *growing in a chemostat* $(D=0.2$ hr.$^{-1}$; *pH* 6.5)

The magnesium, RNA and ribosome contents were determined by the methods described by Tempest & Hunter (1965). The molar ratio of RNA to Mg^{2+} assumes a molecular weight of 340 for RNA-nucleotide.

Temperature	g./100 g. dried bacteria			Molar ratio
(°C.)	Ribosomes	RNA	Magnesium	RNA/Mg^{2+}
	Mg^{2+}-limited organisms			
40	—	9.4	0.14	4.8
35	26.5	10.7	0.15	5.0
30	—	12.4	0.17	5.1
25	33.1	14.9	0.21	5.0
	Glycerol-limited organisms			
40	—	7.6	—	—
35	24.4	10.1	—	—
30	—	11.0	—	—
25	29.9	11.8	—	—

Although the bulk of the magnesium contained in a bacterial cell must be present in the ribosomes, some is associated with substances that form the cell envelope. Thus, Gordon & MacLeod (1966) reported a molar ratio of P:Mg^{2+} of about 7:1 in phospholipid-containing extracts from the envelopes of *Pseudomonas aeruginosa* and of a marine pseudomonad (NCMB 19). That this magnesium is intimately involved in maintaining the structural integrity and functional ability of the envelope is suggested by the opposite effects of EDTA and Mg^{2+} on the permeability of bacteria towards various intracellular and extracellular substances (Strange & Dark, 1962; Strange, 1964; Strange & Postgate, 1964; Leive, 1965; Gray & Wilkinson, 1965; Hamilton-Miller, 1966) and the evidence that association-dissociation of membrane particles from bacteria, *in vitro*, is magnesium-dependent (Brown, 1965). Also of relevance in this connection is the observation of Sykes & Tempest (1965) that washed suspensions of Mg^{2+}-limited *Pseudomonas putida* were not permeable to glucose unless Mg^{2+} was present in the suspending buffer.

As well as the membrane-bound magnesium, much magnesium usually is to be found loosely attached to the surface of organisms that have been separated from media containing an excess of this cation; this 'adsorbed' magnesium is readily removed by suspending the organisms in 0.85%, w/v, NaCl solution (Strange & Shon, 1964). Although no loosely attached magnesium was to be found on the surface of Mg^{2+}-limited bacteria (Tempest & Strange, 1966; Tempest *et al.* 1967) the surface-bound magnesium, when present, may have some functional significance. For example, it was found to stimulate polysaccharide synthesis in *Aerobacter aerogenes* and to increase the resistance of bacteria to stresses such as starvation, heat-accelerated death, substrate-accelerated death and cold-shock (Tempest & Strange, 1966). Furthermore, it is no doubt of some significance that the capacity of bacterial cell walls to bind magnesium was found to vary with the growth condition (D. W. Tempest & J. L. Meers, unpublished observation); growing either *A. aerogenes* or *B. subtilis* in Mg^{2+}-limited environments, or in environments of high salinity (i.e. 2–4 %, w/v, NaCl) resulted in the synthesis of walls with an enhanced binding capacity, and affinity, for Mg^{2+}. This adaptive change, presumably in the cell-wall composition, strongly suggests that surface anionic structures play some role in the uptake of magnesium from the environment.

From a strictly quantitative point of view, the enzyme-bound magnesium of a bacterial cell must be of little significance. Nevertheless, variations in the concentration of intracellular non-ribosomal magnesium may be expected to affect the metabolic activity of the growing cell. Wyatt (1964) proposed a model system for the regulation of some enzyme activities in bacteria in which cations were specific activators and the metabolic end-products chelating agents that competed with the enzymes for the activating ions. Whether such regulatory systems actually operate in growing bacteria is not known, but the findings of Sykes & Tempest (1965) of a substantial effect of magnesium on the oxidation of citrate and acetate by washed suspensions of citrate-grown, Mg^{2+}-limited, *Pseudomonas putida*, but not by washed suspensions of Mg^{2+}-sufficient, citrate-limited organisms, indicates that the enzyme-associated magnesium content of this organism (in contrast to the ribosome-associated magnesium content) varied substantially with the growth condition.

POTASSIUM

As indicated previously (Table 1) much potassium is present in growing bacteria yet the physiological basis for this requirement is not fully understood. Lubin & Kessel (1960) isolated a mutant of *Escherichia coli* that had a decreased ability to concentrate potassium from its environment and found that the growth rate of this mutant, in a defined medium, was proportional to the culture potassium concentration over a wide range. Furthermore, with this mutant Ennis & Lubin

(1961, 1965) showed that when the intracellular potassium concentration was decreased to a low value, the organisms were unable to synthesize protein, though still capable of synthesizing RNA. Ennis & Lubin suggested that these potassium-depleted conditions produced a specific impairment of protein synthesis similar to that imparted by chloramphenicol. Previously Eddy & Hinshelwood (1951) had concluded that potassium was associated with cell metabolism and was not, at any rate in the full amount, among the permanent structural requisites. They suggested that intracellular potassium was associated with certain intermediates involved in the degradation of carbon substrates (see also Pulver & Verzar, 1940; Roberts, Roberts & Cowie, 1949; Perry & Evans, 1961; Blond & Whittam, 1965) and that the specificity for potassium resulted from its being of a size suitable for incorporation into a temporary structure involving phosphorylated intermediates and enzyme surfaces. Recently, however, Harold & Baarda (1968) showed that replacement of the potassium of *Streptococcus faecalis* by sodium did not greatly affect the rate of glycolysis, but protein synthesis was completely stopped.

Since, in the absence of rubidium, bacterial growth can be made dependent on potassium (MacLeod & Snell, 1948; Lester, 1958), chemostat conditions can be arranged in which availability of this cation limits the growth rate. In such a K^+-limited culture the organisms may be expected to contain the minimum concentration of potassium necessary for growth at the imposed rate in the prescribed environment. Thus, a systematic analysis of the effects of changing growth rate (and other parameters) on the physiology of K^+-limited organisms should provide data from which the structural and functional sites that require high intracellular concentrations of potassium may be located and identified.

Studies with Aerobacter aerogenes

As with Mg^{2+}-limited cultures, the concentration of *Aerobacter aerogenes* in a K^+-limited chemostat culture was found to be proportional to the medium potassium concentration. But although a plot of the culture bacterial dry weight versus the culture potassium content was linear (Fig. 2) it did not pass through the origin, suggesting the presence of substance(s) in the environment that were able to replace the bacterial potassium requirement to a small extent (cf. Fig. 1). With a K^+-limited medium (containing 40 μg. K^+/ml.) the concentration of *A. aerogenes* varied with the growth rate in a manner that was very similar to the variation observed with Mg^{2+}-limited cultures of this organism;

thus, the yield value (g. organisms synthesized/g. K^+ consumed) approximately doubled with a decrease in dilution rate from about 0·8 to 0·1 hr.$^{-1}$ (Tempest, Dicks & Hunter, 1966). This change in yield value indicated a variation in the minimum requirement for potassium with changes in growth rate and, since no accumulation of storage-type compounds could be detected, the changes in cell-bound potassium content obviously had some functional significance.

Because the variation in yield value of K^+-limited *Aerobacter aerogenes* cultures with dilution rate was similar to that with Mg^{2+}-limited cultures, and because the magnesium content of *A. aerogenes* appeared to be stoichiometrically related to its RNA content, it was reasonable to suppose that an interdependence between RNA and potassium also existed in growing *A. aerogenes*. This was confirmed by direct analysis of bacteria from K^+-limited *A. aerogenes* cultures grown at several different dilution rates (Table 5). Also included in Table 5 are data for the bacterial phosphorus contents; clearly the changes in cellular magnesium, potassium, RNA and phosphorus, with growth rate, were proportionately almost identical. Thus, the molar ratios of cell-bound magnesium:potassium:RNA (as nucleotide):phosphorus remained almost constant (at 1:4:5:8, respectively), regardless of the growth rate.

Table 5. *Potassium, magnesium, phosphorus and ribonucleic acid contents of K^+-limited* Aerobacter aerogenes, *grown at different dilution rates in a chemostat* (35°; *pH* 6·5)

Dilution rate (hr.$^{-1}$)	g./100 g. dry bacteria			
	Potassium	Magnesium	Phosphorus	RNA
0·1	0·80	0·12	1·31	7·3
0·2	1·08	0·16	1·69	11·7
0·4	1·42	0·19	2·17	15·0
0·6	1·55	0·21	2·31	17·4
0·8	1·68	0·23	2·56	17·6

Data of Tempest *et al.* (1966)

The growth-rate-independent stoichiometry between the magnesium, potassium, RNA and phosphorus contents of K^+-limited *Aerobacter aerogenes* organisms implied a functional inter-relationship between these substances and, since the bulk of the RNA and magnesium probably are contained in the ribosomes, the above findings suggested that this is where the bulk of the potassium also is located in the growing organisms. But in order to establish unequivocally a direct relationship between bacterial potassium and ribosome contents it again was necessary to show that under all conditions where the bacterial ribosome content varied, the cellular potassium content varied proportionately.

Changes in the potassium, magnesium and phosphorus contents of *Aerobacter aerogenes* have been measured as functions of the bacterial RNA content when the latter was caused to vary by altering the incubation temperature at a fixed dilution rate (under K^+-limiting conditions) and by changing the growth rate, at a fixed temperature, under conditions both of Mg^{2+}-limitation and of PO_4^{3-}-limitation. In all cases the changes in RNA content were accompanied by corresponding changes in bacterial magnesium, potassium and phosphorus contents and the above-mentioned molar relationship was maintained throughout (Dicks & Tempest, 1966).

The high degree of correlation between the increases in RNA and potassium in K^+-limited *Aerobacter aerogenes* as temperature was decreased (Table 6) is considered in particular to be evidence of an association between potassium and ribosomes in growing bacteria. This, together with the observation of a general correlation between these substances and magnesium in bacteria, irrespective of the nature of the growth limitation, makes it seem extremely likely that the large cellular requirement for potassium for growth results from a specific requirement for this cation by the functioning ribosomes.

Table 6. *Analyses of K^+-limited* Aerobacter aerogenes, *grown in a chemostat at different temperatures* $(D=0.2 \ hr.^{-1}; \ pH \ 6.5)$

Average values from duplicate samples collected and analysed on different days.

Temperature (°C.)	Yield (g. dry bacteria. g. K^+ consumed)	(g./100 g. dry bacteria)			
		Potassium	Magnesium	Phosphorus	RNA
35	95·3	1·05	0·16	1·76	12·1
30	83·7	1·20	0·17	1·92	13·3
25	71·5	1·40	0·19	2·30	16·4

Data of Dicks & Tempest (1966).

A generalization which has been derived from numerous *in vitro* studies on ribosomes is that the monovalent cation:Mg^{2+} ratio in the suspending medium is of paramount importance in the stabilization of these particles. Thus, high ratios tend to result in dissociation of the ribosomes into subunits, some degree of unfolding of the RNA chain and, possibly, breakdown through the action of adsorbed ribonuclease (Petermann, 1964; Cammack & Wade, 1965; Gavrilova, Ivanov & Spirin, 1966). At the other extreme, low ratios favour tighter cross-

7

linking of the RNA, association of subunits and ribosomal aggregation (Petermann, 1964). It is reasonable to suppose that over the range of possible association-dissociation and configurational states, only within fairly precise limits are these organelles functional. Since, in the growing bacteria, potassium is the predominant monovalent cation (Table 1), its concentration around the ribosomes presumably must be kept constant, relative to magnesium, for the ribosomes to be maintained in a functional state. As stated above, it probably is this particular requirement for potassium that accounts for the high concentration of this cation in growing *Aerobacter aerogenes*.

Contrary to the above conclusion, Epstein & Schultz (1965) showed that sudden changes in the osmolarity of the environment in which *Escherichia coli* was being grown produced rapid changes in the bacterial potassium content. They proposed, therefore, that a major determinant of the intracellular potassium concentration was the medium osmolarity. However, Epstein & Schultz did not equalize the growth rates of organisms in cultures of different osmolarity, consequently differences in bacterial ribosome content may have contributed to the observed differences in bacterial potassium content (though presumably this could not account for the rapid initial changes observed). On the other hand, the experiments of Tempest and coworkers, described above, took no account of the medium osmolarity which was not controlled and probably varied; therefore the parallel changes in cellular RNA and potassium contents that were observed when either growth rate or incubation temperature were varied could have been fortuitous. Experiments to resolve this conflict were carried out by Tempest & Meers (1968). They were able to confirm that even in a K^+-limited chemostat culture (where the organisms may be expected to contain the minimum concentration of potassium necessary for growth in the particular environment) the potassium content of *Aerobacter aerogenes* did indeed vary with medium salinity (Fig. 3); and this also occurred with either glycerol-limited or Mg^{2+}-limited cultures of this organism. But independent of any changes in the medium salinity, the potassium content of *Aerobacter aerogenes* still varied with growth rate. Therefore it must be concluded that the potassium content of an organism such as *Aerobacter aerogenes* is basically prescribed by its cellular ribosome content and this amount of potassium, together with small amounts of other low molecular weight compounds, is sufficient to provide an adequate internal hydrostatic pressure when the organisms are suspended in environments of moderately low ionic strength. However, when exposed to environments of high salinity, plasmolysis can be prevented only by a further uptake of potassium (plus an equivalent amount of low molecular weight anion, presumably) and this occurs. But since there is no corresponding change in the RNA and magnesium contents of the organisms, this 'extra' potassium must be compartmented in a way such that it does not affect the structural configuration, and functioning, of the ribosomes.

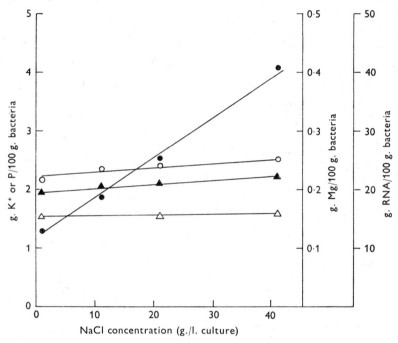

Fig. 3. Influence of medium salinity on the magnesium, potassium, RNA and phosphorus contents of *Aerobacter aerogenes* growing in a K^+-limited culture at a dilution rate of 0·3 hr.$^{-1}$ (35°C.; pH 6·5). (●), Bacterial potassium contents; (○), phosphorus content; (▲), magnesium contents; (△), RNA content. Data from Tempest & Meers (1968).

Studies with Bacillus subtilis var. niger

If, in media of moderate salinity, the bulk of the bacterial potassium really is required for maintenance of ribosomal configuration, then one might expect a stoichiometric relationship between magnesium, potassium, RNA and phosphorus, similar to that found in *Aerobacter aerogenes*, in all other organisms; particularly since bacterial ribosomes do not seem to vary greatly from species to species (Taylor & Storck, 1964). In this connection, no significant differences were found between the intracellular RNA and magnesium contents of *Aerobacter aerogenes* and *Bacillus subtilis* when grown at identical rates in similar environments (Tempest *et al.* 1967). However, a considerable difference is evident (Table 1) between the potassium contents of vegetative bacilli and Gram-negative organisms such as *Escherichia coli* and *Aerobacter aerogenes*. Examination of Mg^{2+}-limited *Bacillus subtilis* var. *niger*, grown in a chemostat in a simple salts medium, showed it to contain about 300 % more potassium and 50 % more phosphorus than Mg^{2+}-limited *Aerobacter aerogenes*, grown at the same rate (Tempest, Dicks & Ellwood, 1968). Growth of this bacillus in a K^+-limited environment (where the organisms may be expected to contain the minimum concentration of potassium) did not cause the cellular potassium and phosphorus contents to be lowered and the proportions of cell-bound magnesium:potassium:RNA (as nucleotide):phosphorus were found to approximate to 1:13:5:13, respectively.

Table 7. *Distribution of phosphorus in* Aerobacter aerogenes *and* Bacillus subtilis *growing at different dilution rates in a chemostat*

The data contained in this table are taken from the paper of Tempest *et al.* (1968).

	Distribution of phosphorus [g. (RNA + DNA)–P/g. total P]				
Dilution rate (hr.$^{-1}$)	0·05	0·1	0·2	0·4	0·8
A. aerogenes (K^+-limited)	—	0·67	0·73	0·73	0·71
B. subtilis (K^+-limited)	0·49	0·45	0·41	0·41	—
B. subtilis (Mg^{2+}-limited)	—	0·40	0·44	0·40	—
B. subtilis (PO_4^{3-}-limited)	—	0·91	0·86	0·74	—

The higher contents of both potassium and phosphorus in *Bacillus subtilis* (when compared with *Aerobacter aerogenes*) suggested the presence of phosphorus-containing compounds other than RNA and DNA which presumably had much potassium associated with them. Indeed, whereas the nucleic acid content of K^+-limited *Aerobacter aerogenes* accounted for over 70 % of the total cell-bound phosphorus, it accounted for no more than 50 % of the cellular phosphorus of either Mg^{2+}-limited or K^+-limited *Bacillus subtilis* (Table 7); but whether these non-nucleic acid phosphorus-containing substances have potassium specifically associated with them is conjectural. If the large potassium content of *Bacillus subtilis* is due to the presence of non-nucleic acid phosphorus-containing anionic substances, then any decrease in their concentration should effect a corresponding decrease in the bacterial potassium content. Growth of *Bacillus subtilis* var. *niger* in a phosphate-limited environment caused the cellular phosphorus content to be lowered to a value similar to that in Mg^{2+}-limited *Aerobacter aerogenes* and the nucleic acids now accounted for over

80 % of the total cell-bound phosphorus, but the potassium content was not correspondingly lowered (Tempest *et al.* 1968). It seemed, therefore, that, in *Bacillus subtilis* at least, there was no direct correlation between potassium, RNA and phosphate contents.

Table 8. *Influence of growth-limiting component of the medium on the composition of the cell walls of* Bacillus subtilis *var.* niger

The data in this table are from the paper of Tempest *et al.* (1968). The organisms were grown at 35°, pH 7·0, in 0·5 l. Porton-type chemostats (Herbert *et al.* 1965).

	Content (g. component/100 g. dry cell walls)	
	Phosphate-limited organisms	Magnesium-limited organisms
TCA-extractable polymer	40	40
Phosphorus	0·2	6·0
Glucose	< 1	28 (27*)
Glucuronic acid	22	< 2
Galactosamine	14	< 2

* Values determined by different methods.

Although the presence of much of the potassium in bacillus organisms could not be explained, the marked fluctuation in the concentration of non-nucleic acid phosphorus in *Bacillus subtilis*, when the organisms were changed from being K^+-limited to being phosphorus-limited, seemed worth investigating further. Since *Bacillus subtilis* is known to contain teichoic acids (Armstrong *et al.* 1958), which are phosphorus-containing polymers, these could have accounted for the extra phosphorus present in both Mg^{2+}-limited and K^+-limited *Bacillus subtilis* var. *niger* cultures; and in that case no such polymers ought to be present in the phosphate-limited organisms. Examination of the walls of Mg^{2+}-limited *Bacillus subtilis* showed them to contain about 6 % phosphorus (sufficient to account entirely for the non-nucleic acid phosphorus present in these bacilli) and a single extraction of the walls with trichloroacetic acid solubilized the bulk of the phosphate in a form that could then be precipitated as a polymer. Examination of this polymer showed it to be a glycerol-containing teichoic acid. On the other hand, walls from phosphate-limited bacilli contained little phosphorus (indicating the absence of teichoic acid) but nevertheless still contained a polymer that could be extracted with trichloroacetic acid, as above. Examination of this polymer (Table 8) showed that it was a teichuronic acid (Janczura, Perkins & Rogers, 1961). Because the teichoic acid present in the Mg^{2+}-limited *Bacillus subtilis* cell walls contained no alanine it would be strongly acidic; so would be the teichuronic acid present in the phosphate-limited cell walls (particularly since the galactosamine was N-acetylated) and both polymers were found to behave as strong anions when examined by paper electrophoresis (Dr D. C. Ellwood, personal communication). Thus, if the presence of teichoic acids in the walls of Mg^{2+}-limited and K^+-limited Bacillus subtilis did account for the presence of the 'extra' potassium, then replacement of teichoic acid by teichuronic acid (another anionic polymer) in the walls of phosphate-limited bacilli could account for the continued presence of the 'extra' potassium in these organisms and the apparent breakdown in the relationship between cell-bound potassium and phosphorus. But, assuming the specificity for potassium accumulation to reside at the plasma membrane, a specific association between potassium and cell-wall anionic polymers would not be expected—particularly when the organisms were grown in media that were high in sodium and low in potassium (i.e., K^+-limited). However, little cell-bound sodium could be detected in rapidly growing Bacillus subtilis organisms, even when they were K^+-limited, suggesting that potassium may indeed be bound selectively within the cell wall.

To get direct evidence for a specific association between cell-bound potassium and the anionic polymers present in the bacillus cell walls would necessitate culturing organisms that had walls which lacked both teichoic acid and teichuronic acid. This could not be accomplished with cultures of Bacillus subtilis var. niger but possibly was achieved with a chemostat culture of Bacillus megaterium, strain KM. Thus, whereas growth of Bacillus megaterium in a Mg^{2+}-limited environment produced organisms that had teichoic acids in their walls, and growth of this bacillus in a phosphate-limited environment produced organisms that had a teichuronic acid in their walls, when grown in a K^+-limited medium Bacillus megaterium cell walls apparently lacked both polymers; at least no anionic polymers were extracted from the walls by treatment with 10% (w/v) trichloroacetic acid for 16 hr. at 37°C. Actively growing K^+-limited Bacillus megaterium contained considerably less potassium than either Mg^{2+}-limited or phosphate-limited bacilli; the ratios of cell-bound K^+/Mg^{2+} and P/Mg^{2+} were close to those observed with K^+-limited cultures of Aerobacter aerogenes and very different from those observed with K^+-limited cultures of Bacillus subtilis var. niger (Table 9). It is not unreasonable to suggest, therefore, that the 'extra' potassium present in actively growing Bacilli (i.e. the potassium over and above that present in similarly grown Aerobacter aerogenes) is indeed associated with anionic polymers (teichoic acid and teichuronic acid) which are present largely in the bacterial cell walls.

Table 9. *The influence of the growth-limiting component of the medium on the potassium and phosphorus contents of* Bacillus megaterium, *strain KM comparison with* Bacillus subtilis *var.* niger *and* Aerobacter aerogenes

Data are from cultures growing in a simple salts medium in a chemostat ($D = 0.3$ hr.$^{-1}$; $35°$; pH 7.0, 7.0 and 6.5, respectively).

Growth	g./100 g. dried bacteria			Molar ratios	
Condition	Magnesium	Potassium	Phosphorus	K^+/Mg^{2+}	P/Mg^{2+}
	B. megaterium				
Mg^{2+}-limited	0·16	3·10	2·38	11·9	11·5
$PO_4{}^{3-}$-limited	0·17	2·19	1·70	7·9	7·8
K^+-limited	0·16	1·20	1·64	4·6	7·9
	B. subtilis var. *niger*				
K^+-limited	0·21	4·30	3·35	12·6	12·3
	A. aerogenes				
K^+-limited	0·18	1·25	1·93	4·3	8·3

CALCIUM

Whether all bacteria have a requirement for calcium, for growth, is not certain but this cation does seem to be necessary for the growth of several species of *Azotobacter* (Norris & Jensen, 1957) and *Leptospira pomona* (Johnson & Gary, 1963). When *Rhizobium trifolii* was grown in a calcium-depleted medium the cells had an abnormal morphology suggesting some impairment of synthesis of the bacterial cell wall (Vincent & Colburn, 1961). Isolation and examination of the walls from calcium-depleted organisms showed them to contain all the organic components typical of a Gram-negative cell wall but with only about 60 % of the expected calcium content (Humphrey & Vincent, 1962). Since nearly all the calcium present in a calcium-depleted culture was found to be contained in the bacterial cell walls, it was concluded that calcium was an essential component of the wall; that is, it could not be functionally replaced by magnesium or by other divalent cations.

There is much general evidence suggesting a preferred accumulation of calcium in the cell walls, and of magnesium in the cytoplasm, of organisms (Williams & Wacker, 1967) but since the bacterial requirement for calcium is quantitatively much less than for magnesium (see Table 1) it is difficult to obtain genuinely calcium-limited growth conditions in a chemostat culture. However, the calcium present in organisms grown in various environments in a chemostat can be determined and any differences correlated with changes in bacterial composition and activity. A preliminary investigation along these lines has been carried out by the author; the results are shown in Table 10. Changes in bacterial calcium content did not correlate with changes in bacterial RNA, magnesium, potassium or phosphorus but were similar to those in total cell-wall content. However, since the composition of bacterial walls may vary with growth condition (Ellwood & Tempest, 1967 a, b, 1968) and the calcium content of organisms probably relates more to the amount of some cell wall component than the cell wall as a whole, the significance of the data contained in Table 10 is difficult to assess. Nevertheless, one point is clear: since Mg^{2+}-limited *Aerobacter aerogenes* contained less calcium than glycerol-limited organisms, calcium obviously did not replace any of the bacterial magnesium requirement.

Table 10. *Effect of growth rate on the calcium content of* Aerobacter aerogenes *and* Bacillus subtilis *and its possible relationship to bacterial cell-wall content*

Organisms were grown in 0·5 l. Porton-type chemostats (Herbert, Phipps & Tempest, 1965) at 35° and pH 6·5 (for *A. aerogenes*) and 7·0 (for *B. subtilis*). The media were as described by Tempest, Hunter & Sykes (1965) with glycerol replaced by glucose for the growth of the *Bacillus*.

Growth condition	Dilution rate (hr.$^{-1}$)	g./100 g. dried bacteria		μg. Ca^{2+}/mg. cell wall
		Cell wall	Calcium	
Aerobacter aerogenes				
Glycerol-limited	0·1	20	0·025	1·25
	0·3	14	0·019	1·35
	0·7	11	0·015	1·36
Mg^{2+}-limited	0·1	19	0·016	0·84
	0·3	15	0·014	0·93
	0·7	13	0·014	1·08
Bacillus subtilis var. *niger*				
Mg^{2+}-limited	0·1	27	0·012	0·44
	0·2	24	0·010	0·42
	0·4	18	0·007	0·39

In contrast to vegetative bacilli, bacterial spores do contain large amounts of calcium (Table 1). Although the calcium present in some spores can be largely replaced by other divalent cations (Mn^{2+}, Zn^{2+}, Ni^{2+}, Co^{2+}, Cu^{2+}) their thermal resistance then is considerably lowered (Slepecky & Foster, 1959). Altering the calcium content of the medium in which *Bacillus cereus* was being grown altered simultaneously the calcium and dipicolinic acid contents of their resultant spores, suggesting some inter-relationship between these substances in the spores (Halvorson & Howitt, 1961). However, in other experiments Levison, Hyatt & Moore (1961) were able to show that the Ca^{2+}: dipicolinic acid ratio could vary considerably, but not without simultaneously affecting their heat tolerance.

OTHER CATIONS

Although potassium, magnesium and calcium are, quantitatively, the most important cations present in vegetative bacteria and spores, bacterial growth may be markedly affected by the presence or absence of other cations. For example, Webb (1968) found that growth and the uptake of magnesium by Mg^{2+}-limited cultures of *Bacillus megaterium* and *Bacillus subtilis* F3 (but not by similarly limited cultures of *Escherichia coli*) was markedly affected by added Mn^{2+}. He concluded that although manganese could not support the growth of organisms in the absence of magnesium, it was assimilated during growth and facilitated in some way the uptake of Mg^{2+} from dilute solutions (that is, those containing less than 1 μg. Mg^{2+}/ml.). Since it is known, however, that manganese is essential for the growth of some Gram-positive organisms (MacLeod & Snell, 1947; Lankford, Kustoff & Sergeant, 1957; Weinberg, 1964), a more probable explanation for the above findings is that *Bacillus megaterium* and *Bacillus subtilis* F3 have a small, but absolute, requirement for manganese that normally is met by contaminating amounts of this cation in the simple salts media. Thus, in the absence of added manganese, growth and the assimilation of nutrients (including magnesium) would cease if the concentration of contaminating manganese happened to be abnormally low and therefore became totally depleted before complete uptake of any of the other essential nutrients. And this situation would be most likely to occur if magnesium salts were the major source of manganese contamination and cultures were grown in low-magnesium media.

The quantitative requirements of bacteria for other cations is generally so small that it is almost impossible to prepare media sufficiently devoid of them to demonstrate an absolute need. However, it is possible to 'deferrate' some media to an extent sufficient to obtain Fe^{3+}-limited growth conditions (Waring & Werkman, 1942; Young *et al.* 1944). Thus it was found that iron was essential for the growth of aerobic bacteria since its availability influenced the synthesis and functioning of iron-containing enzymes and electron carriers (catalase, peroxidase, cytochromes) necessary for an aerobic metabolism (Waring & Werkman, 1944).

CONCLUSIONS

Growth can be defined as the co-ordinated synthesis of all cellular components and since the cellular content of each component, and its composition, may vary with environment (Herbert, 1961; Neidhardt, 1963; Tempest & Herbert, 1965; Dicks & Tempest, 1967) it is essential to take fully into account any progressive change in environment when attempting to analyse quantitatively the processes involved in growth. In this connection it is important to realize that in a 'batch' culture the environment changes continuously throughout the growth cycle, due to assimilation of utilizable substances and excretion of waste end-products by the growing organisms; thus, the physiological properties of organisms in a batch culture change continuously throughout the growth cycle. The environment can be kept constant only by means of a continuous-flow culture technique and the 'chemostat' method is usually preferred to that of the 'turbidostat' because it is operationally more stable (Herbert, 1958; Powell, 1965; Tempest, 1968) and can provide not only controlled environments but a considerable range of environments.

Application of continuous culture techniques to the study of cation metabolism in bacteria has yielded information from which it has been possible to determine, at least in part, the structural and functional bases of the cellular requirements for magnesium, potassium and calcium. It is concluded that the high concentrations of these substances in bacteria are connected mainly with the presence of certain anionic polymers, with which they are functionally associated. Thus, the magnesium content seems to be determined primarily by the cellular ribosome content, and so does the potassium content of *Aerobacter aerogenes* when the organisms are growing in media of moderate salinity. In contrast, the gross potassium content of *Bacillus subtilis* (which generally is much greater than that of *Aerobacter aerogenes*) is not determined largely by the bacterial ribosome content but by the concentration of anionic polymers (teichoic acids and teichuronic acids) that are present in, or associated with, the bacterial cell walls. The concentration of calcium in bacteria also possibly depends upon their cell wall content or, more particularly, upon the content of some cell wall component. In bacterial spores, on the other hand, the calcium is truly intracellular and is probably associated with dipicolinic acid.

Although the techniques used in studying cation involvement in bacterial growth and metabolism may be applied, almost without modification, to elucidating the quantitative requirements of bacteria for specific amino acids, purines, pyrimidines and vitamins, their main contribution in the future must be in unravelling the complex inter-relationships between environment and the mechanisms that control the precise composition, structure and functioning of cellular components such as ribosomes (see Sykes & Tempest, 1965; Tempest et al. 1965), membranes, and cell walls (see Sud & Schaechter, 1964; Collins, 1964). In this latter case much evidence is now accumulating to suggest that the wall may be one of the most dynamic components of the cell (Ellwood & Tempest, 1967 a, b, 1968; Tempest & Ellwood, 1968), and the significance of this to problems of vaccine production, for example, may be enormous.

REFERENCES

ARMSTRONG, J. J., BADDILEY, J., BUCHANAN, J. G., CARSS, B. & GREENBERG, G. R. (1958). Isolation and structure of ribitol phosphate derivatives (teichoic acids) from bacterial cell walls. *J. Chem. Soc.* p. 4344.

BLOND, D. M. & WHITTAM, R. (1965). Effects of sodium and potassium ions on oxidative phosphorylation in relation to respiratory control by a cell-membrane adenosine triphosphatase. *Biochem. J.* **97**, 523.

BROWN, J. W. (1965). Evidence for a magnesium-dependent dissociation of bacterial cytoplasmic membrane particles. *Biochim. biophys. Acta*, **94**, 97.

BURTON, K. (1956). A study of the conditions and mechanism of the diphenylamine reaction for the colorimetric estimation of deoxyribonucleic acid. *Biochem. J.* **62**, 315.

CAMMACK, K. A. & WADE, H. E. (1965). The sedimentation behaviour of ribonuclease-active and -inactive ribosomes from bacteria. *Biochem. J.* **96**, 671.

CHRISTIAN, J. H. B. & WALTHO, J. A. (1962). Solute concentrations within halophilic and non-halophilic bacteria. *Biochim. biophys. Acta*, **65**, 506.

COHEN, P. S. & ENNIS, H. L. (1967). Amino-acid regulation of RNA synthesis during recovery of *Escherichia coli* from Mg^{2+}-starvation. *Biochim. biophys. Acta*, **145**, 300.

COLLINS, F. M. (1964). The effect of the growth rate on the composition of *Salmonella enteritidis* cell walls. *Aust. J. exp. Biol. med. Sci.* **42**, 255.

CURRAN, H. R., BRUNSTETTER, B. C. & MYERS, A. T. (1943). Spectrochemical analysis of vegetative cells and spores of bacteria. *J. Bact.* **45**, 485.

DICKS, J. W. & TEMPEST, D. W. (1966). The influence of temperature and growth rate on the quantitative relationship between potassium, magnesium, phosphorus and ribonucleic acid of *Aerobacter aerogenes* growing in a chemostat. *J. gen. Microbiol.* **45**, 547.

DICKS, J. W. & TEMPEST, D. W. (1967). Potassium-ammonium antagonism in polysaccharide synthesis by *Aerobacter aerogenes* NCTC 418. *Biochim. biophys. Acta*, **136**, 176.

DIXON, M. & WEBB, E. C. (1964). *Enzymes*, 2nd ed. London: Longmans Green.

ECKER, R. E. & SCHAECHTER, M. (1963). Ribosome content and rate of growth of *Salmonella typhimurium*. *Biochim. biophys. Acta*, **76**, 275.

EDDY, A. A. & HINSHELWOOD, C. N. (1950). The utilization of potassium by *Bacterium lactis aerogenes*. *Proc. R. Soc.* B, **136**, 544.

EDDY, A. A. & HINSHELWOOD, C. N. (1951). Alkali-metal ions in the metabolism of *Bacterium lactis aerogenes*. III. General discussion of their role and mode of action. *Proc. R. Soc.* B, **138**, 237.

EDELMAN, I. S., TS'O, P. O. P. & VINOGRAD, J. (1960). The binding of magnesium to microsomal nucleoprotein and ribonucleic acid. *Biochim. biophys. Acta*, **43**, 393.

ELLWOOD, D. C. & TEMPEST, D. W. (1967*a*). Teichoic acid or teichuronic acid in the walls of *Bacillus subtilis* var. *niger*, grown in a chemostat. *Biochem. J.* **104**, 69P.

ELLWOOD, D. C. & TEMPEST, D. W. (1967*b*). Influence of growth condition on the cell wall content and wall composition of *Aerobacter aerogenes*. *Biochem. J.* **105**, 9P.

ELLWOOD, D. C. & TEMPEST, D. W. (1968). The teichoic acids of *Bacillus subtilis* var. *niger* and *Bacillus subtilis* W 23, grown in a chemostat. *Biochem. J.* **108**, 40.

ENNIS, H. L. & LUBIN, M. (1961). Dissociation of ribonucleic acid and protein synthesis in bacteria deprived of potassium. *Biochim. biophys. Acta*, **50**, 399.

ENNIS, H. L. & LUBIN, M. (1965). Pre-ribosomal particles formed in potassium-depleted cells. Studies on degradation and stabilisation. *Biochim. biophys. Acta*, **95**, 605.

EPSTEIN, W. & SCHULTZ, S. G. (1965). Cation transport in *Escherichia coli*, V. Regulation of cation content. *J. gen. Physiol.* **49**, 221.

GAVRILOVA, L. P., IVANOV, D. A. & SPIRIN, A. S. (1966). Studies on the structure of ribosomes. III. Stepwise unfolding of the 50S particles without loss of ribosomal protein. *J. molec. Biol.* **16**, 473.

GOLDBERG, A. (1966). Magnesium binding by *Escherichia coli* ribosomes. *J. molec. Biol.* **15**, 663.

GORDON, R. C. & MACLEOD, R. A. (1966). Mg^{2+} phospholipids in cell envelopes of a marine and terrestrial Pseudomonad. *Biochem. Biophys. Res. Commun.* **24**, 684.

GRAY, G. W. & WILKINSON, S. G. (1965). The action of ethylenediaminetetra-acetic acid on *Pseudomonas aeruginosa*. *J. appl. Bact.* **28**, 153.

HALVORSON, H. O. & HOWITT, C. (1961). The role of DPA in bacterial spores. In *Spores*, vol. II, p. 149. Ed. H. O. Halvorson. Minneapolis: Burgess Publ. Co.

HAMILTON-MILLER, J. M. T. (1966). Damaging effects of ethylene-diamine-tetra-acetate and penicillins on permeability barriers in Gram-negative bacteria. *Biochem. J.* **100**, 675.

HARDER, W. & VELDKAMP, H. (1967). A continuous culture study of an obligately psychrophilic *Pseudomonas* species. *Archiv. Mikrobiol.* **59**, 123.

HAROLD, F. M. & BAARDA, J. R. (1968). Effects of nigericin and monactin on cation permeability of *Streptococcus faecalis* and metabolic capacities of potassium-depleted cells. *J. Bact.* **95**, 816.

HERBERT, D. (1958). Some principles of continuous culture. In *Recent Progress in Microbiology, VIIth int. Congr. Microbiol.* p. 381.

HERBERT, D. (1961). The chemical composition of microorganisms as a function of their environment. *Symp. Soc. gen. Microbiol.* **11**, 391.

HERBERT, D., PHIPPS, P. J. & TEMPEST, D. W. (1965). The chemostat: design and instrumentation. *Lab. Practice*, **14**, 1150.

HUMPHREY, B. & VINCENT, J. M. (1962). Calcium in the cell walls of *Rhizobium trifolii*. *J. gen. Microbiol.* **29**, 557.

JANCZURA, E., PERKINS, H. R. & ROGERS, H. J. (1961). Teichuronic acid: a muco-polysaccharide present in wall preparations from vegetative cells of *Bacillus subtilis*. *Biochem. J.* **80**, 82.

JOHNSON, R. C. & GARY, N. D. (1963). Nutrition of *Leptospira pomona* III. Calcium, magnesium and potassium requirements. *J. Bact.* **85**, 983.

KENNELL, D. E. (1967). Nucleic acid and protein metabolism in *Aerobacter aerogenes* during magnesium starvation. In *Microbial Physiology and Continuous Culture*, p. 76. Ed. E. O. Powell, C. G. T. Evans, R. E. Strange and D. W. Tempest. London: H.M.S.O.

KJELDGAARD, N. O. & KURLAND, C. G. (1963). The distribution of soluble and ribosomal RNA as a function of the growth rate. *J. molec. Biol.* **6**, 341.

LANKFORD, C. E., KUSTOFF, T. Y. & SERGEANT, T. P. (1957). Chelating agents in growth initiation of *Bacillus globigii*. *J. Bact.* **74**, 737.

LEIVE, L. (1965). A non-specific increase in permeability in *Escherichia coli* produced by EDTA. *Proc. natn. Acad. Sci.; U.S.A.* **53**, 745.

LESTER, G. (1958). Requirement for potassium by bacteria. *J. Bact.* **75**, 426.

LEVISON, H. S., HYATT, M. T. & MOORE, F. E. (1961). Dependence of the heat resistance of bacterial spores on the calcium: dipicolinic acid ratio. *Biochem. Biophys. Res. Commun.* **5**, 417.

LUBIN, M. & KESSEL, D. (1960). Preliminary mapping of the genetic locus for potassium transport in *Escherichia coli*. *Biochem. Biophys. Res. Commun.* **2**, 249.

MACLEOD, R. A. & SNELL, E. E. (1947). Some mineral requirements of the lactic acid bacteria. *J. biol. Chem.* **170**, 351.

MACLEOD, R. A. & SNELL, E. E. (1948). The effect of related ions on the potassium requirement of lactic acid bacteria. *J. biol. Chem.* **176**, 39.

McCARTHY, B. J. (1962). Effects of magnesium starvation on the ribosome content of *Escherichia coli*. *Biochim. biophys. Acta*, **55**, 880.

MILITZER, W. E. (1946). Note on the orcinol reagent. *Archs Biochem.* **9**, 85.

NEIDHARDT, F. C. (1963). Effects of environment on the composition of bacterial cells. *Annu. Rev. Microbiol.* **17**, 61.

NORRIS, J. R. & JENSEN, H. L. (1957). Calcium requirements of *Azotobacter*. *Nature, Lond.* **180**, 1493.

PERRY, J. J. & EVANS, J. B. (1961). Role of potassium in the oxidative metabolism of *Micrococcus sodonensis*. *J. Bact.* **82**, 551.

PETERMANN, M. (1964). *The Physical and Chemical Properties of Ribosomes*. Amsterdam: Elsevier.

POWELL, E. O. (1965). Theory of the chemostat. *Lab. Practice*, **14**, 1145.

PULVER, R. & VERZAR, F. (1940). Connexion between carbohydrate and potassium metabolism in the yeast cell. *Nature, Lond.* **145**, 823.

ROBERTS, R. B., ROBERTS, Z. & COWIE, D. B. (1949). Potassium metabolism in *Escherichia coli*. 2. Metabolism in presence of carbohydrates and their derivatives. *J. cell. comp. Physiol.* **34**, 259.

RODGERS, A. (1964). The exchange properties of magnesium in *Escherichia coli* ribosomes. *Biochem. J.* **90**, 548.

ROUF, M. A. (1964). Spectrochemical analysis of inorganic elements in bacteria. *J. Bact.* **88**, 1545.

SCHAECHTER, M., MAALØE, O. & KJELDGAARD, N. O. (1958). Dependency on medium and temperature of cell size and chemical composition during balanced growth of *Salmonella typhimurium*. *J. gen. Microbiol.* **19**, 592.

SCHULTZ, S. G., EPSTEIN, W. & SOLOMON, A. K. (1963). Cation transport in *Escherichia coli*. IV. Kinetics of net K uptake. *J. gen. Physiol.* **47**, 329.

SLEPECKY, R. & FOSTER, J. W. (1959). Alterations in the metal content of spores of *Bacillus megaterium* and the effect on some spore properties. *J. Bact.* **68**, 117.

STRANGE, R. E. (1964). Effect of magnesium on permeability control in chilled bacteria. *Nature, Lond.* **203**, 1304.

STRANGE, R. E. & DARK, F. A. (1962). Effect of chilling on *Aerobacter aerogenes* in aqueous suspension. *J. gen. Microbiol.* **29**, 719.

STRANGE, R. E. & POSTGATE, J. R. (1964). Penetration of substance into cold-shocked bacteria. *J. gen. Microbiol.* **36**, 393.

STRANGE, R. E. & SHON, M. (1964). Effects of thermal stress on viability and ribonucleic acid of *Aerobacter aerogenes* in aqueous suspension. *J. gen. Microbiol.* **34**, 99.

SUD, I. J. & SCHAECHTER, M. (1964). Dependence of the content of cell envelopes on the growth rate of *Bacillus megaterium*. *J. Bact.* **88**, 1612.

SUZUKI, H. & HAYASHI, Y. (1964). The formation of ribosomal RNA in *Escherichia coli* during recovery from magnesium starvation. *Biochim. biophys. Acta*, **87**, 610.

SYKES, J. & TEMPEST, D. W. (1965). The effect of magnesium and of carbon limitation on the macromolecular organisation and metabolic activity of *Pseudomonas* sp., strain C–1 B. *Biochim. biophys. Acta*, **103**, 93.

TAYLOR, M. M. & STORCK, R. (1964). Uniqueness of bacterial ribosomes. *Proc. natn. Acad. Sci., U.S.A.* **52**, 958.

TEMPEST, D. W. (1965). An 8-liter continuous flow culture apparatus for laboratory production of microorganisms. *Biotechnol. Bioengng*, **7**, 367.

TEMPEST, D. W. (1968). The continuous cultivation of micro-organisms. 1. Theory of the chemostat. In *Methods in Microbiology*. Ed. D. W. Ribbons and J. R. Norris. London: Academic Press.

TEMPEST, D. W. & HERBERT, D. (1965). Effect of dilution rate and growth-limiting substrate on the metabolic activity of *Torula utilis* cultures. *J. gen. Microbiol.* **41**, 143.

TEMPEST, D. W. & HUNTER, J. R. (1965). The influence of temperature and pH value on the macromolecular composition of magnesium-limited and glycerol-limited *Aerobacter aerogenes* growing in a chemostat. *J. gen. Microbiol.* **41**, 267.

TEMPEST, D. W. & MEERS, J. L. (1968). The influence of medium NaCl concentration on the potassium content of *Aerobacter aerogenes* and on the inter-relationships between potassium, magnesium, and ribonucleic acid in the growing bacteria. *J. gen. Microbiol.* (in the Press).

TEMPEST, D. W. & STRANGE, R. E. (1966). Variation in content and distribution of magnesium, and its influence on survival, in *Aerobacter aerogenes* grown in a chemostat. *J. gen. Microbiol.* **44**, 273.

TEMPEST, D. W., DICKS, J. W. & ELLWOOD, D. C. (1968). Influence of growth condition on the concentration of potassium in *Bacillus subtilis* var. *niger* and its possible relationship to cellular ribonucleic acid, teichoic acid and teichuronic acid. *Biochem. J.* **106**, 237.

TEMPEST, D. W., DICKS, J. W. & HUNTER, J. R. (1966). The interrelationship between potassium, magnesium and phosphorus in potassium-limited chemostat cultures of *Aerobacter aerogenes*. *J. gen. Microbiol.* **45**, 135.

TEMPEST, D. W., DICKS, J. W. & MEERS, J. L. (1967). Magnesium-limited growth of *Bacillus subtilis* in pure and mixed cultures, in a chemostat. *J. gen. Microbiol.* **49**, 139.

TEMPEST, D. W., HUNTER, J. R. & SYKES, J. (1965). Magnesium-limited growth of *Aerobacter aerogenes* in a chemostat. *J. gen. Microbiol.* **39**, 355.

TISSIERES, A. & WATSON, J. D. (1958). Ribonucleoprotein particles from *Escherichia coli*. *Nature, Lond.* **182**, 778.

VINCENT, J. M. & COLBURN, J. R. (1961). Cytological abnormalities in *Rhizobium trifolii* due to a deficiency of calcium and magnesium. *Aust. J. Sci.* **23**, 269.

WADE, H. E. & MORGAN, D. M. (1957). The nature of the fluctuating ribonucleic acid in *Escherichia coli*. *Biochem. J.* **65**, 321.

WARING, W. S. & WERKMAN, C. H. (1942). Growth of bacteria in an iron-free medium. *Archs Biochem.* **1**, 303.

WARING, W. S. & WERKMAN, C. H. (1944). Iron deficiency in bacterial metabolism. *Archs Biochem.* **4**, 75.

WEBB, M. (1949). The influence of magnesium on cell division. 2. The effect of magnesium on the growth and cell division of various bacterial species in complex media. *J. gen. Microbiol.* **3**, 410.

WEBB, M. (1966). The utilization of magnesium by certain Gram-positive and Gram-negative bacteria. *J. gen. Microbiol.* **43**, 401.

WEBB, M. (1968). The influence of certain trace metals on bacterial growth and magnesium utilization. *J. gen. Microbiol.* **51**, 325.

WEINBERG, E. D. (1964). Manganese requirement for sporulation and other secondary biosynthetic processes in *Bacillus. Appl. Microbiol.* **12**, 436.

WILLIAMS, R. J. P. & WACKER, W. E. C. (1967). Cation balance in biological systems. *J. Am. med. Ass.* **201**, 18.

WYATT, H. V. (1964). Cations, enzymes and control of cell metabolism. *J. Theoret. Biol.* **6**, 441.

YOUNG, E. G., BEGG, R. W. & PENTZ, E. I. (1944). The inorganic nutrient requirements of *Escherichia coli. Archs Biochem.* **5**, 121.

ZARLENGO, M. H. & SCHULTZ, S. G. (1966). Cation transport and metabolism in *Streptococcus faecalis. Biochim. biophys. Acta*, **126**, 308.

THE REGULATION OF ENZYME SYNTHESIS DURING GROWTH

PATRICIA H. CLARKE AND M. D. LILLY

University College London

INTRODUCTION

The third Symposium of the Society in 1953 was devoted to a discussion of the ways in which micro-organisms adapt themselves to their environment. Most of the earlier studies had been concerned with enzyme variation in response to different growth substrates but there were also observations on 'training' cultures to dispense with certain nutrilites and on the development of resistance to drugs and toxic agents. The discussions at this Symposium brought out very clearly the need to distinguish between an adaptation in a culture which was due to a phenotypic change affecting all the cells of the culture, now termed enzyme induction, and an adaptation which resulted from mutation and selection. The discovery that an enzyme could be induced under gratuitous conditions by non-substrate inducers (Monod, Cohen-Bazire & Cohn, 1951) and the demonstration that β-galactosidase was synthesized *de novo* in an exponentially growing culture following the addition of an inducer (Rotman & Spiegelman, 1954; Hogness, Cohn & Monod, 1955), began a period of detailed investigation by many workers on the kinetics of induced enzyme synthesis during growth. This went on at a time when very rapid developments were taking place in microbial genetics and led to the general theory of control of protein synthesis by induction and repression put forward by Jacob & Monod (1961).

The regulation of enzyme synthesis during growth has been investigated in various ways. Some workers have examined the variation of a single enzyme under different growth conditions. Others have set up defined growth conditions and looked for variation in a number of different enzymes. In cases where enzyme measurements were complicated, or when a large number of enzymes were being measured simultaneously, it was often impossible to make measurements at more than one stage of growth. It is no accident that the most intensive investigations have been made on enzymes such as β-galactosidase and alkaline phosphatase for which rapid and simple assays are available. Most of the studies on the kinetics of enzyme synthesis have been

8

made with exponentially growing batch cultures, although some workers have used continuous cultures. Much useful information has been obtained with mutants, especially when it was also possible to employ a system of genetic recombination. In practical terms the experimenter is always concerned with how much enzyme is being produced in his experimental conditions and how active that enzyme is in that particular environment. Before looking at some experimental findings we shall consider some of the theoretical background to these problems.

REGULATION OF ENZYME ACTIVITY

Current ideas on enzyme control are centred around the regulation of enzyme activity and the regulation of enzyme synthesis. Specific regulation of enzyme activity by metabolic products was first observed for biosynthetic enzymes. For example, amino acid biosynthesis is commonly regulated in bacteria by feedback inhibition of the first enzyme of the pathway by the end product. An increase in the concentration of the end product therefore results in the immediate slowing down of the entire pathway. The chemical dissimilarity between the feedback inhibitor and the enzyme substrate was noted by several workers (e.g. Ames, Martin & Garry (1961) reported that histidine inhibits the first enzyme of its pathway, 5-phosphoribosyl-ATP-pyrophosphorylase). Monod, Changeux & Jacob (1963) described inhibition of this type as allosteric inhibition brought about by the combination of the inhibitor with the enzyme at a site other than the active site and resulting in an altered conformation of the enzyme protein. Many examples are now known in which enzyme activity is regulated by an effector molecule which combines with the enzyme in such a way that the enzyme conformation is changed and its affinity for its substrate is altered (Monod, Wyman & Changeux, 1965). Aspartate transcarbamylase, for example, is made up of two types of subunits (Gerhardt & Schachman, 1965). The substrate-binding subunit carries the active site of the enzyme and the regulatory subunit carries the binding site for the feedback inhibitor, CTP, which is the allosteric inhibitor molecule controlling the functioning of the pyrimidine biosynthetic pathway. Gerhardt & Schachman (1968) obtained evidence for the existence of two different conformational states of this enzyme in the presence or absence of the specific ligands.

End-product inhibition of this kind provides a simple and direct method of control of a linear biosynthetic pathway. However, an enzyme may carry out a reaction which is common to more than one pathway. Inhibition by one of the end products could result in the reduction of the concentration of a common intermediate below that required to synthesize sufficient of another metabolic product. This could mean that the addition of one end product to a growth medium would result in inhibition of growth since feedback inhibition would prevent the synthesis of an essential product on another branch of the metabolic pathway. Several mechanisms have evolved to deal with this problem. Aspartokinase is the first enzyme of the branched pathway for the biosynthesis of all the aspartate family of amino acids and *Escherichia coli* produces two distinct aspartokinases. One of these enzymes is inhibited by lysine and another by threonine. In *Rhodopseudomonas capsulatus* a single aspartokinase is subject to concerted feedback inhibition by a combination of threonine and lysine (Datta & Gest, 1964). Such branched pathways are regulated by complex patterns of feedback inhibition since, in addition to inhibition of the first enzyme of the pathway, there is often inhibition of the first enzyme after each of the branch points. In a growing culture this provides a very sensitive and rapidly adjusting control of amino acid biosynthesis. Feedback inhibition of amino acid pathways was reviewed by Cohen (1965).

Several of the key enzymes of the central metabolic pathways are also thought to be regulated by allosteric inhibition by intermediary metabolites. Isocitrate lyase is essential for growth of *Escherichia coli* on acetate and compounds which generate 2-carbon moieties. It is subject to allosteric inhibition by phosphoenolpyruvate (Ashworth & Kornberg, 1963). Phosphoenolpyruvate is also the allosteric inhibitor of phosphofructokinase in *E. coli* (Blangy, Buc & Monod, 1968). This enzyme is essential for glycolysis via the Embden–Meyerhof pathway and is known to be one of the key enzymes in the regulation of glycolysis and gluconeogenesis in animal tissues. Allosteric inhibition, either by end products or by more proximal metabolites, therefore seems to be a very general mechanism for the regulation of enzyme activity. Such control systems are important in maintaining the balance between anabolism and catabolism during growth and end product control of biosynthesis can confer a growth advantage to the organism by ensuring that energy is not expended unnecessarily on biosynthetic processes when an adequate supply of the end product is available.

REGULATION OF ENZYME SYNTHESIS

The regulation of metabolic pathways by end-product inhibition of enzyme activity is of some economic value to the cell, but if this were the only control system enzymes would be synthesized which were not required for cell functioning. The other main control system depends on regulating the rate of enzyme synthesis. This allows those enzymes to be synthesized which are required for growth in a particular chemical environment. Many organisms are able to synthesize very large amounts of certain enzymes. If such enzymes are normally synthesized only at very low levels when not needed for growth, then amino acids and energy will not be squandered on the synthesis of redundant proteins.

The lac operon and the cytoplasmic repressor

Most of the current ideas on the regulation of enzyme synthesis have been developed from studies on the regulation of the synthesis of β-galactosidase in *Escherichia coli*. The operon model of Jacob & Monod (1961) is one of negative control (Fig. 1). A regulator gene (i) is thought to produce a cytoplasmic repressor molecule which prevents the expression of the structural genes (z, y, a) which determine the enzyme β-galactosidase, β-galactoside permease and thiotransacetylase. These three genes are closely linked and are released from repression when an inducer combines with the repressor molecule and removes it from the operator site (o) adjacent to the structural genes. The *lac* operon consists of the three linked structural genes, which are co-ordinately regulated in this way, and includes the operator and an additional genetic element necessary for the transcription of the structural genes called the promoter (p) (Jacob, Ullmann & Monod, 1964; Scaife & Beckwith, 1966). Ippen, Miller, Scaife & Beckwith (1968), suggest that the promoter region serves as the initiation point for transcription possibly by acting as a binding site for the RNA polymerase. One of the most important pieces of evidence for the negative role of the regulator gene product was the finding that i^+/i^- diploids were inducible and that the i^+ gene was *trans* dominant to an i^- mutation. This suggested that in the wild-type inducible strain the regulator gene produced a substance which acted through the cytoplasm and repressed enzyme synthesis. The subsequent isolation of the *lac* repressor by Gilbert & Müller-Hill (1966) confirmed that the i gene product was a protein. It has a very high affinity for the non-substrate inducers of β-galactosidase such as isopropyl-β-D-thiogalactoside (IPTG) and also for the DNA of the *lac* operon (Gilbert & Müller-Hill, 1967). The *lac* repressor protein is a tetramer and the inducer is thought to act as an allosteric effector by changing the conformation in such a way that its affinity for the operator site is decreased. From theoretical considerations it appeared likely that the regulator gene produced only a very small number of repressor molecules. The wild type is thought to produce about 10 molecules of *lac* repressor for each *i* gene. This is an interesting example of severe restriction of the rate of synthesis of a protein and could be due to a very low level of protein synthesis determined by a gene almost com-

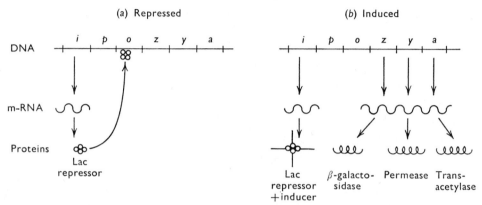

Fig. 1. Model for the regulation of the lactose operon (modified from Jacob & Monod (1961)). (*a*) The regulator gene *i* produces a repressor molecule which combines with the operator *o* to prevent transcription. (*b*) An inducer combines with the repressor removing it from the operator allowing transcription of genes, *z*, *y* and *a*. The promoter *p* is essential for the initiation of transcription.

pletely repressed by a cytoplasmic repressor produced by another regulator gene. There is no evidence for such a regulatory control of the *i* gene and Müller-Hill, Crapo & Gilbert (1968) suggest that this severe restriction of protein synthesis is due to an inefficient promoter. They have isolated mutants (i^q) which produce more *lac* repressor molecules than the wild type and these may be promoter mutations.

Jacob & Monod (1961) suggested that the synthesis of biosynthetic enzymes was also controlled by regulator genes which produced cytoplasmic repressors. For repressible systems the cytoplasmic repressor was thought to be produced in an inactive form and to be able to combine with the operator site only after it had first combined with the appropriate low molecular weight corepressor molecule such as an amino acid. It could be shown that the enzymes required for the biosynthesis of histidine in *Salmonella typhimurium* were coordinately repressed (Ames, Garry & Herzenberg, 1960) and it was later shown that the structural genes formed a single operon of 10 linked genes. However, six other genes have been implicated in the regulation of the synthesis of these enzymes (Roth Antón & Hartman, 1966; Antón, 1968). A single regulator gene was found to control the synthesis of the tryptophan biosynthetic enzymes in *Escherichia coli* (Cohen & Jacob, 1959; Ito & Crawford, 1965). The trytophan structural genes in *E. coli* are also closely linked and coordinately regulated (Imamoto, Ito & Yanofsky, 1966). Other aspects of the operon model for these systems are discussed in a later section.

Regulation by positive control

Most discussion of the regulation of enzyme synthesis has centred round a negative control of the type suggested by Jacob & Monod (1961), but an alternative system of positive control was suggested by Engelsberg, Irr, Power & Lee (1965) for the enzymes required for arabinose metabolism in *Escherichia coli*. A single regulator gene *araC* controls three linked structural genes and an unlinked permease gene. The wild-type strain is inducible by arabinose. Regulator mutants *araC^c* were obtained which were constitutive for the three arabinose enzymes and the permease. Other mutants *araC$^-$* were pleiotropically negative. The significant findings were that in complementation tests with diploids both *araC$^+$* (inducible) and *araC^c* (constitutive) were *trans* dominant to *araC$^-$* (arabinose negative). These results suggested that the regulator gene *C* had a positive role and produced an activator substance which was required for the expression of the structural genes. Sheppard & Englesberg (1967) suggest that the wild type *araC$^+$* produces a substance P1 which in the presence of arabinose is converted into the active form P2 (Fig. 2) and that the constitutive mutants *araC^c* produce altered products (P3 etc.) which can activate without first being combined with arabinose. The inducible *araC$^+$* is trans dominant to *araC^c* in merodiploids and it was suggested that this could be due to interaction of the regulator gene products, possibly by the formation of inactive protein complexes from the two types of subunits. Interaction of repressor subunits was suggested by Müller-Hill *et al.* (1968) to explain some of their results with the *lac* repressors. They showed that although the *i^s* mutation (super-repressible combining firmly with the operator and not released by inducer thereby producing a *lac$^-$* phenotype) is dominant to the wild type *i$^+$* (making an *i^s/i$^+$* diploid *lac$^-$*), when the mutation producing more *lac* repressor was present, the *i^s/i^q* diploid was *lac$^+$*. They thought that with the additional i^q *lac* repressor molecules most of the subunits in the tetramer would be i^q and therefore capable of being removed from the operator by the inducer.

A positive control has also been suggested for the L-rhamnose meta-bolizing enzymes of *Escherichia coli* by Power (1967), and Henning, Denner, Hertel & Shipp (1966) suggested that the pyruvate dehydro-genase complex in *E. coli* might also be subject to positive control by one of the proteins produced by one of the genes. There is less evidence for the positive control model in these two cases than for the arabinose system.

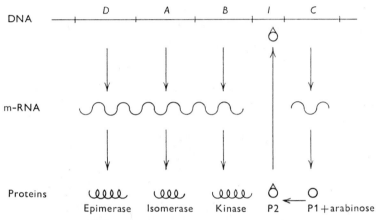

Fig. 2. Model for the regulation of the arabinose operon (modified from Sheppard & Englesberg (1967). The regulator gene *C* produces a repressor P1 which is converted into the activator P2 by combination with arabinose. P2 attaches to the initiator site *I* and activates the transcription of the structural genes *B, A* and *D*.

Transcription and translation

So far we have considered only the over-all results of the series of molecular events which result in enzyme synthesis. The DNA of the structural gene, which determines the sequence of amino acids of the enzyme protein, must first be copied as the RNA sequence of the specific messenger RNA (transcription). The machinery of protein synthesis, including the messenger RNA, ribosomes and transfer RNAs, then enables the sequence of amino acids defined by the gene to be assembled into the unique structure of the enzyme protein (trans-lation). Interference at either the transcription or translation stage would affect the rate of protein synthesis. Jacob & Monod (1961) suggested that the regulation of the *lac* operon was due to the cyto-plasmic repressor preventing the transcription of the *lac* m-RNA rather than by interfering with the translation of the m-RNA into protein. The properties of the *lac* repressor isolated by Gilbert & Müller-Hill (1966, 1967) would fit this hypothesis. Other workers have suggested that the regulation of enzyme synthesis also involves control of the translation process and that transcription and translation may be very tightly bound (Stent, 1964; Brenner, 1965; Cline & Bock, 1966). These conclusions are partly based on the properties of polarity mutants. Polarity mutations in structural genes are defined as those which result in reduced levels of the enzymes on the operator-distal side of the mutated structural gene. Polarity effects have been observed with frame-shift mutations and with nonsense mutations of the amber and ochre type, but not with missense mutations (Newton, Beckwith, Zipser & Brenner, 1965; Martin, Whitfield, Berkowitz & Voll, 1966; Yanofsky & Ito, 1966). Polarity appears therefore to be due to a translational effect resulting from the presence of a chain-terminating codon. Imamoto & Yanofsky (1967) showed, however, that some polar mutations in the tryptophan operon were also accompanied by the production of less tryptophan m-RNA and therefore resulted in altered transcription as well as altered translation.

Mutation in several different genes leads to derepression of the histidine operon and some of these genes are concerned with the synthesis of histidinyl-t-RNA. Roth *et al.* (1966) consider that control in this system acts at the level of m-RNA translation in which histidinyl-t-RNA, or a derivative, acts directly to block translation (and transcription) of the histidine m-RNA. Several genes are concerned in the regulation of the methionine enzymes so that this operon may be regulated in a similar way to the histidine operon (Lawrence, Smith & Rowbury, 1968). It has also been suggested that valine-t-RNA is the repressor of the isoleucine-valine enzymes (Freundlich, 1967). On the other hand, the aminoacyl-t-RNA does not appear to repress the tryptophan operon. Mutants of *Escherichia coli* with an altered tryptophanyl-t-RNA synthetase were not derepressed and the analogue 5-methyltryptophan, which represses the tryptophan enzymes, was not a substrate for the synthetase (Doolittle & Yanofsky, 1968). It seems therefore unlikely that tryptophanyl-t-RNA is involved in the repression of the tryptophan operon.

When an inducer of β-galactosidase is added to a culture there is a lag of about 3 min. before enzyme synthesis can be detected which is thought to be partly due to the time taken to synthesize *lac* m-RNA (Attardi *et al.* 1963; Képès, 1967). Nakada & Magasanik (1964) showed that enzyme potential, usually interpreted as specific m-RNA synthesis, developed during the induction lag and in the apparent absence of any synthesis of protein. However, Mehdi & Yudkin (1967) have pointed out that the concentration of chloramphenicol used by these workers (50 μg./ml.) was not sufficient to inhibit all protein synthesis. Mehdi & Yudkin (1967) used 500 μg./ml. chloramphenicol and showed that at this concentration, with a strain which was wild-type for the *lac* operon but insensitive to catabolite repression, there was no synthesis of the *lac* m-RNA. They concluded, as had been suggested by workers using other methods to study this problem, that transcription of m-RNA is normally linked to translation and movement of ribosomes along the nascent m-RNA strand.

The operon and gene linkage

The simple model of the operon has a single regulator gene producing a repressor molecule which combines with the operator (o) and controls several structural genes linked in an operon. Striking support for this model of control was provided by the isolation of deletion mutants in which the synthesis of the lactose enzymes had become controlled by adenine. The deletions had joined part of the *lac* operon, regulated by induction by β-galactosides, to part of the operon for the biosynthesis of purines controlled by repression by adenine (Jacob, Ullmann & Monod, 1965). In these mutants the deletion extended on one side into the z gene and on the other side into one cistron of the *Purβ* gene region. They were still able to make β-galactoside permease and thio-transacetylase but the synthesis of these proteins was no longer induced by β-galactosides and was now repressed by adenine. The deletion of a large chromosomal segment had therefore resulted in the formation of a new operon in which the purine regulator gene and operator controlled the lactose enzymes.

In both *Escherichia coli* and *Salmonella typhimurium* linkage of structural genes for related enzymes is common. The genes for histidine biosynthesis in *S. typhimurium* are linked in a single operon (Roth *et al.* 1966) with the operator adjacent to the gene for the first enzyme of the pathway, but several other unlinked genes are also involved in the regulation of this operon. The genes for the tryptophan enzymes in *E. coli* form a single operon with the operator adjacent to the first of five structural genes (Imamoto *et al.* 1966). In *S. typhimurium* while the tryptophan genes are also linked into a single linkage group and form a single regulation group, it appears to be made of two parts with respect to expression, and Margolin & Bauerle (1966) suggest that it forms an operon with a single operator but two promotors.

Ramakrishnan & Adelburg (1965) found that the linked genes for the isoleucine-valine pathway in *E. coli* were not all expressed in the same way and suggested that there were at least two operator sites. The arginine structural genes in *E. coli* are not linked but are scattered around the chromosome although there is a single regulator gene. Baumberg, Bacon & Vogel (1965) suggest that the regulator gene produces a common repressor and that each single structural gene or small gene cluster has its own recognition site.

Linkage of genes for related enzymes is uncommon in higher organisms and appears to be infrequent in *Pseudomonas aeruginosa* (Fargie & Holloway, 1965). Mee & Lee (1967) found that the structural genes for histidine biosynthesis in this organism appeared to fall into five distinct linkage groups. Similarly, three groups of tryptophan requiring mutants formed separate linkage groups (Fargie & Holloway, 1965). Some gene linkage could be shown with isoleucine-valine mutants (Pearce & Loutit, 1965). In view of the complex regulatory relationships in pseudomonads the absence of a high degree of genetic linkage is of considerable interest. The general significance of the operon as a unit of regulation was discussed by Ames & Martin (1964).

Catabolite repression

Many inducible enzymes are also subject to catabolite repression (Magasanik, 1961) manifested by the repression of enzyme synthesis when various carbon compounds are added to a culture in the presence of the inducer. This type of repression is still often referred to as the 'glucose effect', since glucose is known to repress the synthesis of many enzymes. However, it is now generally thought that catabolite repression is more specific and it has been suggested that the effective repressing molecule may be a common intermediate of the metabolism of both the substrate inducer and the compound producing catabolite repression (McFall & Mandelstam, 1963).

The search for a specific catabolite repressor molecule for β-galactosidase has implicated many glucose metabolites as well as the pyridine nucleotide coenzymes and ATP. Observations that catabolite repression of β-galactosidase synthesis during growth in a glucose medium could be relieved by making the culture anaerobic (Cohn & Horibata, 1959; Okinaka & Dobrogosz, 1967) suggested that catabolite repression of this enzyme was associated with a high growth rate. Although at one time it was thought that catabolite repression was not affected by the presence of inducer it is now known that the effect is dependent on the relative concentrations of inducer and catabolite repressor. The transient stage of glucose repression of β-galactosidase synthesis can be reversed by a sufficiently high concentration of a powerful inducer such as IPTG (Clark & Marr, 1964; Moses & Prevost, 1966). Nakada & Magasanik (1964) suggested that the catabolite repressor, like the enzyme inducer, acted at the stage of transcription of the specific m-RNA. The i gene product was thought not to be involved, since constitutive (i^-) strains are also sensitive to catabolite repression by glucose (McFall & Mandelstam, 1963). Loomis & Magasanik (1964) showed that glucose also repressed β-galactosidase synthesis during the brief period of constitutivity in a merozygote following the introduction of i^+z^+ genes into an i^-z^- recipient. These findings do not necessarily exclude the participation of the i gene product in catabolite repression, since many i^- strains produce mutant i gene products (Jayaraman, Müller-Hill & Rickenberg, 1966). The period of constitutivity in the merozygote appears to be due to the time taken to reach a sufficient number of cytoplasmic repressor molecules to prevent gene expression. Müller-Hill et al. (1968) showed that when an i^qz^+ mutant, producing more lac repressor molecules than the wild-type i^+z^+, was used in the cross with an i^-z^- strain then the synthesis of β-galactosidase by the merozygote was much more effectively repressed. If the effector molecule for catabolite repression were to increase the affinity of the wild-type lac repressor for the operator, this could also result in an increase in repression in the merozygote. The experimental findings are also compatible with the catabolite repressor acting at the translation stage.

Palmer & Moses (1967) showed that deletion of the o gene abolished acute transient repression of β-galactosidase by glucose and concluded that a functional operator region was necessary for catabolite repression and thought that the i gene might also be involved. McFall (1964, 1967) isolated mutants of *Escherichia coli* which were constitutive for serine deaminase. Two of these had altered catabolite repressibility and the mutation sites mapped in the regulator gene for this enzyme. Loomis & Magasanik (1967) isolated a catabolite irrepressible mutant (CR$^-$) for β-galactosidase. The CR$^-$ mutation was not in the lac operon but mapped near the tryptophan locus. They suggested that the wild-type produced another cytoplasmic repressor which combined with a genetic region continuous to the lac operon. Catabolite repression of the aliphatic amidase of *Pseudomonas aeruginosa* is discussed in a later section.

ENZYME REGULATION AND GROWTH CONDITIONS

Induction and repression during growth

During growth the over-all metabolic pattern depends on the inter-action of the factors controlling enzyme activity and those controlling induction and repression of enzyme synthesis. It is possible to disturb these systems by adding inducing or repressing compounds to the growth medium, by adding different growth substrates or by altering the growth rate of the culture. Much useful information may be obtained from studying mutants with altered regulatory properties even in organisms where a detailed genetic analysis is not yet possible. Most variation is observed for catabolic enzymes which may vary several thousand-fold under different growth conditions. On the whole, biosynthetic enzymes tend to vary less, but differences of a hundred-fold have been encountered between repressed and derepressed states. Much less information is available about regulation of the enzymes of the central metabolic pathways, although it is known that the enzymes of the tricarboxylic acid cycle in *Escherichia coli* and *Bacillus subtilis* are sensitive to repression by glucose, organic nitrogen compounds and anaerobic conditions (Gray, Wimpenny & Mossman, 1966; Hanson & Cox, 1967; Wimpenny, this Symposium).

In discussing the mechanism of control of enzyme synthesis it was necessary to distinguish between inducible enzymes such as β-galacto-sidase, repressible enzymes such as those for the biosynthesis of amino acids and the phenomenon of catabolite repression of induced or constitutive enzyme synthesis. If catabolite repression is due to the repressing activity of specific metabolic intermediates as suggested by Magasanik (1961) and McFall & Mandelstam (1963) it may well be that the mechanism of action of these three types of regulation is essentially similar. The terms 'repression' and 'catabolite repression' are often used indiscriminately. For instance, isocitrate lyase is re-pressed during growth on succinate and the effective co-repressing molecule is thought to be phosphoenolpyruvate (Ashworth & Korn-berg, 1963). The isolation of mutants which are constitutive for iso-citrate lyase at higher temperatures suggests that repression may be effected via a protein determined by a regulator gene (cf. β-galacto-sidase). There is no evidence that isocitrate lyase can be induced and it can reasonably be described as an enzyme controlled by repression (Kornberg, 1966). When *Pseudomonas aeruginosa* is growing on acet-amide as carbon source it requires an acetamidase to hydrolyse acet-amide and also isocitrate lyase to assimilate acetate (Skinner & Clarke,

1968). Amidase is induced by acetamide in the wild-type strain and both enzymes are repressed by the presence of succinate in the growth medium. If we take an amidase constitutive mutant, then the synthesis of both amidase and isocitrate lyase appears to be regulated by repression by products of succinate metabolism. In this case we shall refer to the effect of succinate on amidase synthesis as catabolite repression mainly to emphasize that this enzyme is normally under dual control.

Under experimental conditions it may be found that particular compounds increase the rate of synthesis of an enzyme. In most cases it is correct to assume that this represents enzyme induction but an alternative explanation is that it is due to the removal of repressing intermediary metabolites from the metabolic pool, as for example in the apparent induction of isocitrate lyase by acetate (Kornberg, 1966). In the following discussion of experimental findings an increase in the rate of synthesis of an enzyme will be referred to as enzyme induction unless there is evidence to the contrary. Similarly, any decrease in the rate of synthesis of an enzyme resulting from the presence of particular compounds will be referred to as repression. The term catabolite repression will be used in certain cases to imply repression by unknown metabolites of the synthesis of enzymes which may or may not be subject to regulation by induction. In the absence of detailed studies of the kinetics of enzyme synthesis and the genetics of the systems concerned, these terms are related only to empirical observations.

Multiple control of enzyme synthesis is probably much more common than has been generally realized. Urease in *Proteus rettgeri* is induced by urea, repressed by ammonia and is also subject to catabolite repression (Magana-Plaza & Ruiz-Herrera, 1967). Nitrate reductase is induced by nitrate and repressed by ammonia in *Neurospora* (Kinsky, 1961) and in *Chlorella* (Morris & Syrett, 1965). Under certain conditions synthesis of ornithine transcarbamylase in *Escherichia coli* is induced by glutamic acid and repressed by arginine (Gorini, 1963). Aryl sulphatase in *Aerobacter aerogenes* is repressed by sulphur compounds and induced by tyramine (Harada & Spencer, 1964). During growth the rate of synthesis of such enzymes will depend on the way in which the opposing controls of induction and repression interact under the particular environmental conditions.

Continuous culture

Surprisingly little attention has been paid to continuous culture as a tool for investigating enzyme regulation. In batch culture the chemical environment is continually changing thereby introducing another varying parameter into what is already a complex situation and it is difficult to vary the growth rate except by altering the carbon and/or energy source or by changing the temperature. In a chemostat the growth rates of cultures can be limited by any of the nutrients essential for growth under exactly defined conditions. The culture can be set at steady-state conditions at a particular concentration of limiting nutrient and the growth rate varied by changing the flow rate. Additional information may be obtained by studying enzyme changes during the transition period from one steady state to another following a change in the flow rate of the culture or a change in the incoming medium (Boddy, Clarke, Houldsworth & Lilly, 1967). At low substrate concentrations chemostat conditions may result in a change in the microbial population due to the selection of mutants and in long-term experiments this possibility must always be borne in mind. However, this may provide a useful method for the selection of mutants. Constitutive mutants for inducible enzyme systems may be selected by growing the wild-type in continuous culture at very low concentrations of a substrate inducer. This has been used successfully to isolate constitutive mutants for β-galactosidase in *Escherichia coli* (Horiuchi, Tomizawa & Novick, 1962) and for the enzymes of the mandelate pathway in *Pseudomonas putida* (Hegeman, 1966c).

The dual control of an enzyme by induction and repression may result in a growth diauxie in batch culture. If *Escherichia coli* is inoculated from a glucose medium into a medium containing glucose and lactose the resultant growth will be diauxic and the culture does not utilize the lactose until the glucose concentration has fallen to a very low level. It is quite clear that during the first growth phase glucose represses the synthesis of β-galactosidase; when the glucose has almost disappeared, lactose induces β-galactosidase and after a lag period growth continues on lactose as the carbon source. In continuous culture in media containing mixed substrates it is frequently observed that the substrates are not equally available for growth. Baidya, Webb & Lilly (1967) studied the utilization of glucose and lactose in continuous culture by a strain of *Klebsiella aerogenes* which had a very marked diauxic growth lag in batch culture on these two sugars. They found that glucose was utilized very rapidly at all dilution rates but when the culture was grown in steady-state conditions with glucose as the sole carbon source and the incoming medium was changed to one containing both glucose and lactose there was a pronounced lag before lactose was utilized at low dilution rates, and at very high dilution rates lactose was not metabolized at all. Similar results were obtained by Harte &

Webb (1967) with glucose and maltose as the mixed growth substrates. Mateles, Chian & Silver (1967) showed that with mixtures of glucose and fructose both sugars were completely utilized for growth by *E. coli*, *Pseudomonas fluorescens* and *Saccaromyces cerevisiae* in continuous culture at low dilution rates, but at high dilution rates much of the fructose was not assimilated and remained in the medium. These experiments did not include enzyme measurements, but it is probable that the incomplete utilization of the second substrate was due to repression by glucose of some of the enzymes required for growth.

REGULATION OF ENZYME SYNTHESIS IN PSEUDOMONADS

The pseudomonads are biochemically very versatile organisms (Stanier, Palleroni & Doudoroff, 1966). To be able to grow on such a wide variety of organic compounds they must be able to produce many rather uncommon enzymes which are required to transform these growth substrates into suitable intermediates of the central metabolic pathways. Very many enzymes and permease systems are known to be inducible and many of these appear also to be regulated by catabolite repression. Complex regulatory patterns have evolved for the regulation of the synthesis of groups of enzymes required for these metabolic pathways. Enzyme regulation in pseudomonads has been studied in continuous culture as well as in batch culture and regulator mutants have been isolated for several systems including the mandelate pathway enzymes in *Pseudomonas putida* (Hegeman, 1966c) and the aliphatic amidase of *P. aeruginosa* (Brammar, Clarke & Skinner, 1967). The genetic study of pseudomonads is much less developed than that of *Escherichia coli* and *Salmonella typhimurium*, but in *P. aeruginosa* transfer of genetic material by conjugation has been known for some time (Holloway, 1955; Holloway & Fargie, 1960; Loutit & Marinus, 1967). Several transducing bacteriophages have been isolated and genetic linkage has been studied by transduction (Fargie & Holloway, 1965; Pearce & Loutit, 1965; Mee & Lee, 1967; Brammar *et al.* 1967).

We have chosen three different regulatory systems in pseudomonads for more detailed consideration; the enzymes of the aromatic pathways; the glucose enzymes and the aliphatic amidase. In each case the growth substrates are used both for energy production and biosynthesis and the metabolic pathways can be traced to the points where intermediates of the tricarboxylic acid cycle can be identified. These systems have been studied by different workers and between them they have employed many different methods of approach including the use of mutants and continuous culture techniques. The studies on the regulation of the aromatic pathways illustrate the ways in which convergent metabolic pathways may be regulated. The regulatory patterns which have developed also provide for economy of enzyme synthesis when certain intermediates of these pathways are available as growth substrates, and induction of enzyme synthesis by substrates appears to be balanced by repression by later metabolic products. The study of enzyme synthesis during growth on glucose and citrate provides a relatively simple example of the effect of one growth substrate in inducing its own enzymes and repressing those of another pathway. At least in pseudomonads, glucose is not always the dominant metabolite and while glucose represses some of the enzymes required for citrate metabolism, repression by citrate of the enzymes required for glucose metabolism is much more striking. In our own work we have been investigating the regulation of a single enzyme, the inducible amidase which allows *Pseudomonas aeruginosa* to grow on acetamide and a few other aliphatic amides. It has been possible to isolate mutants with very different growth patterns towards amides and some of these have been shown to be regulator mutants, while others have mutations in the amidase structural gene. For this enzyme the way in which the rate of synthesis is poised between induction and repression shows up very clearly in growth in continuous culture.

Growth on aromatic compounds

The aromatic compounds tryptophan, mandelic acid and *p*-hydroxy-mandelic acid are metabolized by a series of enzymes which eventually convert them into compounds which can enter the tricarboxylic acid cycle (as succinyl coenzyme A and acetyl coenzyme A). These enzymes fall into several groups which are separately regulated (Fig. 3). Hegeman (1966*a*,*b*) showed that the first groups of enzymes of the mandelate pathway in *Pseudomonas putida* are coordinately regulated and form a single regulation group (group 1). These enzymes are inducible by D- and L-mandelate (or benzoyl formate) and mutants were obtained which were constitutive for all these enzymes (Hegeman, 1966*c*). Figure 4 shows that DL-mandelate immediately induced the synthesis of two enzymes of the first regulation group, mandelate racemase and L-mandelate dehydrogenase, but two enzymes belonging to a later enzyme regulation group were synthesized only after a considerable lag.

The second regulation group consists of a single enzyme, benzoate oxidase, oxidizing benzoate to catechol, induced by benzoate (group 2). The enzyme for the following step was found to be induced, not by catechol, but by its product *cis-cis*-muconate, which was also the inducer for the next two enzymes (Ornston, 1966). These two enzymes,

Enzyme group	Reaction	Regulation

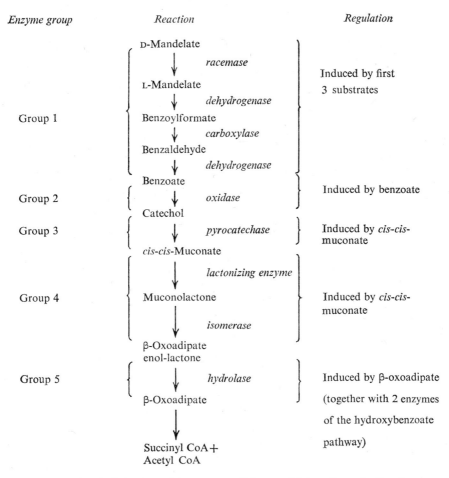

Fig. 3. Regulation by induction of the enzymes of the mandelate pathway in *Pseudomonas putida*. Data from Hegeman (1966*a*) and Ornston (1966).

but not catechol oxygenase, were coordinately induced so that they formed two more regulation units (groups 3 and 4). Further metabolism depends on the induction of the next enzyme by its product *β*-oxoadipate but, at the same time and coordinately, it also induces two earlier enzymes of the *p*-hydroxymandelate-*p*-hydroxybenzoate-protocatechuate pathway (group 5). These two enzymes are required for

9

growth on the hydroxy compounds but are produced gratuitously when the organism is growing on mandelate or benzoate.

The catabolic pathway for tryptophan converges with that for mandelate and benzoate (Fig. 5). The first two enzymes of the tryptophan pathway are induced by kynurenine, which is the product of the

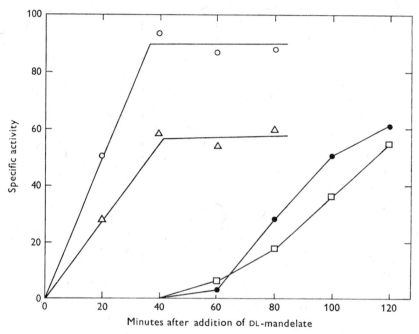

Fig. 4. Time course of the appearance of four enzymes of the mandelate pathway after the addition of DL-mandelate to a culture of *Pseudomonas putida*. Specific activities are expressed as μmoles substrate converted/mg. protein/min. for all enzymes except muconolactone isomerase, which is expressed as 10^{-4} μmoles/mg. protein/min. Two enzymes of the first regulation group were induced without lag. ○—○, mandelate racemase; △—△, L-mandelate dehydrogenase. Two enzymes occurring later in the pathway were induced after about 40 min. lag; □—□, muconolactone isomerase; ●—●, lactonizing enzyme (Hegeman, 1966a).

action of these two enzymes, and successive steps lead to catechol (Palleroni & Stanier, 1964). These regulatory patterns for the breakdown of the aromatic compounds are very interesting. The enzymes at the beginning of a pathway (belonging to the first regulation group) are induced by the substrate (or product) of the first enzymes and as a result of their activities the inducer is formed for the enzymes of the next regulation group. The entire pathway is thus regulated by a rather complex form of sequential induction. This allows various intermediates of the pathway to be used as growth substrates since they are able to induce the necessary enzymes for their own metabolism. Also, where the catabolic pathways for different starting substrates converge, a new induction sequence begins so that a general economy is achieved in the number of enzymes required for metabolizing diverse growth substrates. The genes corresponding to the enzymes of each regulation group may also be associated in separate linkage groups (Stanier, 1968).

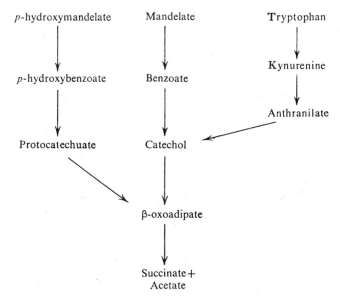

Fig. 5. Convergent metabolic pathways for mandelate, p-hydroxymandelate and tryptophan in pseudomonads.

A different regulatory pattern for the enzymes metabolizing these aromatic compounds has been described by Canovas & Stanier (1967) in *Moraxella calcoacetica*. The metabolic pathways are chemically homologous but the enzyme regulation groups are different.

The inducible enzymes of the aromatic pathways are also controlled by very complex catabolite (or end-product) repression. Mandelstam & Jacoby (1965), using the same strain of *Pseudomonas putida* as Hegeman (1966a) (although they then described it as *P. fluorescens*), found that the synthesis of the first regulation group of enzymes was not significantly affected by glucose but was repressed by benzoate, catechol and succinate. In each case the repression could be overcome by the addition of more inducer so that the rate of enzyme synthesis was determined by the balance between induction and repression. Stevenson & Mandelstam (1965) showed that the enzyme of the next regulation group, benzoate oxidase, was repressed by catechol, succinate and acetate and that the enzymes of the following groups were repressed by succinate and acetate. There was a parallel system of repression of the enzymes of the pathway for the corresponding hydroxy compounds. The over-all effect was that the enzymes of each regulation group were repressed by their immediate products and also by the products of the succeeding groups of enzymes (Fig. 6).

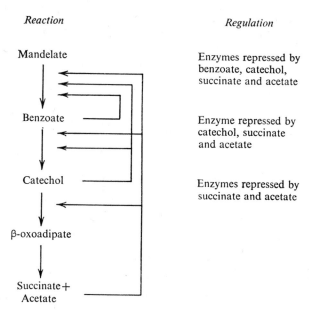

Fig. 6. Regulation by repression of the enzymes of the mandelate pathway in *Pseudomonas putida*. Data from Mandelstam & Jacoby (1965) and Stevenson & Mandelstam (1965).

Growth on glucose and citrate

Hamilton & Dawes (1959) observed a growth diauxie when *Pseudomonas aeruginosa*, grown on citrate, was inoculated into a medium containing both citrate and glucose. This was particularly interesting, since, in contrast to observations made with other bacteria growing in media containing glucose together with an alternative carbon source, this pseudomonad utilized citrate in preference to glucose. Further experiments with batch cultures showed that glucose utilization was not completely repressed when citrate-grown bacteria were transferred to a medium containing a mixture of citrate and glucose but, when added to a culture growing exponentially on citrate, glucose was not utilized at all until the citrate was almost exhausted. When a medium containing citrate and glucose was inoculated with glucose-grown bacteria both substrates were utilized for growth. Hamilton & Dawes (1960, 1961) showed that glucose uptake required a specific permease, which was induced by growth in the presence of glucose, and also that bacteria grown on succinate or citrate had very low levels of the glucose enzymes. In this organism glucose is metabolized mainly by the Entner–Doudoroff pathway (see Fig. 7) and the activities of these enzymes are

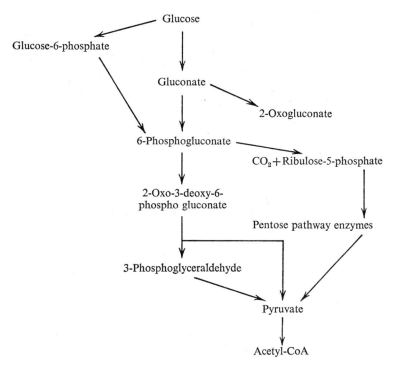

Fig. 7. Pathways of glucose metabolism in *Pseudomonas aeruginosa*.

very dependent on the nature of the growth substrates. Hamlin, Ng & Dawes (1967) showed that in batch culture the activities of the enzymes required for glucose metabolism, particularly glucose 6-phosphate and 6-phosphogluconate dehydrogenases, and the combined activities of the specific Entner–Doudoroff enzymes, were very much higher in cultures grown on glucose than in those grown on citrate or peptone (Table 1). Measurements were also made on two tricarboxylic acid cycle enzymes and it was found that peptone repressed both isocitrate dehydrogenase and aconitase while glucose appeared to repress only isocitrate dehydrogenase under these conditions.

Table 1. *Comparison of specific activities of enzymes from cells of* Pseudomonas aeruginosa *grown in batch culture on various carbon sources* (*Hamlin* et al. *1967*)

Enzyme	Carbon source for growth		
	Glucose	Citrate	Peptone
Hexokinase	38·1	27·5	10·0
Glucose 6-phosphate dehydrogenase	126·3	4·4	15·0
6-Phosphogluconate dehydrogenase	31·3	1·3	3·8
6-Phosphogluconate dehydratase + 2-oxo-3-deoxy-6-phosphogluconate aldolase	30·0	4·4	9·4
Glucose dehydrogenase	4·4	1·3	4·4
Gluconate dehydrogenase	8·2	3·2	6·3
Isocitrate dehydrogenase	228·8	240·0	128·1
Aconitase	23·8	46·3	11·9

Specific activities are recorded as μmoles of substrate metabolized per hour per mg. nitrogen.

In continuous culture it was shown that at a constant flow rate and a constant level of citrate, an increase in the glucose concentration of the incoming medium resulted in an increase in glucose enzymes. When the enzyme activities were measured during the transition period from one steady state to another, following a change in the incoming medium to a higher glucose concentration, the period during which glucose 6-phosphate and 6-phosphogluconate dehydrogenases were steadily increasing lasted for several hours. In the same way, when the incoming medium was maintained at a constant glucose concentration and changed from a higher to a lower citrate concentration, there was again a transition period of several hours before these enzymes reached the new steady-state level (Hamlin *et al.* 1967). Some of these experiments were complicated by changes from carbon to nitrogen limitation as the concentration of the carbon source was increased but in later experiments it was arranged that at all times the cultures would be nitrogen-limited. We are grateful to Professor E. A. Dawes for permission to refer to the following unpublished results. Table 2 gives some of the results in an experiment in which the dilution rate was set at 0·25 hr.$^{-1}$, the citrate concentration was kept constant and the ratio of glucose to citrate was increased from 0·5:75 mM to 30:75 mM. The specific activities of all the glucose enzymes increased as the glucose to citrate ratio rose. The aconitase activity also increased and there was some increase in the activity of isocitrate dehydrogenase although much less than for aconitase.

In addition to studying the effect on enzyme synthesis of alterations in the concentrations of the growth substrates, F. M-W. Ng & E. A. Dawes (personal communication) also investigated the effect of different dilution rates on the synthesis of these enzymes in cultures

Table 2. *Effect of relative citrate: glucose concentrations in inflowing medium of chemostat on enzymic activities of* Pseudomonas aeruginosa (*F. M.-W. Ng & E. A. Dawes, personal communication*)

(Dilution rate 0·25 hr.$^{-1}$)

Specific activities

Citrate: glucose ratio (mM) in medium	G 6–P de-hydro-genase	Hexo-kinase	6–P G de-hydro-genase	Entner–Doudor-off enzymes	Glucose de-hydro-genase	Gluco-nate de-hydro-genase	Iso-citric de-hydro-genase	Aconi-tase
75 : 0·5	16·5	13·1	0·4	3·8	1·7	16·0	269·4	30·0
75 : 1	14·7	9·4	0·6	4·3	0·9	18·7	295·5	33·2
75 : 2	19·7	16·0	0·6	5·2	1·9	23·6	274·6	31·5
75 : 4	22·4	15·4	2·7	5·4	1·6	30·7	290·6	27·2
75 : 8	73·6	29·8	18·6	16·8	2·7	38·9	332·9	35·8
75 : 15	94·2	37·7	20·6	21·8	2·5	74·9	352·6	40·4
75 : 30	101·4	39·2	24·7	30·3	3·1	86·9	338·2	44·0

Specific activities are expressed as μmoles of substrate utilized/mg. protein nitrogen/hr.

Table 3. *Effect of dilution rate on enzymic activities of citrate-grown* Pseudomonas aeruginosa (*F. M.-W. Ng & E. A. Dawes, personal communication*)

Specific activities

Dilution rate (hr.$^{-1}$)	G 6-P de-hydro-genase	Hexo-kinase	6–P G de-hydro-genase	Entner–Doudor-off enzymes	Glucose de-hydro-genase	Glu-conate de-hydro-genase	Iso-citric de-hydro-genase	Aconi-tase
0·125	27·3	13·8	0·7	0·7	0·6	4·7	400·7	41·4
0·170	22·5	12·0	0·8	1·6	0·3	3·5	392·6	41·2
0·200	19·8	12·1	0·9	1·8	0·7	3·8	363·7	43·5
0·250	20·1	9·9	0·7	1·5	0·9	7·4	346·3	29·8
0·500	12·5	11·2	0·7	1·2	1·1	4·0	381·5	27·8
Batch culture								
Glucose	126·3	38·1	31·3	30	4·4	8·2	228	23·8
Citrate	4·4	27·5	1·3	4·4	1·3	3·2	240	46·3

Specific activities are expressed as μmoles of substrate utilized/mg. protein nitrogen/hr.

growing at a constant level of either citrate or a mixture of citrate and glucose (Tables 3, 4). When the cultures were growing on citrate as the sole carbon source there was a marked repression of hexokinase and glucose, gluconate, glucose 6-phosphate and 6-phosphogluconate dehydrogenases at all dilution rates (0·125–0·50 hr.$^{-1}$) and in most cases the activities were lower than those previously reported for batch cultures grown on citrate (Table 1). On the other hand, the isocitrate dehydrogenase activity was higher than that in batch culture. Several enzymes, particularly aconitase and glucose 6-phosphate dehydrogenase, showed increased repression as the growth rate increased. When similar experiments were carried out with a mixture of citrate and glucose in the incoming medium it was clear that the presence of glucose resulted in induction or derepression of all the glucose enzymes (Table 4). For several of these enzymes, particularly glucose 6-phosphate dehydrogenase and gluconate dehydrogenase, there was an increase in the enzyme activities as the growth rate increased. Isocitrate dehydrogenase was more repressed as the dilution rate increased, but, while the aconitase activity was considerably higher than that in citrate alone at the lower dilution rates, it was repressed to about the same extent at the higher rates.

Table 4. *Effect of dilution rate on enzymic activities of* Pseudomonas aeruginosa *growing on a mixture of citrate and glucose* (F. M.-W. Ng & E. A. Dawes, *personal communication*)

Dilution rate (hr.⁻¹)	Specific activities							
	G 6-P de-hydro-genase	Hexo-kinase	6-P G de-hydro-genase	Entner–Doudor-off enzymes	Glucose de-hydro-genase	Glu-conate de-hydro-genase	Iso-citric de-hydro-genase	Aconi-tase
0·125	35·6	14·2	9·3	3·6	0·7	21·0	449·2	79·0
0·170	33·5	15·2	9·5	4·3	2·0	30·6	451·7	65·5
0·200	27·4	19·3	2·6	5·8	4·0	32·0	378·8	51·9
0·250	24·9	17·1	3·0	6·0	1·8	34·1	322·7	30·0
0·500	62·0	25·0	9·1	5·4	4·3	37·4	321·0	21·6

Specific activities are expressed as μmoles of substrate utilized/mg. protein nitrogen/hr. Concentrations of residual citrate were 27–30 mM and the concentrations of inflowing glucose were 4 mM at all dilution rates.

These experiments suggest that the glucose enzymes are regulated by induction by glucose (or its metabolites) and by repression by products of citrate metabolism. The chemostat experiments, in which the ratio of citrate to glucose was varied, demonstrate that the synthesis of the glucose enzymes can be increased, either by increasing the amount of glucose or by decreasing the amount of citrate in the medium. These experiments have also shown that aconitase and isocitrate dehydrogenase, which are required for both citrate utilization and the complete oxidation of glucose, are under regulatory control. None of these experiments showed coordinate repression of these two enzymes. The tricarboxylic acid cycle has a dual role and is important for biosynthesis as well as for terminal oxidation and the very marked repression of both aconitase and isocitrate dehydrogenase in complex medium suggests that nitrogen compounds may be more important than glucose in the regulation of these enzymes. It is well known that the tricarboxylic acid enzymes are repressed both by glucose and organic nitrogen compounds in facultative anaerobes (Gray *et al.* 1966) and it is interesting to observe comparable effects in an aerobic organism.

Growth on amides

The aliphatic amidase produced by *Pseudomonas aeruginosa* 8602 is induced by growth in the presence of several substrate or non-substrate amides (Kelly & Clarke, 1962) and is also subject to catabolite repression. A specific catabolite repressor molecule has not been identified, but induced enzyme synthesis by an exponentially growing culture is repressed by acetate, propionate, succinate and other compounds. This can be seen in batch culture where the rate of induced enzyme synthesis is very dependent on the growth substrate and is much higher in pyruvate medium than in succinate medium. The effectiveness of the carbon compound used as a growth substrate in repressing amidase synthesis is also reflected in the induction lag. In succinate medium there is a lag of about one generation after the inducer is added before amidase can be detected but in pyruvate medium the lag is only 0·25 of a generation (Brammar & Clarke, 1964). Amidase induction in succinate medium can be repressed further by the addition of propionate or acetate and this can be relieved by increasing the concentration of acetamide. This enzyme is therefore under dual control and is induced by amides and repressed by one or more intermediary metabolites.

When *Pseudomonas aeruginosa* grows in a medium containing acetamide as the sole source of carbon and nitrogen the physiological role of the amidase is amide hydrolysis. The mechanism of enzyme action appears to be that of acyl group transfer and this provides a convenient assay method using hydroxylamine as the acyl acceptor. We used the transferase method (Brammar & Clarke, 1964) for all routine enzyme assays.

Hydrolase $\quad CH_3CONH_2 + H_2O \rightleftharpoons CH_3COO' + NH_4^+$

Transferase $\quad CH_3CONH_2 + NH_2OH + H^+ \rightleftharpoons CH_3CONHOH + NH_4^+$

The enzyme exhibits fairly high substrate specificity. Of the aliphatic amides, acetamide and propionamide are the best substrates both for the hydrolase and transferase reactions and are also good inducers.

Table 5. *Comparison of amides as substrates and inducers of*
Pseudomonas aeruginosa *amidase*

Amide		Substrate activity	Inducer activity
Formamide	$HCONH_2$	14	6
Acetamide	CH_3CONH_2	100	100
Propionamide	$CH_3CH_2CONH_2$	280	50
Butyramide	$CH_3CH_2CH_2CONH_2$	2	0
Lactamide	$CH_3CHOHCONH_2$	9	50
N-acetylacetamide	$CH_3CONHCOCH_3$	0	100

Substrate activity was measured as the rate of amide hydrolysis compared with acet-amide = 100. Inducer activity was measured as the differential rate of amidase synthesis in succinate medium compared with acetamide = 100.

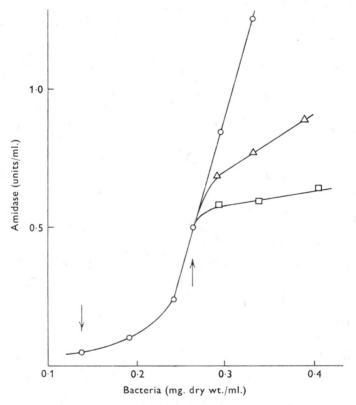

Fig. 8. Induction and repression of amidase synthesis in *Pseudomonas aeruginosa* 8602 (wild-type). N-acetylacetamide was added at the first arrow to give a concentration of 10 mM. At the second arrow the culture was divided into three parts; O—O, control flask no additions. Cyanoacetamide was added to the other two flasks to give the following concentrations: □—□, 10 mM; △—△, 0·1 mM. Amidase activity was measured by the transferase method and is expressed as μmoles hydroxamate formed/min. (Brammar & Clarke, 1964).

Glycollamide is an effective substrate and inducer but lactamide, although a good inducer, is a very poor substrate (Table 5). Several *N*-substituted amides, e.g. *N*-methylacetamide and *N*-acetylacetamide, are effective as inducers although they are not enzyme substrates. Cyanoacetamide acts as a repressor of amidase induction by both *N*-acetylacetamide and acetamide. The consequence of this is that

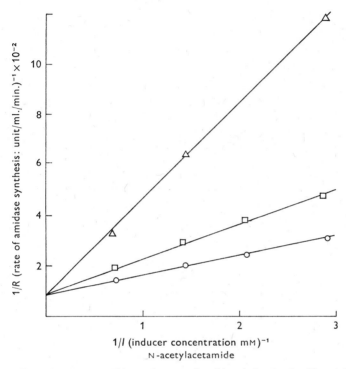

Fig. 9. Effect of cyanoacetamide on the rate of amidase induction by *N*-acetylacetamide in carbon-starved cells of *Pseudomonas aeruginosa* 8602. △—△, 1.4×10^{-1} mм cyanoacetamide; □—□, 3.4×10^{-2} mм cyanoacetamide; ○—○, no cyanoacetamide (Brammar, 1965).

cyanoacetamide is able to inhibit growth of *Pseudomonas aeruginosa* on acetamide unless the culture has been previously induced. Figure 8 shows a typical amidase induction and repression experiment with a culture growing exponentially in succinate medium. Amidase was induced by 10 mм *N*-acetylacetamide and after synthesis was well established the culture was divided into three parts. 10 mм cyanoacetamide was added to repress synthesis in one of the flasks and 0·1 mм cyanoacetamide was added to a second flask. A much higher concentration of cyanoacetamide was required to repress amidase induction if acetamide was used as the inducer and these results suggested that the cyanoacetamide repression was due to competition for an amide binding site. This was confirmed by measuring amidase induction and cyanoacetamide repression at several different inducer concentrations in carbon-starved suspensions in which catabolite repression was minimal (Fig. 9). With N-acetylacetamide as inducer, the value obtained for K_{Ind} from a double reciprocal plot of 1/Rate of amidase synthesis against 1/Inducer concentration (equivalent to the Lineweaver–Burk method for determining the Michaelis constant for enzyme and substrate) was about 10^{-3}M and for acetamide 7×10^{-6}M and the 'inhibitor constant' for cyanoacetamide was 5×10^{-5}M (Clarke & Brammar, 1964; Brammar, 1965). If the assumption is made that this is a regulation system of the *lac* type these values could represent the ratios of the affinities of the amides for the amidase cytoplasmic repressor. It must be pointed out that these experiments were carried out with whole cells but the results do not appear to be due to permeability effects (Brammar, McFarlane & Clarke, 1966). In a similar way, 2-nitrophenyl-β-D-fucoside is a specific inhibitor of induced synthesis of β-galactosidase in *Escherichia coli* and there is good evidence that it interacts with the repressor (Jayaraman *et al.* 1966).

Amidase regulator mutants

The wild-type strain does not grow on formamide as a carbon source but can use it as a nitrogen source. It is however unable to grow on succinate formamide agar (S/F) medium since insufficient enzyme is induced in this medium in which succinate acts as a catabolite repressor and formamide as a very poor inducer. Two types of regulator mutants have been isolated from S/F plates; (*a*) constitutive (C) mutants which include both magno-constitutive mutants producing enzyme at the same or a somewhat higher rate than a fully induced wild-type and semi-constitutive mutants which can be induced further by the addition of an inducing amide to the growth medium; (*b*) formamide inducible (F) mutants which are wholly inducible but have altered response to amide induction. Figure 10 *a* and *b* show the rates of induction of amidase in the wild-type and in mutant F6 by formamide, acetamide and propionamide. The C and F characters were always co-transduced with the amidase positive character, suggesting that the amidase regulator gene and the amidase structural gene were closely linked (Brammar *et al.* 1967).

It has not been possible to construct diploids to examine whether or not inducibility is dominant to constitutivity. The assumption is made in the following discussion that the regulation is of the *lac* type, with a regulator gene producing a cytoplasmic repressor molecule which represses the synthesis of the enzyme specified by the amidase structural gene, but most of the results could also be fitted to a positive control model of the *ara* type. One of the magno-constitutive mutants was repressed both by the amide analogue repressor cyanoacetamide and by *N*-acetylacetamide which is an inducer for the wild-type. This is

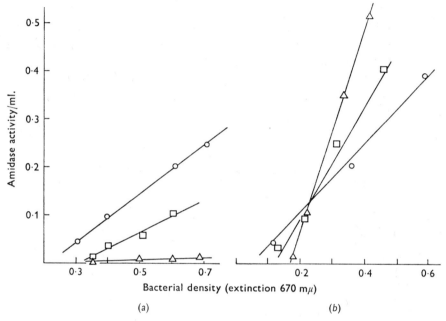

Fig. 10. Induction of amidase synthesis in *Pseudomonas aeruginosa* 8602 by amides: △—△, formamide; O—O, acetamide; □—□, propionamide; (*a*) wild-type; (*b*) mutant F6; 10 mM amides in succinate medium (Brammar *et al.* 1967).

probably due to the production of an altered cytoplasmic repressor which responds in a different way to combination with amides and is reminiscent of the arginine regulation pattern in which arginine represses the synthesis of the enzymes required for its biosynthesis in *Escherichia coli* K12 but acts as an inducer in *E. coli* B (Jacoby & Gorini, 1967).

The wild-type strain does not grow on butyramide as a carbon source. Butyramide is a very poor substrate for the wild-type enzyme and like cyanoacetamide represses amidase induction by *N*-acetylacet-amide. Some of the constitutive mutants isolated from *S/F* plates were also unable to grow on butyramide and it was found that they were as sensitive as the wild-type to repression by butyramide. Other constitutive mutants from S/F plates, and a further series of constitutive mutants isolated directly on butyramide plates, were all much less sensitive to butyramide repression. The enzyme produced was indistinguishable from the wild-type so that the essential difference which allowed the second group of constitutive mutants to grow was the absolute amount of enzyme produced on butyramide medium (Brown, 1969).

Structural gene mutants

A very different mutational change which would allow growth on a poor substrate would be an alteration in the substrate specificity of the enzyme. From the constitutive strain C 11, which like the wild-type

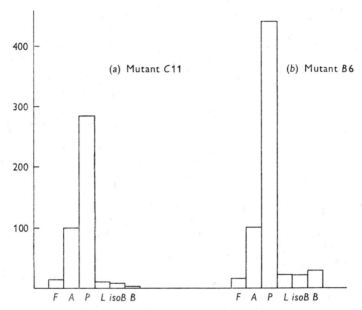

Fig. 11. Relative rates of amide hydrolysis by whole cells of *Pseudomonas aeruginosa* 8602 mutants: (*a*) C11, (*b*) B6. Rate of hydrolysis of acetamide assigned a value of 100. *F*, Formamide; *A*, acetamide; *P*, propionamide; *L*, lactamide; *isoB*, isobutyramide; *B*, butyramide.

is strongly repressed by butyramide, several mutants (B mutants) were isolated which grew on butyramide and produced an amidase with different electrophoretic mobility from that produced by the wild-type (Brown & Clarke, 1967). An examination of the substrate specificity for amide hydrolysis showed that these B mutants produced an enzyme with a significantly higher rate of butyramide hydrolysis. Figure 11 compares the substrate profile of the enzyme produced by C11 (A amidase) and the enzyme produced by mutant B6 (B amidase). The mutants of the B class are all as sensitive to repression of amidase synthesis by butyramide as the constitutive strain C11 from which they were derived. They have evolved to grow on butyramide by producing a more efficient enzyme for this substrate rather than by producing more of the less efficient enzyme produced by the wild-type and the constitutive mutants. There must be a critical value relating the amount of amidase synthesized and the activity of the enzyme towards butyramide above which growth can take place and below which the hydrolysis of butyramide is too slow. If the amount of enzyme synthesized in unit time is expressed as E and the rate of butyramide hydrolysis per enzyme molecule in unit time as B, then there is a critical value for P in the equation $E \times B = P$ which determines whether or not growth will occur.

Mutants with altered catabolite repressibility

The wild-type strain grows well on 1 % lactamide as a source of carbon and nitrogen but hardly grows at all if lactamide 0.02–0.1 % is used as the nitrogen source and succinate (1 %) as the carbon source (S/L agar). Mutants isolated from these plates produced wild-type enzyme but were all found to be less sensitive than the wild-type to catabolite repression by succinate (L mutants). Some of the L mutants were inducible and others were constitutive. Transduction analysis showed that the character determining decreased sensitivity to catabolite repression in this class of mutants was not associated with the amidase structural gene or the linked regulator gene. The constitutive L mutants appeared to have been double mutants since in crosses with amidase-negative recipients they all gave constitutive transductants (able to grow on S/F agar) but none of the transductants were able to grow on S/L agar. No specific gene defects have yet been identified but these mutants probably have alterations in the genes for one or more metabolic enzymes which alter the rate at which the compound (or compounds) which cause catabolite repression of the amidase is synthesized or used up.

The growth characteristics of the amidase mutants are summarized in Table 6.

Amidase synthesis in continuous culture

The balance between induction and catabolite repression can be shown more strikingly in chemostat continuous culture than in batch culture. We have examined the effect of growth rate on amidase induction

Table 6. *Properties of amidase mutants of* Pseudomonas
aeruginosa *strain 8602*

Mutant	Selection medium	Phenotype
C11	Succinate/formamide	Magno-constitutive, repressed by butyramide, repressed by succinate
C17	Succinate/formamide	Semi-constitutive, repressed by butyramide, repressed by succinate
C1	Succinate/formamide	Magno-constitutive, reduced repression by butyramide, grows on butyramide medium, repressed by succinate
F6	Succinate/formamide	Induced rapidly by formamide, does not grow on butyramide medium, repressed by succinate
B6	Butyramide	Magno-constitutive, isolated from C11, repressed by butyramide, produces amidase with altered substrate specificity
L9	Succinate/lactamide	Magno-constitutive, reduced repression by butyramide, grows on butyramide medium, reduced repression by succinate
L11	Succinate/lactamide	Inducible, does not grow on butyramide, reduced repression by succinate

when acetamide is added to a steady-state culture limited by succinate. Figure 12 shows a typical experiment with a culture growing at a dilution rate of 0·48 hr.$^{-1}$ on 10 mM succinate for which the incoming medium was changed to 10 mM succinate + 20 mM acetamide. After a transition period of several hours the culture reached the new steady-state level and the bacterial density was approximately doubled. There was a considerable lag before an appreciable rate of amidase synthesis could be detected in this experiment and the rates of amidase synthesis during the transition period were therefore compared for cultures growing at several different dilution rates. Figure 13 shows that at low dilution rates ($D = 0·22$ hr.$^{-1}$) amidase synthesis occurred rapidly with little or no lag, but at higher dilution rates the lag became longer until at a dilution rate of 0·76 hr.$^{-1}$ almost no amidase had been synthesized during the first 4 hr. after the change of medium. A similar increase in repression of amidase synthesis with dilution rate occurred under gratuitous conditions when the incoming medium was changed from 10 mM succinate to 20 mM succinate + 10 mM *N*-acetylacetamide (Fig. 14). In these experiments the increase in the concentration of the growth substrates in the incoming medium following the changeover was sufficient in all cases to allow a doubling in the mass of the culture if the substrates had been fully utilized. This would have entailed a temporary spurt in growth rate before the culture became established at the new steady-state level. We considered that the severe repression of amidase synthesis during the transition period was related to this temporary growth spurt. Figure 14 also shows the rate of amidase synthesis during the transition period when the gratuitous inducer

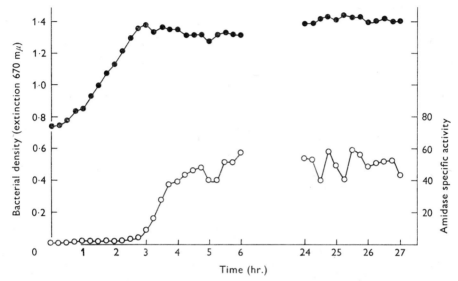

Fig. 12. Amidase synthesis by *Pseudomonas aeruginosa* 8602 in chemostat continuous culture. At zero time the ingoing medium was changed from 10 mM succinate to 10 mM succinate + 20 mM acetamide. ●—●, Bacterial concentration; ○—○, amidase specific activity μmoles acethydroxamate formed/mg. dry wt. bacteria/min.; $D = 0.48$ hr.$^{-1}$ (Boddy *et al.* 1967).

N-acetylacetamide was introduced into the incoming medium at a constant growth rate by keeping the succinate concentration constant at 10 mM. There was no immediate severe repression of amidase induction and the brief lag was due to the time taken to build up an effective concentration of inducer in the culture vessel. The differences in specific activities after 4 hr. of the cultures growing at three different dilution rates are probably significant and will be discussed later. This experiment showed that with a carbon-limited continuous culture there was no lag period for amidase induction with a gratuitous inducer although in batch culture the induction lag had been significant in all growth media. Marr & Marcus (1962) obtained similar results for mannitol dehydrogenase in *Azotobacter agilis*. They found that there was no lag in the gratuitous induction of mannitol dehydrogenase in carbon-limited continuous culture, although there was a long induction lag for this enzyme in batch culture.

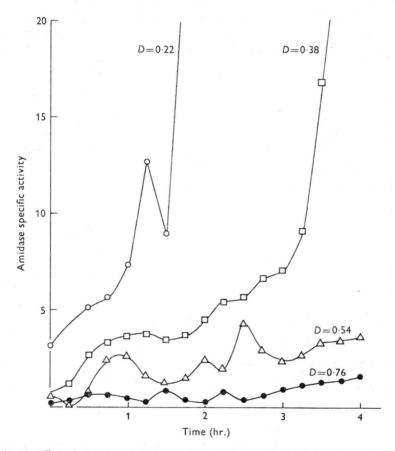

Fig. 13. Effect of dilution rate (D=hr.$^{-1}$) on amidase synthesis by *Pseudomonas aeruginosa* 8602 following a change of the ingoing medium at zero time from 10 mM succinate to 10 mM succinate + 20 mM acetamide.

It was also possible to demonstrate a balance between amidase induction and repression under steady-state conditions (Clarke, Houldsworth & Lilly, 1968). When the wild-type was grown in continuous culture under carbon limitation either in a medium containing 20 mM acetamide, or 20 mM acetamide + 10 mM succinate, there was a sharp peak of amidase activity at a dilution rate of about 0·3 hr.$^{-1}$ (Fig. 15). It was tempting to assume that the rise in amidase activity as the dilution rate increased was due to the inducer control and that the fall at the higher dilution rates was due to control by the catabolite repressor. Thus, as the dilution rate increased, at the lower end of the range the effective concentration of the inducer at the inducer binding site increased and, as the growth rate increased still further, the concentration of the catabolite repressor molecules also increased until the point was reached at which catabolite repression became dominant to induction. Since acetamide is rapidly hydrolysed to acetate it could

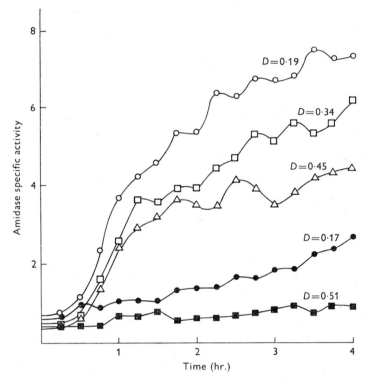

Fig. 14. Effect of dilution rate (D=hr.$^{-1}$) on amidase synthesis in the presence of the non-substrate inducer N-acetylacetamide. At zero time the ingoing medium was changed from 10 mM succinate either to 10 mM succinate+10 mM N-acetylacetamide (upper three curves), or to 20 mM succinate+10 mM N-acetylacetamide (bottom two curves).

provide the source of both inducer and catabolite repressor. In the acetamide + succinate medium the part of the curve thought to correspond to induction is the same but repression appears at a lower dilution rate.

If these curves reflected the induction-repression controls for this enzyme it seemed likely that different curves would be obtained for the mutants which were known by their growth on plates and in batch culture to have altered amidase regulation. Figure 16 shows that the fully constitutive mutant C11 produced most enzyme at the lowest dilution rate ($D = 0.06$). In batch growth the rate of amidase synthesis by this fully constitutive mutant is unaffected by the presence of the non-substrate inducer N-acetylacetamide and any substrate inducer is

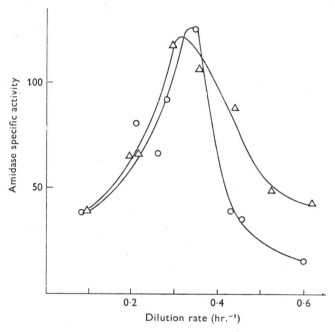

Fig. 15. Effect of dilution rate on amidase synthesis by *Pseudomonas aeruginosa* 8602 (wild-type) growing in continuous culture under steady-state conditions. O—O, Minimal salts medium containing 10 mM succinate + 20 mM acetamide; △—△, minimal salts medium containing 20 mM acetamide (Clarke *et al.* 1968).

rapidly hydrolysed. In effect, as can be seen from Fig. 16, the curve obtained for amidase synthesis by C11 in continuous culture is similar to the falling side of the wild-type curve but displaced to a lower dilution rate. The rate of synthesis of amidase by this mutant would seem to be wholly controlled by catabolite repression. Mutant L9 was one of the group of constitutive mutants isolated from S/L plates which appeared to have a double mutation and has both constitutivity and altered catabolite repressibility. Figure 16 shows that in continuous culture it behaves in a similar way to C11. The curve for L9 is displaced to higher dilution rates compared with that for C11, which is consistent with the view that this decrease in the rate of amidase synthesis at the higher dilution rates is due to the onset of catabolite repression.

Similar comparisons were made with a mutant which was semi-constitutive but had wild-type catabolite repression and a mutant which had altered catabolite repression but had wild-type inducibility. Mutant C17, which is semi-constitutive, gave a curve which was intermediate between that for the wild-type and C11. This mutant presumably produces an altered cytoplasmic repressor which differs from the wild-type in its affinities for the inducer and the gene region which it controls (or it may be an operator mutant). The balance between induction and repression is again shown by the sharp peak of maximum production (Fig. 17). Mutant L11, which is inducible but is less sensitive to catabolite repression, was induced to form amidase in batch culture by 10 mM N-acetylacetamide at a slightly higher rate in succinate medium than the wild-type strain and the induction lag was markedly decreased. In the continuous culture experiment these differences showed up in an earlier onset of induction, i.e. at a lower dilution rate, and a much more gradual decline in the rate of amidase synthesis as the dilution rate was increased.

It was not possible to carry out the steady-state experiments under gratuitous conditions since N-acetylacetamide is not sufficiently stable for long-term experiments. It was however possible to continue the transition experiments for about 4 hr. without any measurable spontaneous hydrolysis of the N-acetylacetamide and the amidase levels at three dilution rates were compared after 4 hr. as the culture approached steady state conditions (Fig. 14). It is clear that the highest amidase activities reached were at the lower dilution rates. So that when the inducing amide is N-acetylacetamide there is a steady decline in the rate of amidase synthesis by the wild-type strain as the dilution rate is increased. There are two important differences between adding acetamide and N-acetylacetamide. Acetamide is not only an inducer but is also a source of the catabolite repressor so that it would tend both to increase the rate of synthesis of the enzyme and to repress it. At very low dilution rates as soon as any enzyme is synthesized acetamide tends to be destroyed too fast to be very effective as an inducer. N-acetylacetamide, on the other hand, acts only as an inducer and is not destroyed by the enzyme but it is much less effective as an inducer than acetamide. The experiments with carbon-starved suspensions had indicated that acetamide was about 100 times more effective as an inducer than N-acetylacetamide. We suggest that it is in fact so weak an inducer that in our experimental conditions its inducing action under steady-state conditions is manifest as resistance to catabolite repression at low dilution rates, but this is overcome by catabolite repression as the dilution rate is increased. Whether or not inducible enzymes will exhibit a sharp peak of enzyme synthesis at a particular dilution rate will therefore depend on the relative importance of the growth substrate as inducer or as source of a catabolite repressor.

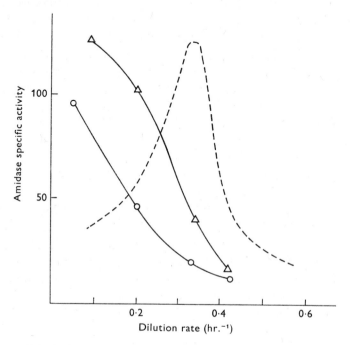

Fig. 16. Effect of dilution rate on amidase synthesis by magno-constitutive mutants growing in continuous culture under steady-state conditions in minimal medium containing 10 mM succinate + 20 mM acetamide. O—O, Mutant C11; △—△, mutant, L9; - - - -, wild-type on the same medium (Clarke *et al.* 1968).

During the transition period from one steady state to another, in the experiments in which changes were made in the incoming medium, we observed marked oscillations in amidase activity (Boddy *et al.* 1967). We concluded that these were due to incomplete balance in the control of the level of rate of enzyme synthesis by the opposing effects of induction and catabolite repression. The oscillations continued into the new steady state (see Fig. 12) although they were then much less marked. The significance of regular oscillations in enzyme synthesis is discussed in more detail by Goodwin (this Symposium).

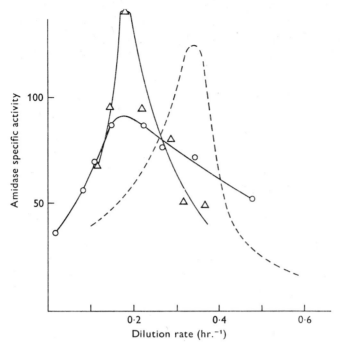

Fig. 17. Effect of dilution rate on amidase synthesis by two mutants growing in continuous culture under steady-state conditions in minimal medium containing 10 mM succinate + 20 mM *N*-acetylacetamide. △—△, Mutant C17, semi-constitutive; ○—○, mutant L11, fully inducible, reduced catabolite repressibility; - - - -, wild-type on the same medium.

ENZYME REGULATION AND EVOLUTION

During evolution bacteria have evolved biochemical economies of several kinds. Those living in environments providing complex nutrients have tended to become biochemically exacting and have lost various biosynthetic pathways. The biochemically versatile organisms have evolved control mechanisms which ensure that enzymes are metabolically active only when required, or are synthesized only under conditions when they are needed for growth. It has been widely assumed that biochemical economies of this kind would convey to the mutants a selective growth advantage over the corresponding wild-type strains.

Zamenhof & Eichhorn (1967) have recently shown that mutants which have lost the ability to produce certain biosynthetic enzymes have a growth advantage over the wild-type in media containing the appropriate end product. In one of their experiments a histidine-requiring mutant (*his⁻*) of *Bacillus subtilis*, strain 168, was grown in continuous culture with a histidine-independent back-mutant (*his⁺*)

in the presence of 0·32 mM histidine. The *his⁻* strain had a marked growth advantage over the *his⁺* strain and in 48 generations the ratio of *his⁺/his⁻* had decreased 4000-fold (Fig. 18*a*). Similar results were obtained with an indole-requiring mutant *ind⁺* and its back-mutant

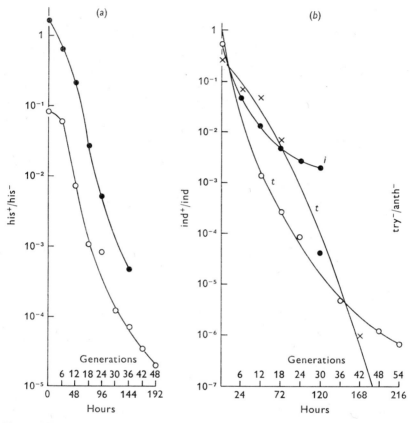

Fig. 18. Change in the ratio of cell numbers of *Bacillus subtilis* strains during growth in continuous culture in minimal medium. (*a*) A histidine-requiring mutant (his⁻) was grown with a spontaneous back-mutant (his⁺) in the presence of 0·32 mM histidine. (*b*) An indole-requiring auxotroph (ind⁻) was grown with a back-mutant (ind⁺) in the presence of 0·25 mM tryptophan (curves *t*) or indole (curve *i*) (Zamenhof & Eichhorn, 1967).

ind⁻ grown in continuous culture in the presence of 0·25 mM-tryptophan (Fig. 18*b*). They also compared a mutant which was derepressed for the tryptophan enzymes, and excreted tryptophan into the medium, with the wild-type strain in which the tryptophan enzymes are under tryptophan repression. A mixture of equal numbers of the two strains was grown in continuous culture in glucose-limited basal medium and the proportion of the derepressed mutant decreased 10⁵-fold in 26 generations.

Baich & Johnson (1968) showed that mutants which have lost feedback control of a metabolic pathway are also at a growth disadvantage compared with the wild-type controlled strain. *Escherichia coli* W was compared with a mutant strain which had lost the feedback control of proline biosynthesis. They mixed equal numbers of the bacteria of the two strains and transferred subcultures to fresh medium at regular intervals. They found that the time taken to establish a change in ratio of mutant/wild-type of 1/10 was very dependent on the growth medium. The presence of proline at low concentrations enhanced the growth advantage of the wild-type, but even in a complex medium the mutant was at a disadvantage. In this case the enzymes were present and the biochemical economy was only in the actual operation of the proline pathway.

CONCLUSIONS

Our discussion has centred around the ways in which growth depends on the balance of enzyme synthesis. We have emphasized that especially for catabolic enzymes there is a dual control by induction and repression and that in a particular environment the amount of enzyme formed depends on the balance of these opposing forces. We would like to know much more about the relative rates of enzyme synthesis in cultures growing on mixed substrates metabolized by enzymes which are regulated by both induction and repression. It is well known that individual enzymes can be synthesized at such rates that they represent about 10 % of the total protein of the bacterial cell. One would like to find out how many enzymes could be derepressed simultaneously, and to what extent. Many workers have noticed that enzymes become markedly derepressed towards the end of the exponential growth phase. This cannot obviously be the case for all enzymes but since most investigators look at only one enzyme at a time there is almost no information on this.

The model on which most discussion is based is still that for the lactose operon suggested by Jacob & Monod (1961) with subsequent modifications. We all have cause to be grateful for this powerful hypothesis but the very success of most of its major predictions have made it difficult to envisage other forms of control. It is by no means certain that all catabolic enzymes are regulated in this way (cf. Englesberg *et al.* 1965 on the arabinose operon) and detailed studies of biosynthetic enzymes suggest that several variations may occur in control of enzyme synthesis at the levels of transcription and translation.

The biosynthetic enzymes and the enzymes of the central metabolic pathways appear to be under much tighter control than the catabolic enzymes and there is no clear-cut evidence that they can be induced. The biosynthetic enzymes such as those for amino acid biosynthesis are probably repressed by specific molecules (amino acids or aminoacyl-t-RNAs), but whether these act in combination with a regulator gene product of the *lac* type is far from clear in all cases. Does catabolite repression of enzymes, such as those for the aromatic pathways, and glucose metabolism and the TCA enzymes also involve highly specific metabolic interactions? It could well be that repression of all or some of the enzymes concerned with energy metabolism is dependent on the concentration of one or more key cell metabolites. Obvious candidates for a rather general repression of this type would be ATP, the nucleotide coenzymes, some of the phosphate compounds such as phosphoenolpyruvate, fructose-1,6-diphosphate or other phosphate esters.

A rather pleasing economy of control is the feedback inhibition of an enzyme, or metabolic pathway, by the compound which represses enzyme synthesis. This appears to occur with the inhibition and repression of isocitrate lyase by phosphoenolpyruvate. Tryptophan again appears to be both the feedback inhibitor and the corepressor of the enzymes of the tryptophan operon. It is possible to imagine that this type of regulation developed from gene duplication at an earlier stage of evolution. In the same way one could imagine the regulation of an amino acid biosynthetic pathway by an amino acid derivative (e.g. repression of the histidine operon by histidinyl-t-RNA) as a further evolutionary stage in the development of cellular control mechanisms.

Growth on organic compounds is one of the standard criteria for identifying and classifying bacterial species. When mutant strains are found which do not grow on a particular compound it is usually assumed that they have suffered a loss mutation and no longer synthesize one or more enzymes. From our amidase studies it was clear that growth is a function of both the properties of an enzyme and also of its regulatory system. During evolution strains must have acquired the ability to utilize new substrates for growth either by mutations which produced changes in the specificity of the enzyme protein or by mutations which altered the specificity of the product of the regulator gene and allowed induction or derepression by new compounds.

The capacity to produce an enzyme which makes a substrate available for growth by changing it into a common metabolite would obviously give a species an advantage over organisms which could not carry out this reaction. However, when this compound was absent the first species would be wasting energy on synthesizing this enzyme at the expense of essential central metabolic enzymes. It is easy to see that a growth advantage would accompany the development of a regulatory control system which allowed both a high rate of enzyme synthesis when required and an almost complete shut off when not required. For this type of enzyme the wild-type inducible strain would be expected to outgrow a constitutive mutant in the same way in which the wild-type strain repressible for biosynthetic enzymes outgrows the derepressed mutants. Wide variation in the central metabolic enzymes on the other hand might result in metabolic inbalance and it is unlikely that mutants would be viable if they produced a thousand times as much of these enzymes as the wild-type.

REFERENCES

AMES, B. N., GARRY, B. & HERZENBERG, L. A. (1960). The genetic control of the enzymes of histidine biosynthesis in *Salmonella typhimurium. J. gen. Microbiol.* **22**, 369.

AMES, B. N. & MARTIN, R. G. (1964). Biochemical aspects of genetics: The Operon. *Annu. Rev. Biochem.* **33**, 235.

AMES, B. N., MARTIN, R. G. & GARRY, B. J. (1961). The first step in histidine biosynthesis. *J. biol. Chem.* **236**, 2019.

ANTON, D. N. (1968). Histidine regulatory mutants in *Salmonella typhimurium. J. molec. Biol.* **33**, 533.

ASHWORTH, J. M. & KORNBERG, H. L. (1963). Fine control of the glyoxylate cycle by allosteric inhibition of isocitrate lyase. *Biochim. biophys. Acta*, **73**, 519.

ATTARDI, G., NAONO, S., ROUVIÈRE, J., JACOB, F. & GROS, F. (1963). Production of messenger RNA and regulation of protein synthesis. *Cold Spring Harb. Symp. quant. Biol.* **28**, 363.

BAICH, A. & JOHNSON, M. (1968). Evolutionary advantage of a control of a biosynthetic pathway. *Nature, Lond.* **218**, 464.

BAIDYA, T. K. N., WEBB, F. C. & LILLY, M. D. (1967). The utilization of mixed sugars in continuous fermentation. I. *Biotech. Bioeng.* **9**, 195.

BAUMBERG, S., BACON, D. F. & VOGEL, H. J. (1965). Individually repressible enzymes specified by clustered genes of arginine synthesis. *Proc. natn. Acad. Sci. U.S.A.* **53**, 1029.

BLANGY, D., BUC, H. & MONOD, J. (1968). Kinetics of the allosteric interactions of phosphofructokinase from *Escherichia coli. J. molec. Biol.* **31**, 13.

BODDY, A., CLARKE, P. H., HOULDSWORTH, M. A. & LILLY, M. D. (1967). Regulation of amidase synthesis by *Pseudomonas aeruginosa* 8602 in continuous culture. *J. gen. Microbiol.* **48**, 137.

BRAMMAR, W. J. (1965). *The control of amidase synthesis in Pseudomonas aeruginosa by induction and repression mechanisms.* Ph. D. Thesis, University of London.

BRAMMAR, W. J. & CLARKE, P. H. (1964). Induction and repression of *Pseudomonas aeruginosa* amidase. *J. gen. Microbiol.* **37**, 307.

BRAMMAR, W. J., CLARKE, P. H. & SKINNER, A. J. (1967). Biochemical and genetic studies with regulator mutants of the *Pseudomonas aeruginosa* 8602 amidase system. *J. gen. Microbiol.* **47**, 87.

BRAMMAR, W. J., MCFARLANE, N. D. & CLARKE, P. H. (1966). The uptake of aliphatic amides by *Pseudomonas aeruginosa. J. gen. Microbiol.* **44**, 303.

BRENNER, S. (1965). Theories of gene regulation. *Brit. med. Bull.* **21**, 244.

BROWN, J. E. (1969). *Biochemical and genetic studies on amidase synthesis in Pseudomonas aeruginosa.* Ph.D. Thesis, University of London.

BROWN, P. R. & CLARKE, P. H. (1967). Mutants of *Pseudomonas aeruginosa* 8602 producing an amidase with altered substrate specificity. *J. gen. Microbiol.* **46**, x.

CANOVAS, J. L. & STANIER, R. Y. (1967). Regulation of the enzymes of the β-ketoadipate pathway in *Moraxella calcoacetica.* 1. General aspects. *European J. Biochem.* **1**, 289.

CLARK, D. J. & MARR, A. G. (1964). Studies on the repression of β-galactosidase in *Escherichia coli. Biochim. biophys. Acta,* **92**, 85.

CLARKE, P. H. & BRAMMAR, W. J. (1964). Regulation of bacterial enzyme synthesis by induction and repression. *Nature, Lond.* **203**, 1153.

CLARKE, P. H., HOULDSWORTH, M. A. & LILLY, M. D. (1968). Catabolite repression and the induction of amidase synthesis by *Pseudomonas aeruginosa* 8602, in continuous culture. *J. gen. Microbiol.* **51**, 225.

CLINE, A. L. & BOCK, R. M. (1966). Translational control of gene expression. *Cold Spring Harb. symp. quant. Biol.* **31**, 321.

COHEN, G. N. (1965). Regulation of enzyme activity in microorganisms. *Annu. Rev. Microbiol.* **19**, 105.

COHEN, G. N. & JACOBS, F. (1959), Sur la répression de la synthèse des enzymes intervenant dans la formation du tryptophane chez *Escherichia coli. C. r. hebd. Séanc. Acad. Sci. Paris,* **248**, 3490.

COHN, M. & HORIBATA, K. (1959). Physiology of the inhibition by glucose of the induced synthesis of the β-galactosidase-enzyme system of *Escherichia coli. J. Bact.* **78**, 624.

DATTA, P. & GEST, H. (1964). Alternative patterns of end-product control in biosynthesis of amino-acids of the aspartic family. *Nature, Lond.* **203**, 1259.

DOOLITTLE, W. F. & YANOFSKY, C. (1968). Mutants of *Escherichia coli* with an altered tryptophanyl-transfer ribonucleic acid synthetase. *J. Bact.* **95**, 1283.

ENGLESBERG, E., IRR, J., POWER, J. & LEE, N. (1965). Positive control of enzyme synthesis by gene *C* in the L-arabinose system. *J. Bact.* **90**, 946.

FARGIE, B. & HOLLOWAY, B. W. (1965). Absence of clustering of functionally related genes in *Pseudomonas aeruginosa. Genet. Res.* **6**, 284.

FREUNDLICH, M. (1967). Valyl-transfer RNA role in repression of the isoleucine-valine enzymes in *Escherichia coli. Science, N.Y.* **157**, 828.

GERHARDT, J. C. & SCHACHMAN, H. K. (1965). Distinct subunits for the regulation and catalytic activity of aspartic transcarbamylase. *Biochemistry,* **4**, 1054.

GERHARDT, J. C. & SCHACHMAN, H. K. (1968). Allosteric interactions in aspartic transcarbamylase. II. Evidence for different conformational states of the protein in the presence and absence of specific ligands. *Biochemistry,* **7**, 538.

GILBERT, W. & MÜLLER-HILL, B. (1966), Isolation of the Lac repressor. *Proc. natn. Acad. Sci. U.S.A.* **56**, 1891.

GILBERT, W. & MÜLLER-HILL, B. (1967) The Lac operator is DNA. *Proc. natn. Acad. Sci. U.S.A.* **58**, 2415.

GORINI, L. (1963). Control by repression of a biochemical pathway. *Bact. Rev.* **27**, 182.

GRAY, C. T., WIMPENNY, J. W. T. & MOSSMAN, M.R. (1966). Effects of aerobiosis, anaerobiosis and nutrition on the formation of the Krebs cycle enzymes in *Escherichia coli. Biochim. biophys. Acta,* **117**, 33.

HAMILTON, W. A. & DAWES, E. A. (1959). A diauxic effect with *Pseudomonas aeruginosa. Biochem. J.* **71**, 25P.

HAMILTON, W. A. & DAWES, E. A. (1960). The nature of the diauxic effect with glucose and organic acids in *Pseudomonas aeruginosa. Biochem. J.* **76**, 70P.

HAMILTON, W. A. & DAWES, E. A. (1961). Further observations on the nature of the diauxic effect with *Pseudomonas aeruginosa*. *Biochem. J.* **79**, 25 P.

HAMLIN, B. T., NG, F. M.-W. & DAWES, E. A. (1967). Regulation of enzymes of glucose metabolism in *Pseudomonas aeruginosa* by citrate. In *Microbial Physiology and Continuous Culture*, p. 211. Ed. E. O. Powell, C. G. T. Evans, R. E. Strange and D. W. Tempest. London: H.M.S.O.

HANSON, R. S. & COX, D. P. (1967). Effect of different nutritional conditions on the synthesis of tricarboxylic acid cycle enzymes. *J. Bact.* **93**, 1777.

HARADA, T. & SPENCER, B. (1964). Repression and induction of arylsulphatase synthesis in *Aerobacter aerogenes*. *Biochem. J.* **93**, 373.

HARTE, M. J. & WEBB, F. C. (1967). Utilization of mixed sugars in continuous fermentation. II. *Biotech. Bioeng.* **9**, 205.

HEGEMAN, G. D. (1966a). Synthesis of the enzymes of the mandelate pathway by *Pseudomonas putida*. I. Synthesis of enzymes by the wild type. *J. Bact.* **91**, 1141.

HEGEMAN, G. D. (1966b). Synthesis of the enzymes of the mandelate pathway by *Pseudomonas putida*. II. Isolation and properties of blocked mutants. *J. Bact.* **91**, 1155.

HEGEMAN, G. D. (1966c). Synthesis of the enzymes of the mandelate pathway by *Pseudomonas putida*. III. Isolation and properties of constitutive mutants. *J. Bact.* **91**, 1161.

HENNING, U., DENNER, G., HERTEL, R. & SHIPP, W. S. (1966). Translation of the structural genes of the *E. coli* pyruvate dehydrogenase complex. *Cold Spring Harb. Symp. quant. Biol.* **31**, 227.

HOGNESS, D. S., COHN, M. & MONOD, J. (1955). Studies on the induced synthesis of β-galactosidase in *Escherichia coli*: The kinetics and mechanism of sulfur incorporation. *Biochim. biophys. Acta*, **15**, 99.

HOLLOWAY, B. W. (1955). Genetic recombination in *Pseudomonas. J. gen. Microbiol.* **13**, 572.

HOLLOWAY, B. W. & FARGIE, B. (1960). Fertility and genetic linkage in *Pseudomonas aeruginosa*. *J. Bact.* **80**, 362.

HORIUCHI, T., TOMIZAWA, J. & NOVICK, A. (1962). Isolation and properties of bacteria capable of high rates of β-galactosidase synthesis. *Biochim. biophys. Acta*, **55**, 152.

IMAMOTO, F., ITO, J. & YANOFSKY, C. (1966). Polarity in the tryptophan operon of *E. coli*. *Cold Spring Harb. Symp. quant. Biol.* **31**, 235.

IMAMOTO, F. & YANOFSKY, C. (1967). Transcription of the tryptophan operon in polarity mutants of *Escherichia coli*. *J. molec. Biol.* **28**, 1.

IPPEN, K., MILLER, J. H., SCAIFE, J. & BECKWITH, J. (1968). New controlling element in the lac operon of *E. coli*. *Nature, Lond.* **217**, 825.

ITO, J. & CRAWFORD, I. P. (1965). Regulation of the enzymes of the tryptophan pathway in *Escherichia coli*. *Genetics*, **52**, 1303.

JACOB, F. & MONOD, J. (1961). Genetic regulatory mechanisms in the synthesis of proteins. *J. molec. Biol.* **3**, 318.

JACOB, F., ULLMAN, A. & MONOD, J. (1964). Le promoteur élément génétique nécessaire à l'expression d'un opéron. *C. r. hebd. Séanc. Acad. Sci., Paris*, **258**, 3125.

JACOB, F., ULLMAN, A. & MONOD, J. (1965). Délétions fusionnant l'opéron lactose et un opéron purine chez *Escherichia coli*. *J. molec. Biol.* **13**, 704.

JACOBY, G. A. & GORINI, L. (1967). Genetics of control of the arginine pathway in *Escherichia coli* B. & K. *J. molec. Biol.* **24**, 41.

JAYARAMAN, K., MÜLLER-HILL, B. & RICKENBERG, H. V. (1966). Inhibition of the synthesis of β-galactosidase in *Escherichia coli* by 2-nitrophenyl-β-D-fucoside. *J. molec. Biol.* **18**, 339.

KELLY, M. & CLARKE, P. H. (1962). An inducible amidase produced by a strain of *Pseudomonas aeruginosa. J. gen. Microbiol.* **27**, 305.

KÉPÈS, A. (1967). Sequential transcription and translation in the lactose operon of *Escherichia coli. Biochim. biophys. Acta*, **138**, 107.

KINSKY, S. C. (1961). Induction and repression of nitrate reductase in *Neurospora crassa. J. Bact.* **82**, 898.

KORNBERG, H. L. (1966). The role and control of the glyoxylate cycle in *Escherichia coli. Biochem. J.* **99**, 1.

LAWRENCE, D. A., SMITH, D. A. & ROWBURY, R. (1968). Regulation of methionine synthesis in *Salmonella typhimurium. Genetics*, **58**, 473.

LOOMIS, W. F. & MAGASANIK, B. (1964). The relation of catabolite repression to the induction system for β-galactosidase in *Escherichia coli. J. molec. Biol.* **8**, 417.

LOOMIS, W. F. & MAGASANIK, B. (1967). The catabolite repression of the *lac*-operon in *Escherichia coli. J. molec. Biol.* **23**, 487.

LOUTIT, J. S. & MARINUS, M. G. (1967). Polarised chromosome transfer at relatively high frequency in *Pseudomonas aeruginosa. Proc. of the University of Otago Medical School*, **45**, 18.

MAGANA-PLAZA, I. & RUIZ-HERRERA, J. (1967). Mechanisms of regulation of urease biosynthesis in *Proteus rettgeri. J. Bact.* **93**, 1294.

MAGASANIK, B. (1961). Catabolite repression. *Cold Spring Harb. Symp. quant. Biol.* **26**, 249.

MANDELSTAM, J. & JACOBY, G. A. (1965). Induction and multi-sensitive end-product repression in the enzymic pathway degrading mandelate in *Pseudomonas fluorescens. Biochem. J.* **94**, 569.

MARGOLIN, P. & BAUERLE, R. H. (1966). Determinants for regulation and initiation of expression of tryptophan genes. *Cold Spring Harb. Symp. quant. Biol.* **31**, 311.

MARR, A. G. & MARCUS, L. (1962). Kinetics of induction of mannitol dehydrogenase in *Azotobacter agilis. Biochim. biophys. Acta*, **64**, 65.

MARTIN, R. G., WHITFIELD, H. J., JUN., BERKOWITZ, D. B. & VOLL, M. J. (1966). A molecular model of the phenomenon of polarity. *Cold Spring Harb. Symp. quant. Biol.* **31**, 215.

MATELES, R. I., CHIAN, S. K. & SILVER, R. (1967). Continuous culture on mixed substrates. In *Microbial Physiology and Continuous Culture*, p. 232. Ed. E. O. Powell, C. G. T. Evans, R. E. Strange and D. W. Tempest. London: H.M.S.O.

MCFALL, E. (1964). Pleiotropic mutations in the D-serine deaminase system of *Escherichia coli. J. molec. Biol.* **9**, 754.

MCFALL, E. (1967). Mapping of the D-serine deaminase region in *Escherichia coli* K12. *Genetics*, **55**, 91.

MCFALL, E. & MANDELSTAM, J. (1963). Specific metabolic repression of three induced enzymes in *Escherichia coli. Biochem. J.* **89**, 391.

MEDHI, Q. & YUDKIN, M. D. (1967). Coupling of transcription to translation in the induced synthesis of β-galactosidase. *Biochim. biophys. Acta*, **149**, 288.

MEE, B. J. & LEE, T. O. (1967). An analysis of histidine requiring mutants in *Pseudomonas aeruginosa. Genetics*, **55**, 709.

MONOD, J., CHANGEUX, J. P. & JACOB, F. (1963). Allosteric proteins and cellular control systems. *J. molec. Biol.* **14**, 290.

MONOD, J., COHEN-BAZIRE G. & COHN, M. (1951). Sur la biosynthèse de la β-galactosidase (lactase) chez *Escherichia coli*. La specificité de l'induction. *Biochim. biophys. Acta*, **7**, 585.

MONOD, J., WYMAN, J. & CHANGEUX, J-P. (1965). On the nature of allosteric transitions: A plausible model. *J. molec. Biol.* **13**, 88.

MORRIS, I. & SYRETT, P. J. (1965). The effect of nitrogen starvation on the activity of nitrate reductase and other enzymes in *Chlorella. J. gen. Microbiol.* **38**, 21.

MOSES, V. & PREVOST, C. (1966). Catabolite repression of β-galactosidase synthesis in *Escherichia coli*. *Biochem. J.* **100**, 336.

MÜLLER-HILL, B., CRAPO, L. & GILBERT, W. (1968). Mutants that make more *lac* repressor. *Proc. natn. Acad. Sci. U.S.A.* **59**, 1259.

NAKADA, D. & MAGASANIK, B. (1964). The roles of inducer and catabolite repressor in the synthesis of β-galactosidase by *Escherichia coli*. *J. molec. Biol.* **8**, 105.

NEWTON, W. A., BECKWITH, J. R., ZIPSER, D. & BRENNER, S. (1965). Nonsense mutants and polarity in the *lac* operon of *Escherichia coli*. *J. molec. Biol.* **14**, 290.

NG, F. M.-W. & DAWES, E. A. (1967). Regulation of enzymes of glucose metabolism by citrate in *Pseudomonas aeruginosa*. *Biochem. J.* **104**, 48 P.

OKINAKA, R. T. & DOBROGOSZ, W. J. (1967). Catabolite repression and pyruvate metabolism in *Escherichia coli*. *J. Bact.* **93**, 1644.

ORNSTON, L. N. (1966). The conversion of catechol and protocatechuate to β-ketoadipate by *Pseudomonas putida*. *J. biol. Chem.* **241**, 3800.

PALLERONI, N. J. & STANIER, R. Y. (1964). Regulatory mechanisms governing synthesis of the enzymes for tryptophan oxidation by *Pseudomonas aeruginosa*. *J. gen. Microbiol.* **35**, 319.

PALMER, J. & MOSES, V. (1967). Involvement of the *lac* regulatory genes in catabolite repression in *Escherichia coli*. *Biochem. J.* **103**, 358.

PEARCE, L. E. & LOUTIT, J. S. (1965). Biochemical and genetic grouping of isoleucine-valine mutants of *Pseudomonas aeruginosa*. *J. Bact.* **89**, 58.

POWER, J. (1967). The L-rhamnose genetic system in *Escherichia coli* K 12. *Genetics* **55**, 557.

RAMAKRISHNAN, T. & ADELBERG, E. A. (1965). Regulatory mechanisms in the biosynthesis of isoleucine and valine. *J. Bact.* **89**, 654,

ROTH, J. R., ANTÓN, D. N. & HARTMAN, P. E. (1966). Histidine regulatory mutants in *Salmonella typhimurium*. I. Isolation and general properties. *J. molec. Biol.* **22**, 305.

ROTH, J. R., SILBERT, D. F., FINK, G. R., VOLL, M. J., ANTÓN, D., HARTMAN, P. E. & AMES, B. N. (1966). Transfer RNA and the control of the histidine operon. *Cold Spring Harb. Symp. quant. Biol.* **31**, 383.

ROTMAN, B. & SPIEGELMAN, S. (1954). On the origin of the carbon in the induced synthesis of β-galactosidase in *Escherichia coli*. *J. Bact.* **68**, 419.

SCAIFE, J. & BECKWITH, J. R. (1966). Mutational alteration of the maximal level of *lac* operon expression. *Cold Spring Harb. Symp. quant. Biol.* **31**, 403.

SHEPPARD, D. E. & ENGLESBERG, E. (1967). Further evidence for positive control of the L-arabinose system by gene *araC*. *J. molec. Biol.* **25**, 443.

SKINNER, A. J. & CLARKE, P. H. (1968). Acetate and acetamide mutants of *Pseudomonas aeruginosa* 8602. *J. gen. Microbiol.* **50**, 183.

STANIER, R. Y. (1968). Control of the enzymes of the β-oxoadipate pathway in bacteria. *Biochem. J.* **106**, 25 P.

STANIER, R. Y., PALLERONI, N. J. & DOUDOROFF, M. (1966). The aerobic pseudomonads: a taxonomic study. *J. gen. Microbiol.* **43**, 159.

STENT, G. S. (1964). The operon: on its third anniversary. *Science, N.Y.* **144**, 816.

STEVENSON, I. L. & MANDELSTAM, J. (1965). Induction and multi-sensitive end-product repression in two converging pathways degrading aromatic substances in *Pseudomonas fluorescens*. *Biochem. J.* **96**, 354.

YANOFSKY, C. & ITO, J. (1966). Nonsense codons and polarity in the tryptophan operon. *J. molec. Biol.* **21**, 313.

ZAMENHOF, S. & EICHHORN, H. H. (1967). Study of microbial evolution through loss of biosynthetic function: Establishment of 'defective' mutants. *Nature, Lond.* **216**, 456.

OXYGEN AND CARBON DIOXIDE AS REGULATORS OF MICROBIAL GROWTH AND METABOLISM

J. W. T. WIMPENNY

Medical Research Council Group for Microbial Structure and Function,
Department of Microbiology, University College,
Cathays Park, Cardiff

INTRODUCTION

The reactions of oxygen and carbon dioxide occupy a central position in microbial physiology. Oxygen can be assimilated, reduced as a terminal electron acceptor or evolved as a product of the photosynthetic cleavage of water. Carbon dioxide can be assimilated by many fixation reactions, reduced as a terminal electron acceptor—in, for example, the methane-forming bacteria—or evolved as a final end product of carbon metabolism.

Both the presence and absence and the relative concentrations of oxygen and CO_2 may regulate growth and metabolism in many micro-organisms. For example, the concentration of oxygen in a growth medium determines whether mitochondria are present in many yeasts. Chromatophore synthesis in some photolithotrophic bacteria is similarly controlled. Regulation by oxygen of cytochrome pigments and aerobic and anaerobic respiratory enzymes has been extensively investigated in oxygen facultative bacteria. Although CO_2 has long been known to affect the growth of micro-organisms, very little work has been done on its role as a metabolic regulator: most heterotrophic and autotrophic micro-organisms fix CO_2 and an increasing number of 'auto-heterotrophic' organisms are now known to reduce CO_2 by an adaptive Calvin carbon cycle or alternatively metabolize organic carbon sources by adaptive heterotrophic pathways.

The questions that remain to be answered before a clearer picture of oxygen–CO_2 control appears include the following: Are oxygen or CO_2 directly involved at the molecular level as regulators? If not, is the regulator molecule a single metabolite controlling the appearance and disappearance of large groups of enzymes?—or are there many distinct regulator metabolites perhaps controlling each enzyme separately? This question might be rephrased: Is formation of groups of enzymes coordinately or sequentially regulated? How does oxygen in particular regulate the formation of new structures? Are the regulatory mechanisms the same as those reported for other inducible-repressible systems of catabolism and anabolism?

OXYGEN AS A REGULATOR OF GROWTH AND METABOLISM

Micro-organisms show many degrees of reaction to molecular oxygen: strict aerobes show an absolute dependence on it, strict anaerobes cannot tolerate it, while facultative bacteria can grow in the presence or absence of it. Yet other groups can be defined which fall in the twilight zone of near anaerobiosis; for example, the micro-aerophilic bacteria. But these definitions are not precise. Many 'strict' anaerobes can tolerate a small amount of oxygen, whereas growth of strictly aerobic and facultative bacteria is inhibited by excess oxygen. Variation in oxygen concentration leads to changes in the physiology of most microbes, the clearest of these changes being seen in facultative organisms which have to adapt to high or low concentrations of oxygen.

The effect of oxygen on growth
Oxygen provision

The solubility of oxygen in water is low (about $2 \cdot 5 \times 10^{-4}$ M at ambient temperatures) and is reduced by dissolved salts. Whilst air contains 21 %, v/v, oxygen, a nutrient broth used in cultivating micro-organisms may only contain 5–7 μg./ml. at physiological temperature and one atmosphere pressure. Oxygen requirements of microbes in the nutrient medium may, on the other hand, be very high. Lineweaver (1933) measured respiration rates (μl. O_2 utilized/hr./mg. dry weight cells) of approximately 4000 for *Azotobacter vinelandii*. One gram dry weight of this bacterium can use all the oxygen in 1 l. of air-saturated medium in about 7 sec. Clearly the problem of maintaining full aerobiosis is great. Much recent work has demonstrated physiological changes in micro-organisms as a result of changes in oxygen provision. These changes certainly occur in a growing population which uses oxygen more rapidly than it can be provided. The ratio of cell weight to oxygen concentration, in a batch culture, is usually changing continuously, leading to a parallel series of small metabolic changes within the cell.

Arnold & Steele (1958) have listed the various resistances to diffusion of oxygen from the gas phase to respiratory enzyme active centres in the organism. Reduction of the over-all resistance is a biological engineering problem, but obviously much can be done to maintain full aerobiosis at all stages in a growth cycle. For example: (*a*) keep the cell population low by ensuring that another nutrient limits growth; (*b*) ensure a sufficient flow rate of air or possibly oxygen (oxygen toxicity is considered later); (*c*) reduce the bubble size, thus increasing the surface to volume ratio of each bubble and exposing the largest possible diffusion surface; (*d*) ensure the longest possible bubble path in the culture vessel (vigorous stirring, the provision of baffles, a sparger generating small bubbles and a high aspect ratio fermenter assist in (*c*) and (*d*)); (*e*) prevent cell clumping. The solution of these problems is outside the scope of this review, but the reader is referred to recent articles by Finn (1967), Calderbank (1967) and Webb (1964) among many others.

Another important question concerns the concentration of oxygen that limits growth. The oxygen concentration at half maximal uptake rate is, if it represents the kinetics of the oxygen-reducing enzyme, a classical Michaelis constant or K_m. At low concentrations of oxygen, diffusion through cell material may restrict oxygen uptake, whilst at higher concentrations the electron transport system itself may limit oxygen utilization. These are the findings of Johnson (1967) using *Candida utilis* grown in a chemostat at low oxygen tensions. Apparent K_m values for oxygen have been determined in different organisms by a number of different workers (Table 1). The oxygen utilizing systems in these organisms, especially in mammalian mitochondria, have a very high affinity for oxygen. In microbial cultures the oxygen concentration which limits respiration rate is of the order of 1 % of its air saturated level at S.T.P.

Table 1. *Apparent K_m values for oxygen uptake by different cells or organelles*

Organism	Apparent K_m: (μM)	Reference
Candida utilis	1·34	Johnson (1967)
Saccharomyces cerevisiae	2·84	Winzler (1941)
Saccharomyces cerevisiae	0·645	Longmuir (1954)
Saccharomyces cerevisiae	0·73–3·4	Terui, Konno & Sase (1960)
Saccharomyces cerevisiae	0·41	Chance (1965)
Haemophilus parainfluenzae	2·2–11·1	White (1963)
Rat heart particles	0·052	Chance (1965)
Rat liver particles	0·042	Chance (1965)
Pigeon heart particles	0·019	Chance (1965)

Excess oxygen

It is an over-simplification to state, as may have been suggested, that the main problem in cultivating aerobic and facultative bacteria is in providing sufficient oxygen. Oxygen in excess can be toxic and in most cases 'enough is as good as a feast'. Oxygen excess, defined as any concentration above that which leads to maximum growth rate, is of course a very variable factor depending on the organism involved. It can be a very low figure in strictly anaerobic bacteria.

In growth tests using a number of different bacteria Wiseman, Violago, Roberts & Penn (1966) showed that growth of certain bacteria was not inhibited by 4 atm. pure oxygen, others showed maximal growth between 1 and 3 atm. oxygen, whilst still others were not affected by oxygen pressure in the range tested. These effects were due to oxygen and not simply to hydrostatic pressure and in some strains they could be partially reversed by growing the cells on richer media. It is possible that this reversal could involve *p*-aminobenzoic acid (PABA) metabolism, as there appeared to be synergism between PABA analogues and oxygen (Gottlieb & Pakman, 1968). Growth of several fungi and bacteria has been shown to be reversibly inhibited by several weeks' exposure to 10 atm. of pure oxygen although all strains tested grew well in 2 % oxygen at 10·5 atm. pressure. It was suggested that the inhibition might result from the sensitivity of SH group enzymes to elevated oxygen tensions (Caldwell, 1965).

High oxygen solution rates can be shown to inhibit bacterial growth in continuous culture vessels (Dalton & Postgate, 1967; Khmel & Ierusalemski, 1967) and in some experiments (Zobell & Hittle, 1967) it was possible to distinguish clearly between the effects of oxygen and of hydrostatic pressure. Air-saturated water contains approximately 7 μg. oxygen/ml. and *Escherichia coli*, *Bacillus subtilis* and *B. mega-therium* survived 100 atm. hydrostatic pressure as long as the compressed culture liquid contained only this amount of oxygen per ml., but 35 μg. oxygen/ml. at the same pressure was bactericidal. In bakers yeast too one can distinguish between hydrostatic pressure and oxygen inhibition. Oxygen is lethal at 100 atm. pressure, while nitrogen at the same pressure is not (Stuart, Gerschman & Stannard, 1962). The effect of oxygen is apparently to cause increased potassium leakage and this leakage can be partially reversed by glucose. It may be that substrates which can keep intracellular constituents reduced also protect the cell against oxygen toxicity. Comparable evidence has been reported in mammalian systems where succinate reduces oxygen toxicity in rats

(Sanders, Hall & Woodhall, 1965). However, the factors reducing oxygen toxicity vary from one organism to another. In *Euglena* oxygen toxicity is reduced by glucose and also by a high phosphate concentration (Blum & Begin-Heick, 1967), but in *Achromobacter* strain P it was unchanged by glucose and prevented by succinate, glutamate, malate or fumarate (Gottlieb, 1966).

The inhibitory effects of oxygen may be limited to particular enzymes or enzyme systems. For example, oxygen inhibits bacteriochlorophyll synthesis in obligate anaerobes such as *Chromatium* but it has no effect on the motility or viability of the cells (Hurlbert, 1967). In a nitrifying bacterium *Nitrocystis oceanus*, growth is inhibited by elevated partial pressures of oxygen while respiration is actually increased (Gundersen, 1966). In this organism *initiation* of growth is more sensitive than subsequent division (Gundersen, Carlucci & Borström, 1966). One clear case where oxygen toxicity can be explained in enzymic terms has been described by Pichinoty & d'Ornano (1961). They showed that growth of a number of organisms on nitrate as sole nitrogen source is inhibited by pure oxygen and not by air. This is because the enzyme nitrate reductase is inhibited or repressed at high oxygen tensions as I shall discuss later. Nitrogen fixation is also oxygen sensitive. Oxygen concentrations above 25 %, v/v, at atmospheric pressure inhibit root nodule symbiotic nitrogen fixation (Bond, 1961).

It is difficult to explain these toxic efforts of oxygen. Gundersen *et al.* (1966) suggest on the basis of their work and that of Schon (1965) that in bacteria there is competition for reductant between terminal electron acceptor and carbon assimilation, so that in the presence of excess oxygen insufficient reductant is present for CO_2 fixation. This 'short-circuiting' theory may well apply to observations of Dalton & Postgate (1967) and those of Khmel & Ierusalemski (1967) with *Azotobacter vinelandii* cited earlier. This organism can reduce oxygen at a very great rate and may simply run out of reductant necessary for heterotrophic biosynthetic reactions in the presence of excess oxygen. A more remote possibility exists that a 'short circuit' operates in strictly anaerobic bacteria in the presence of oxygen, as some of these have flavoprotein NADH oxidase enzymes whose functions are in doubt (Dolin, 1961).

Whilst mechanisms of oxygen toxicity are not clearly understood, some of the following factors may be important. First, it may involve sensitivity of SH groups to oxidation by oxygen (Haugaard, 1946, 1955, 1968). This includes non-protein SH groups such as glutathione, coenzyme A and lipoic acid as well as SH-group-containing proteins, particularly flavoproteins, pyridoxal-containing enzymes and electron transport proteins such as ferredoxin. Other factors include free radical formation (see Heden & Malmborg, 1961), peroxide formation, repression of enzyme formation and endogenous reductant diverted to reduce excess oxygen leaving inadequate supplies for biosynthesis —the 'short-circuit' theory. Superimposed on these toxicity factors are modifications induced by nutritional factors. These may include agents giving rise to endogenous reductant which may partially or wholly reverse oxygen toxicity. Oxygen toxicity in micro-organisms is discussed by Zobell & Hittle (1967) and in living organisms generally in a comprehensive review by Haugaard (1968). A valuable list of enzymes sensitive to oxygen was compiled by Davies & Davies (1965).

Cell yield

Oxygen clearly influences growth yield. This effect is absolute in the case of strictly aerobic or anaerobic bacteria in the sense that no growth occurs in the absence or presence of oxygen respectively: and relative in facultative bacteria, where any increase in yield in the presence of oxygen is a measure of the greater efficiency of energy conversion. The cell yield and ATP yield of various organisms under aerobic and anaerobic growth conditions has been compared in Table 2 and the results pose interesting questions from the energetic point of view. In aerobically grown cells both the cell yields and the ATP yields are greater than those in anaerobically cultured organisms: the difference in cell yield ranges from two- to sixfold, while the amount of energy made available as ATP shows a much greater range, even if the P:O ratios in microbial systems are lower than those calculated from mammalian mitochrondria. ATP yield per mole glucose is calculated for anaerobic growth and aerobic growth assuming different P:O ratios for oxidative phosphorylation in the oxidation of NADH. For a more detailed consideration of microbial energetics see Bauchop & Elsden (1960), Gunsalus & Shuster (1961) and Forrest (this Symposium). It seems clear that aerobic growth is not energy limited. This conclusion is given weight by the fact that the steady-state ATP pool in at least one facultative bacterium, *Escherichia coli* K12, is highest when the cells are grown aerobically on a simple glucose salts medium. Under these conditions a relative increase in pool size during batch culture can be observed (Cole, Wimpenny & Hughes, 1967). Anaerobically, growth *may* be energy limited although still fast, due to the rapid turnover of substrate made possible by the large surface to volume ratio of small micro-organisms.

Table 2. *A comparison of cell yield (grams dry weight cells/gram substrate) and calculated energy yield in some micro-organisms growing aerobically and anaerobically*

Organism	Growth condition	Aerobic yield	Anaerobic yield	Ratio: Aerobic/Anaerobic	Reference
Aerobacter aerogenes	Complex + glucose	0·72	0·12	6·0	Hernandez & Johnson (1967 a, b)
	Carbon limited, synthetic + glucose	0·38	0·11	3·4	Necklen (personal communication)
Klebsiella aerogenes	Carbon limited, synthetic + glucose	0·4	0·067	6·0	Harrison & Pirt (1967)
	Nitrogen limited synthetic + glucose	0·005	0·002	2·5	
Escherichia coli	Complex + glucose	0·65	0·13	5·0	Hernandez & Johnson (1967 a, b)
	Carbon limited, synthetic + glucose	0·30	0·12	2·5	Wimpenny (unpublished)
	Carbon limited complex + glucose	0·36	0·18	2·0	

ATP yield (moles/mole glucose) assuming P:O ratio for oxidation of $NADH_2$					Assumed P:O ratio for succinate dehydrogenase
3		38	2	19	2
2		28	2	14	2
1		16	2	8	1

Electron acceptors other than oxygen can also increase the growth yield. *Aerobacter aerogenes* grown anaerobically produces 21·8 g. dry weight of cells per mole mannitol compared to 50·6 with added nitrate. The figures for cells grown on glucose anaerobically, aerobically or with nitrate are 26·1, 72·7 and 45·5 respectively. No additional yield was observed with nitrite as electron acceptor, although nitrite was reduced (Hadjipetrou & Stouthamer, 1965). Hadjipetrou, Gray-Young & Lilly (1966) showed that *Escherichia coli* can reduce ferricyanide anaerobically: no increase in yield was observed but there was a change in fermentation products. Thus there appears to be ATP synthesis by oxidative phosphorylation with oxygen or nitrate but not with nitrite or ferricyanide. In some cases optimum cell yield is obtained in less than fully aerobic conditions; for example, with *Bacillus stearothermophilus* (Downey, 1966) with *E. coli* and *A. aerogenes* (J. W. T. Wimpenny & D. K. Necklen, unpublished observations) and *Azotobacter vinelandii* (Dalton & Postgate, 1967; Khmel & Ierusalemsky, 1967). There is a fine distinction here between oxygen toxicity and supraoptimal aeration conditions.

Product formation

The wide range of products formed anaerobically by fermentation reactions in different organisms reflects the need of the cell to produce electron acceptors to balance oxidation-reduction reactions in the absence of oxygen or other inorganic electron acceptors. At the same time endogenously produced acceptor must have a sufficiently positive potential to allow energy released to be coupled to ATP synthesis. At the other extreme, under fully aerobic conditions, a carbon source may be completely oxidized to CO_2 and water. Facultative bacteria show a gradual change in product formation as the oxygen tension changes. Anaerobically *Aerobacter aerogenes* forms ethanol, formic acid, butanediol, acetoin, acetic acid and CO_2. As the oxygen tension rises, ethanol and formic acid then butanediol and acetoin and finally acetic acid disappear: at the same time the amount of CO_2 formed increases (Pirt, 1957; Harrison & Pirt, 1967). The yield of penicillin from *Penicillium* batch cultures is related to aeration efficiency (Bartholomew, Karow, Sfat & Wilhelm, 1950), as is the yield of ustilagic acid from *Ustilago zeae* (Roxburgh, Spencer & Salans, 1954). On the other hand, there is an optimum oxygen concentration for vitamin B_{12} formation in *Propionibacterium freundreichii* (Bartholomew, 1960). Lowered yield at high aeration rates is observed in the hydrolysis of tannin to gallic acid by *Aspergillus niger* (Foster, 1949)—although growth is favoured by high aeration, it is at the expense of gallic acid, which is used as an oxidizable carbon source. Glutamic acid production by *Brevibacterium lactofermentum* is optimal at intermediate oxygen tensions and Hirose, Sonoda, Kinoshita & Okada (1967) attribute this to altered terminal electron-transport enzyme levels, since *in vitro* glutamate production from citrate appeared to be unaffected by oxygen tension.

The effect of oxygen on structures

Yeast mitochondria

Many yeasts grow in the presence or absence of oxygen. Slonimski (1953) showed that anaerobically grown yeast cells have greatly reduced cytochrome pigment levels compared with those grown aerobically. Linnane, Vitols & Nowland (1962), Polakis, Bartley & Meek (1964), Wallace & Linnane (1964) and Linnane (1965) in electron microscope studies showed that mitochondrial structure is absent in anaerobically grown yeast but returns with aeration. It appears that formation of mitochrondrial structure is sensitive to catabolite repression and respiratory enzymes associated with mitochondria are inversely related to the concentration of glucose in the medium (Polakis *et al.* 1964). Use of a non-repressing carbon source such as galactose for anaerobic growth led to the formation of mitochondria and respiratory activity including cytochrome oxidase (Tustanoff & Bartley, 1964). However Somlo & Fukuhara (1965) repeated these experiments, taking rigorous precautions to eliminate oxygen. Gassing with 'pure' nitrogen (containing *ca.* 0·001 % oxygen), with galactose as a carbon source, led to aerobic cytochrome spectra. If this nitrogen was first passed through alkaline pyrogallol, the levels of respiratory dehydrogenase enzymes and cytochromes were greatly reduced. A further reduction in aerobic systems was achieved by inhibiting the cells with actidione before harvesting to prevent induction by oxygen during handling. Thus molecular oxygen is clearly important for induction of respiratory apparatus in yeast. Roodyn (1966) has demonstrated that, *in vitro*, yeast mitochondrial protein synthesis requires molecular oxygen. It may be that respiratory adaptation can only take place in the presence of oxygen because steps in cytochrome biosynthesis are oxygen dependent.

There is a distinct difference among even procaryotic cells in their dependence on molecular oxygen for haem biosynthesis (Goldfine, 1965). For example, under anaerobic conditions *Escherichia coli* can synthesize its 'aerobic' cytochromes as well as an anaerobically induced *c*-type cytochrome (Gray, Wimpenny, Hughes & Mossman, 1966). However *Staphylococcus aureus* and *Bacillus cereus* require oxygen and accumulate coproporphyrin when incubated anaerobically (Lascelles, 1964). Since yeast synthesizes aerobic cytochromes through a step *requiring* molecular oxygen, it is also likely that over-all regulation of mitochondrial formation requires either oxygen or an equivalent of oxygen in terms of redox potential. It may be that catabolite repression exerts an over-all control on mitochondrion synthesis whilst molecular oxygen is required only to complete the biosynthesis of cytochrome pigments. Light with an action spectrum of neutral porphyrin inhibits respiratory adaptation in yeast and appears to block protoporphyrin synthesis but at present this inhibition is not clearly understood (Guerin & Jacques, 1968).

Having said all this, the role of oxygen in mitochondria biogenesis remains almost as mysterious now as it has always done! For a more detailed discussion of this subject see Lloyd (this Symposium).

Bacterial chromatophores

The non-sulphur purple photosynthetic bacteria (*Athiorhodaceae*) can grow autotrophically in the light or heterotrophically with oxygen in the dark. Photosynthetic pigment production and chromatophore synthesis are apparently regulated by oxygen and light. Cohen-Bazire & Kunisawa (1963) and Drew & Giesbrecht (1963) have demonstrated what are apparently cytoplasmic-membrane-derived vesicles formed at the same time as pigment is synthesized in *Rhodospirillum rubrum* and *Rhodopseudomonas spheroides* grown at medium light intensities. Oxygen, or high light intensities, depresses vesicle and pigment production. Anaerobically, at very low light intensities, high pigment levels and cells packed with vesicles are apparent. *Chromatium* D—a strictly anaerobic photosynthetic purple sulphur bacterium—can metabolize and remain viable in the presence of oxygen for some time: but cannot grow because photosynthetic pigment production is absolutely inhibited. Heterotrophic aerobic metabolism is apparently absent in this organism, and *Chromatium* is entirely dependent on its ability to synthesize bacteriochlorophyll (Hurlbert, 1967).

The synthesis of the photosynthetic pigments and structures are controlled by light intensity and oxygen, and it seems likely that the actual regulatory mechanism is related to the internal redox state of the cell. High light intensity is equivalent to a high level of photo-oxidant and a more positive E_h. Similarly, oxygen raises the E_h and a low level of 'anaerobically' induced pigments and vesicle structures results.

Oxygen and aerobic respiratory enzymes

Structures for aerobic respiration in procaryotic cells are far simpler than the mitochondria of eucaryotic organisms, and are confined to the cytoplasmic membrane or in some cases to structures of various degrees of complexity derived therefrom. The formation of these bacterial energy organelles is discussed in detail by Lloyd (this Symposium) and will not therefore be considered here.

Over-all respiratory activity

Oxygen-mediated changes in respiratory activity have been reported in many organisms. Transition to aerobiosis in yeast leads to an increase in aerobic respiration that parallels increase in mitochondrial cytochromes (Slonimski, 1953). The same sort of effect was seen in *Candida utilis*, where there was a 3- to 4-fold increase in Q_{O_2} as oxygen concentration in a continuous culture medium was raised from 0·234 to

28·4 μM (Johnson, 1967). In procaryotic organisms similar observations have been made. The changes from anaerobic to aerobic growth leads to a threefold increase in glucose-oxidizing ability in *Pasteurella pestis* (Engelsberg, Gibor & Levy, 1954); the Q_{O_2} of *Staphylococcus epidermidis* increases 10-fold on aeration (Jacobs & Conti, 1965); and in *Escherichia coli* K12 aerobiosis leads to increased respiratory activity which ranges from 1·8-fold for glucose to 10- and 15-fold for succinate and acetate respectively (Hino & Maeda, 1966). In another facultative anaerobe, *Aerobacter aerogenes*, maximum aerobic respiration is observed at low oxygen tensions (Harrison & Pirt, 1967), and White (1963) made similar observations with *Haemophilus parainfluenzae* grown in batch cultures with poor aeration.

Clearly oxygen induces aerobic respiration capacity in many organisms: often maximally, however, at low aeration levels.

Cytochromes associated with aerobic electron transport

Respiratory adaptation in yeasts leads to the formation of a classical cytochrome system comparable to the mammalian system and consisting of cytochromes a, a_3, b, c and c_1 (Slonimski, 1953; Mackler *et al.* 1962). Other haem-containing compounds present aerobically are cytochrome b_2, catalase and cytochrome c peroxidase (Sherman, 1965). Under anaerobic conditions Chaix & Labbe (1965) reported spectra of a b_1 type cytochrome together with zinc protoporphyrin and a free protoporphyrin.

The effects of oxygen on respiratory pigment synthesis can be seen in the staphylococci. *Staphylococcus epidermidis* is facultative towards oxygen, but cannot synthesize haem anaerobically (Jacobs & Conti, 1965; Jacobs, McClosky & Jacobs, 1967). It appears that haem biosynthesis in this organism requires molecular oxygen as does haem biosynthesis in higher organisms (Goldfine & Bloch, 1963). Under anaerobic conditions it accumulates coproporphyrinogen but has no cytochrome a and only traces of cytochromes b_1 and o, all of which are present in aerobically grown cells. Addition of haem anaerobically allows formation of b_1 and o but not a, and changing to aerobic growth causes increases in cytochromes b_1 and o and finally a (Frerman & White, 1967). It is not clear whether or not anaerobically grown *S. epidermidis* contains the apoenzymes of the cytochrome system but, using a haem-requiring strain of *S. aureus*, Chang & Lascelles (1963) were able to demonstrate the presence of the cytochrome apoenzyme for nitrate reductase in anaerobically grown cells.

Other facultative bacteria *can* form cytochrome pigments anaerobically. When *Escherichia coli* is grown under anaerobic conditions, it still possesses many activities which are characteristic of aerobically grown cells including the cytochromes (Gray, Wimpenny, Hughes & Mossman, 1966). In this and some other facultative anaerobes the maximum levels of cytochromes are often found under conditions of reduced oxygen availability. This was first demonstrated in chemostat cultures of *E. coli* and *Aerobacter aerogenes* where the optimum concentration of oxygen for cytochrome a_2 biosynthesis was 1 μM while formation of cytochrome b_1 remained approximately constant between 1 and 1000 μM oxygen (Moss, 1952, 1956). Similar results have been observed with batch cultures of *Haemophilus parainfluenzae* (White, 1962) and *Spirillum itersonii* (Clarke-Walker, Rittenberg & Lascelles, 1967).

Among denitrifying bacteria, low oxygen concentrations lead to high cytochrome levels, the highest levels being found with *Pseudomonas fluorescens* aerated with 1 % as opposed to 20 % oxygen during growth (Lenhof, Nicholas & Kaplan, 1956). Using nitrate as hydrogen acceptor anaerobically even greater amounts of cytochromes are formed. A 4- to 5-fold increase in cytochromes (especially of the *c* type) is observed when *Pseudomonas aeruginosa*, *P. stutzeri* and *Micrococcus denitrificans* are cultured under these conditions (Verhoefen & Takeda, 1956; Higashi, 1960). In our experiments with *Escherichia coli* grown in a chemostat, nitrate or nitrite or oxygen at certain concentrations induce maximal amounts of cytochrome b_1 at an approximately constant E_h in the medium (Wimpenny, 1967). It is possible that control of cytochrome b_1 formation at low oxygen concentrations and with alternative electron acceptors is directly related to the redox state of the cell. This conjecture is substantiated by the data of Arima & Oka (1965) and Oka & Arima (1965): reduced aeration or nitrate increased the levels of cytochromes a_1, a_2 and b_1 in *Achromobacter*, but cytochrome *o* was not affected. Especially interesting was the observation that mM cyanide in cells resistant to its lethal effects caused high levels of cytochromes a_1, a_2 and b_1 aerobically. Clearly it is not oxygen *per se* but its *availability* to the cell that dictates the synthetic pattern of cytochromes in this organism at least.

There appear to be fundamental differences between facultative organisms in their ability to synthesize cytochromes anaerobically: most aerobic cytochromes of bakers yeast are absent anaerobically as are respiratory pigments in staphylococci and facultative bacilli. In contrast, the Enterobacteriaceae appear to synthesize cytochromes in the absence of oxygen. Reduced oxygen tension often leads to maximum synthesis of cytochromes even in strict aerobes such as *Haemophilus parainfluenzae* and denitrifiers such as the pseudomonads and *Micrococcus denitrificans*. In many of these organisms, however, the molar ratios of the different components of the cytochrome system vary rather widely during adaptation to particular conditions of aeration or anaerobiosis (see, for example, *H. parainfluenzae*, White (1962), and *Staphylococcus epidermidis*, Frerman & White (1967)). The functional respiratory system thus appears like a loosely knit jigsaw puzzle or mosaic functional in a wide variety of configurations. It is interesting to speculate on the way in which electron transport could occur in such a mosaic. The degrees of freedom open to both electron and speculator are at this point enormous.

Oxygen and other electron carriers

Quinones increase in some facultative organisms on changing to aerobic growth conditions. For example, air-grown cells of *Saccharomyces cerevisiae* have 350 times as much coenzyme Q_6 (CoQ_6) as anaerobically grown cells (Lester & Crane, 1959) and, even under non-growing conditions, aeration increases the anaerobic level 5- to 20-fold (Sugimura & Rudney, 1960; Sugimura, Okabe & Rudney, 1964). Similarly, on changing staphylococci from anaerobic to aerobic growth increased levels of vitamin K_2 ranging from 1·6-fold in *Staphylococcus aureus* (Frerman & White, 1967) to 100-fold in *S. albus* (Bishop, Pandya & King, 1962) have been reported. There does not however appear to be much change in quinone levels of the Gram negative bacteria under different conditions of aeration (Bishop *et al.* 1962; Kashket & Brodie, 1960). Despite this, quinones are important aerobically in these organisms: Wyn-Jones & Lascelles (1967) and Wyn-Jones (1967) have shown that mutants of *Escherichia coli* blocked in ubiquinone synthesis could grow normally both aerobically and anaerobically, although aerobically they required either lysine and methionine or succinate, presumably for the synthesis of succinyl CoA. In *Haemophilus parainfluenzae* aeration increased the levels of 2-demethyl vitamin K, but the increase was not coordinate with that of flavoproteins or cytochromes and the way it is integrated into electron transport structures is hard to assess (White, 1965).

Oxygen and terminal respiratory dehydrogenases

Lactate and NADH-ferricyanide reductase activities are present in anaerobically grown yeast (Slonimski, 1956; Lindenmayer & Smith, 1964), but the existence of succinic dehydrogenase has been the subject of conflicting reports. Lindenmayer & Smith (1964) found no succinate–cytochrome c-oxidoreductase activity in anaerobically grown yeast, while Lukins, Tham, Wallace & Linnane (1966) showed that succinate dehydrogenase, measured with phenazinemethosulphate/ dichlorophenol indophenol, was present in detectable amounts in galactose- but not glucose-grown cells. It appears that the flavo-protein succinate dehydrogenase may be present in anaerobically grown yeast, although it cannot couple with cytochrome c. Furthermore its production is under the control of both oxygen and catabolite repressors.

Similar control systems may exist in bacteria. White (1964), as part of his immaculate dissection of the respiratory pathways of *Haemophilus parainfluenzae*, showed that growth with poor aeration and glucose as carbon source caused an increased activity of $D(+)$- and $L(-)$-lactate, succinate, formate and NADH: ferricyanide or cytochrome c reductases over the level in lactate-grown cells with good aeration. It would have been interesting to know the relative effects of glucose and oxygen in this system in view of the results with yeast referred to above. Membrane-bound dehydrogenases in *Escherichia coli* also increase with aeration (Gray, Wimpenny, Hughes & Mossman, 1966).

The effect of oxygen on lactate dehydrogenase apparently depends on its function. If it is acting as an oxidase, that is linking to oxygen, it is present aerobically, as for example in yeast (Lindenmayer & Smith, 1964), *Haemophilus parainfluenzae* (White, 1964) and *Streptococcus faecium* (London, 1968). In the absence of oxygen, lactate dehydrogenase is used as an organic electron acceptor reductase which may be the function of the yeast $D(+)$-lactate reductase described by Roodyn & Wilkie (1968) and of the lactate dehydrogenase in *Staphylococcus aureus* (Collins & Lascelles, 1962).

Oxygen and tricarboxylic acid (TCA) cycle enzymes

Because the TCA cycle is primarily an oxidative mechanism it might be expected to vary significantly in activity on changing oxygen levels. Hirsch (1952) and Slonimski & Hirsch (1952) showed that aconitase, fumarase and isocitrate dehydrogenase increased in activity during adaptation of yeast to aerobic conditions. Similar observations have been made in facultative bacteria. Oxygen induces TCA cycle enzymes in *Bacillus cereus* (Schaeffer, 1952) and in *Pasteurella pestis* (Englesberg & Levy, 1955). The latter workers showed an increase in all TCA cycle enzymes as well as pyruvate oxidase, phosphotransacetylase and acetokinase in aerobic cultures; at the same time levels of two key glycolytic enzymes, 3-phosphoglyceraldehyde dehydrogenase and phosphofructokinase, fell to about half their anaerobic level. Investigations of a series of other nutritional conditions on levels of TCA cycle enzymes in *Escherichia coli* led us to suggest that both catabolite repression and oxygen control the level of these enzymes (Gray, Wimpenny, Hughes & Mossman, 1966; Gray, Wimpenny & Mossman, 1966). Speculation at the moment centres on the possibility that a single mechanism might play a part in both types of regulation. Results of Forget & Pichinoty (1967) showed that in *Aerobacter aerogenes* there is the same pattern of changes in the TCA cycle enzymes during the growth cycle as those reported for *Pasteurella pestis* and *E. coli*. In continuous culture experiments with *E. coli* and *A. aerogenes* we have shown that synthesis of these TCA cycle enzymes is maximal at high aeration rates and at an E_h that is more positive than the optimum figure for cytochrome formation (J. W. T. Wimpenny & D. K. Necklen, unpublished). Apparently the syntheses of cytochromes and TCA cycle enzymes are controlled separately in facultative anaerobes if redox criteria are a true picture of intracellular regulation.

Oxygen and anaerobic respiratory systems

Since an accepted classification of microbial physiological types is based on electron acceptors, I shall consider the effect of oxygen on the inorganic electron acceptor-reducing enzymes (characteristic of *anaerobic respiration*) first, and secondly, the organic electron acceptor-reducing enzymes, those involved in *fermentation*. Both types of electron transport occur in facultative micro-organisms.

Oxygen and enzymes reducing inorganic electron acceptors

The more important inorganic compounds capable of reduction in place of oxygen by various microbes are nitrate, nitrite, nitrous oxide, tetrathionate, thiosulphate, sulphate, chlorate, ferricyanide and hydrogen ions, and many of these reactions are repressed and/or inhibited by oxygen.

Nitrate reduction has been investigated in a series of thorough studies by Pichinoty and his colleagues. Their original findings are summarized and discussed by Pichinoty (1965a) in terms of oxygen regulation. In *Aerobacter aerogenes* less nitrite is produced from nitrate aerobically than anaerobically because nitrate reductase is both repressed and reversibly inhibited by oxygen. In the absence of any other nitrogen source, nitrate can be reductively assimilated aerobically, though not at elevated oxygen tensions. In *Micrococcus denitrificans* (which can liberate nitrogen from nitrate, nitrite or nitrous oxide), oxygen inhibits nitrogen production from each of these substrates, as well as repressing nitrate-, nitrite- or nitrous oxide-reducing enzymes (Pichinoty, 1965*b*). Pichinoty (1966) has distinguished two types of nitrate reductase: type A, which can utilize chlorate in place of nitrate as a substrate, and type B, which is inhibited by chlorate. Some organisms possess A, some B, others A + B as shown in Table 3. Oxygen inhibits the *in vivo* activity of all these enzymes, possibly by competition for electrons. Though complex control is involved the message is clear: where nitrate is acting energetically as a poor relation of oxygen, oxygen represses respiratory nitrate reductases. When the cell must reduce nitrate for biosynthetic reasons, assimilatory nitrate reductase enzyme is unaffected by oxygen unless the partial pressure of oxygen is high.

Table 3. *Bacterial nitrate reductase enzymes (from Pichinoty, 1966)*

Nitrate reductase type	Chlorate is:	Oxygen and enzyme formation	Example of organism	Respiratory (Resp.) or Assimilatory (Assim.)
A	Substrate	Represses	*Aerobacter aerogenes*	Resp.—sometimes Assim.
B	Inhibitor	No effect	*Pseudomonas putida P*	Assim.
B	Inhibitor	Represses	*Providencia alcalifaciens* 4/€0	Resp.
A	Substrate	Represses ⎫	*Micrococcus*	⎰ Resp.
B	Inhibitor	No effect ⎭	*denitrificans*	⎱ Assim.

Nitrite reductase is also repressed by oxygen in denitrifying bacteria (Sacks & Barker, 1949; Pichinoty, 1965*a*), and in *Escherichia coli* (Cole, 1967). There is some evidence that it is related to a low redox potential soluble *c*-type cytochrome produced anaerobically by *E. coli* grown under certain conditions (Fujita & Sato, 1967) and Cole (1967) considers this cytochrome to be nitrite reductase, since the partially purified pigment is readily reoxidized by nitrite. It is cytochrome c_{552} first reported by Gray, Wimpenny, Hughes & Ranlett (1963) and shown to be present in all enteric organisms tested and to be repressed by oxygen (Wimpenny, Ranlett & Gray, 1963). Its presence, purification and properties were described by Fujita (1966) and although O'Hara, Gray, Puig & Pichinoty (1967) have suggested it may function as a carrier in the formic hydrogenlyase reaction, its appearance and control correlates better with nitrite reductase activity than with the formic hydrogenlyase complex.

The enzyme hydrogenase catalyses the following reaction:

$$2H^+ + 2e \rightleftharpoons H_2$$

and can be regarded as an inorganic electron acceptor reducing enzyme. Whilst facultative bacteria have long been known to produce hydrogen gas anaerobically but not aerobically, Pichinoty (1962, 1965a) showed clearly that both hydrogenase itself and the formate hydrogenlyase complex of which hydrogenase is a part are repressed and inhibited by oxygen. Gray, Wimpenny, Hughes & Mossman (1966) demonstrated hydrogenase in both soluble and particulate preparations of anaerobically grown *Escherichia coli* though only traces were present in aerobically grown bacteria. Nitrate and nitrite as well as oxygen repress this activity in *E. coli* (Wimpenny, 1967) and the repression can be correlated with culture redox potential.

Tetrathionate reductase, first described in some facultative anaerobes by Pollock & Knox (1943), is also repressed and inhibited by oxygen (Pichinoty & Bigliardi-Rouvier, 1962) as is the thiosulphate reductase of *Proteus vulgaris* (Pichinoty, 1965a). A number of other inorganic reductases have been described, but as no details are available of their oxygen response they will not be considered here.

Oxygen and enzymes reducing organic electron acceptors

Dehydrogenase enzymes coupling through terminal electron transport enzymes and carriers to oxygen have already been discussed. Another group of dehydrogenases is responsible for reducing organic compounds which are then released from the cells as the fermentation products of anaerobic metabolism.

The control of lactate dehydrogenase has been examined in *Staphylococcus aureus* where anaerobiosis leads to a 10-fold increase in NAD-linked lactate dehydrogenase and aerobic fermentation ability is one-third of that in anaerobically grown cells (Collins & Lascelles, 1962). Only one species of lactate dehydrogenase is present in *S. aureus*. Aerobic levels of this enzyme are constant but low and highest specific activities appear in anaerobic pyruvate-grown cells. Some fascinating observations have been made with haeminless mutants of *S. aureus*. In the absence of haemin, levels of lactate dehydrogenase are equally high aerobically or anaerobically, but addition of haemin to aerobic mutants leads to repression of lactate dehydrogenase (Garrard & Lascelles, 1968). The absence of terminal respiratory pathways under aerobic conditions in this case seems likely to cause reducing conditions inside the cell which may induce anaerobic levels of relevant enzymes.

This system seems to have interesting possibilities—again oxygen, though present, does not interact with the cell owing to the absence of cytochromes.

The situation seems more complex in *Aerobacter aerogenes* (Pascal, 1966). This organism possesses two stereospecific lactate dehydrogenase enzymes: a D-lactate dehydrogenase which appears to be constitutive and an L-isomer dehydrogenase which is inducible by oxygen. Yeast possesses three lactate dehydrogenases: aerobically it has an oxygen-inducible D- (Nygaard, 1961) and an L- (Slonimski, 1953) lactate cytochrome *c* reductase, but it also has a D-lactate dehydrogenase like that of *A. aerogenes* which is not under the control of oxygen (observations of Somlo, quoted by Pascal, 1966).

In *Escherichia coli* anaerobic conditions induce high levels of both lactate and ethanol dehydrogenases, but, in a mutant unable to synthesize ubiquinone, levels of these two enzymes are high even under aerobic conditions (Wyn-Jones & Lascelles, 1967). Since the mutant cannot oxidize NADH by the normal aerobic route, it presumably has to use these two dehydrogenases for NADH-oxidation even when oxygen is present. We have in this case another example of a cell grown in the presence of oxygen behaving as an anaerobe.

A number of dehydrogenases required for anaerobic growth have been shown to be repressed by oxygen. They include the ethanol and glycerol dehydrogenases of *Aerobacter aerogenes* (McPhedran, Sommer & Lin, 1961; Lin, Levin & Magasanik, 1960) and the acetoin and fumarate reductases of the same organism (Pichinoty, 1965*a*). In *Escherichia coli* fumarate reductase is also repressed by oxygen (Hirsch, Rasminsky, Davis & Lin, 1963; Wyn-Jones & Lascelles, 1967). However, formate dehydrogenase can exist in two forms—a membrane bound state linking the cytochrome system to oxygen, in which case the aerobic level is higher than the anaerobic one—and a soluble one which functions in the anaerobic formic hydrogenlyase enzyme complex and which is controlled by oxygen, nutrition and pH (Gray, Wimpenny, Hughes & Mossman, 1966).

While it is possible to extend this catalogue of observations it is obvious that two classes of dehydrogenase enzymes are controlled in opposite ways by oxygen. The aerobic TCA cycle dehydrogenase enzymes and flavoprotein dehydrogenases linking directly to the aerobic terminal electron transport pathway are increased by oxygen. On the other hand, dehydrogenase enzymes with Michaelis constants favouring substrate reduction operate during anaerobic fermentation reactions and are repressed by oxygen.

Oxygen and redox potential

Oxygen has a profound effect on oxidation-reduction reactions in micro-organisms. It is only natural to consider the over-all redox state of living cells both in terms of their external environment and in terms of the steady-state levels of oxidized and reduced metabolites within the cell. The redox potential (E_h) of a system of oxidized and reduced compounds is given by the equation

$$E_h = E_0' + \frac{RT}{nF} \ln \frac{[\text{Oxidant}]}{[\text{Reductant}]},$$

where the E_0' is the standard potential relative to the potential of the hydrogen electrode at pH 7·0 with molar activities of each reactant, n is the number of electrons involved in the reaction, R is the gas constant, T is absolute temperature, F is the Faraday. Early work on the physiology of microbial cultures concentrated on the E_h during batch culture of a variety of bacteria and demonstrated the effect of aerobiosis and anaerobiosis on the E_h values. This work has been summarized by Hewitt (1950). A 'typical' anaerobic culture might grow in the region -150 mV. down to -420 mV. (E_h of the hydrogen electrode); whilst aerobic cultures starting at $+300$ mV. might fall to -200 mV. or, if they produced hydrogen (as many facultative anaerobes do), to -420 mV. Data of this type (using a measurable yardstick, E_h) defined the type of electron-donating or electron-accepting environment in which different physiological types of micro-organism could survive. In addition, carefully controlled experiments of Knight & Fildes (1930) demonstrated the upper limit of E_h at which spores of anaerobic bacteria were able to germinate. Since then measurements of E_h appear to have fallen into disrepute mainly, I suspect, because of the difficulty of determining the various redox couples in the system that contribute to the potential. We have reintroduced E_h measurements for the main reason that use of continuous culture techniques permits examination under far more controlled conditions. Although the factors in each culture responsible for the measured E_h have not been analysed as yet, certain physiological changes induced in cells by various electron acceptors can now be correlated with E_h. When oxygen concentration is varied in chemostat cultures of *Escherichia coli* K.12, the E_h is observed to change whilst oxygen is barely detectable with a sensitive Mackereth type oxygen electrode (Fig. 1). As E_h rises hydrogenase activity falls to zero, at between $+100$ and $+200$ mV. using oxygen nitrate or nitrite as electron acceptors. Cytochrome b_1 is synthesized

12-2

maximally at between $+50$ and $+150$ mV. with nitrate or oxygen (Fig. 2). Similar observations have been made using nitrite as electron acceptor. Cytochrome a_2 appears to be regulated by oxygen and not by nitrate (Fig. 2). With oxygen TCA cycle enzymes are synthesized maximally at around $+250$ to $+300$ mV. This potential cannot be achieved with nitrate or nitrite as electron acceptor (Fig. 3). Results with another facultative anaerobe, *Aerobacter aerogenes*, are almost identical.

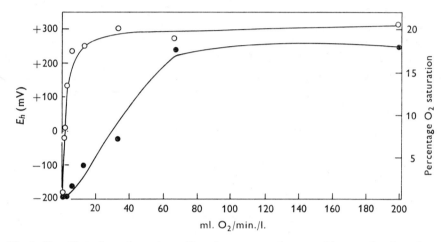

Fig. 1. The effect of aeration rate on E_h and on measured oxygen % saturation in carbon limited chemostat cultures of *Escherichia coli*. ●—●, % oxygen saturation; ○—○, E_h.

Several interesting points emerge from these experiments. First, E_h changes at low oxygen concentrations may be a convenient parameter for measuring and controlling oxygen addition. Secondly and more important: distinctions in cell regulation between different groups of enzymes are clear. Hydrogenase is repressed at a relatively low E_h, then, as the E_h rises, maximum cytochrome synthesis occurs, and finally at an E_h at least 200 mV. higher TCA cycle enzymes are formed in greatest amounts. These data also show that oxygen itself may not be necessary to repress hydrogenase or induce high amounts of cytochrome b_1 in *Escherichia coli* and *Aerobacter aerogenes*.

The whole question of the validity of E_h measurements is still in doubt. It seems likely, though not proven, that the extracellular components of the culture which generate the E_h figure are in equilibrium with the metabolic pool inside the cell, and that changes in external E_h are reflections of, or are reflected by, the redox state of this pool.

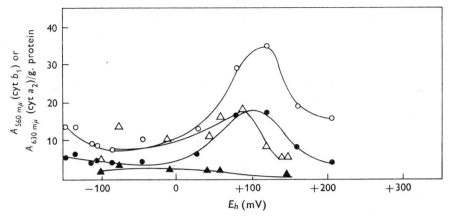

Fig. 2. Effect of E_h on cytochromes b_1 and a_2 in particulate preparations of *E. coli* grown in carbon limited chemostat cultures. O—O, cyt b_1, O_2 as acceptor; ●—●, cyt a_2, O_2 as acceptor; △—△, cyt b_1, NO^-_3 as acceptor; ▲—▲, cyt a_2, NO^-_3 as acceptor.

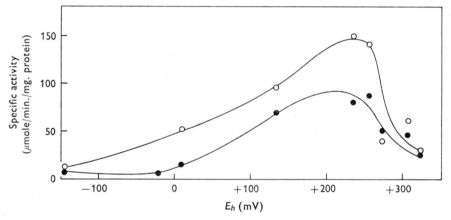

Fig. 3. The effect of E_h on tricarboxylic acid cycle enzyme synthesis in carbon limited chemostat cultures of *Escherichia coli*. ●—●, Isocitrate dehydrogenase; O—O, fumarase.

Summary and discussion

The effects, then, that oxygen has on growth and metabolism, although extensive, seem capable of classification. Most aerobic and facultatively anaerobic bacteria have a high affinity for oxygen but all bacteria are sensitive to excess. In general the effects of oxygen are either on induction or repression. Oxygen can induce the synthesis of the respiratory structures themselves and also of the component cytochromes. This synthesis may or may not take place anaerobically depending on whether the cell requires molecular oxygen for cytochrome formation.

In many organisms there is an optimum oxygen concentration for cytochrome synthesis which is much lower than that of air. In *Escherichia coli* and *Aerobacter aerogenes*, at least, culture E_h rather than concentration of a particular electron acceptor seems to regulate cytochrome synthesis. Furthermore, oxygen induces quinone formation in procaryotic and eucaryotic microbes, and the TCA cycle enzymes and dehydrogenase enzymes whether they link directly, or through, terminal electron transport systems to oxygen. Repression by oxygen is equally important. It represses chromatophore formation and bacteriochlorophyll synthesis in obligate or facultative photosynthetic bacteria. It also represses the formation of many enzymes reducing alternative inorganic electron acceptors, certain cytochromes (e.g. cytochrome c_{552} in enteric bacteria), and enzymes reducing organic electron acceptors.

These are briefly the observations. How are these changes regulated? Pichinoty (1965a) has suggested three alternative working hypotheses for oxygen repression; (i) that oxygen itself is the corepressor; (ii) that a low molecular weight metabolite whose presence depends on the redox state of the cells is the corepressor; (iii) that oxygen inhibits the removal of these enzymes from protein synthesizing sites. Obviously these theories may equally well apply to oxygen-inducible systems. Hypothesis (i) seems unlikely, at least as far as cytochrome induction and hydrogenase repression in *Escherichia coli* is concerned, for two reasons. First, in continuous culture experiments these systems are controlled at oxygen tensions that are undetectable using a sensitive Mackereth electrode. Hence the regulation site would need to be highly sensitive to minute concentrations of oxygen (of the order of 10^{-8} to 10^{-7} M) in the medium. Secondly, similar regulation occurs using electron acceptors other than oxygen, and correlates well with culture E_h. Hypothesis (iii) can be neither proved or disproved as yet, leaving hypothesis (ii) as the most likely explanation at present. The difficulty here remains that of deciding whether a single metabolite or a range of metabolites are responsible for the observed changes, and then identifying each regulator. Certain molecules spring to mind as potential candidates. Intuitively the couple $NADH + H^+ \rightleftharpoons NAD$ has several advantages as a regulator. First, of all molecules in the metabolic pool its oxidation and reduction state is the most important, since the function of most electron transport enzymes is geared to reoxidizing reduced pyridine nucleotide. Secondly, the dependence of synthesis of certain enzymes on environmental E_h, rather than oxygen on its own, points to an ubiquitous redox couple within the cell such as

NAD/NADH. Thirdly, these cofactors seem ideally suited as allosteric effectors in view of their ability to bind to active centres of enzymes. A further and even more outrageous extension of this idea may be that NAD/NADH ratios are responsible for catabolite repression, but detailed consideration of the evidence is outside the scope of this review.

Whilst all this is pure speculation at present it is certainly easy to test for positive correlation under a variety of conditions. However, positive correlation is no proof and further more sophisticated experiments must be designed. Techniques exist for the extraction and measurement of intracellular pool constituents. Total pyridine nucleotides have been measured in a variety of bacteria (London & Knight, 1966; London, 1968); adenosine triphosphate can be measured in very young liquid cultures of micro-organisms (Cole, Wimpenny & Hughes, 1967) and pyridine nucleotides, adenine nucleotides and a variety of other metabolites can now be measured by very sensitive fluorometric methods (Maitra & Estabrook, 1964; Estabrook & Maitra, 1962). Using these methods, the change from aerobiosis to anaerobiosis in yeast has been shown to cause oscillations in the levels of nucleotides and glycolytic intermediates. Analysis of these systems has thus thrown much light on short-term regulation due to feed-back inhibition and stimulation (Hommes, 1964; Ghosh & Chance, 1964; Betz & Chance, 1965). These experiments are exciting but probably concern events that occur an order of magnitude or more faster than regulatory mechanisms responsible for induction and repression of enzymes. Further work on oxygen regulation must now be directed to a detailed analysis of changes in metabolic flux on changing from aerobiosis to anaerobiosis and investigation into changes in components of the actual protein synthesizing system. Already Ning Kwan, Apirion & Schlessinger (1968) have shown changes in isoleucyl transfer RNA and the 50S ribosomes of *Escherichia coli* induced by anaerobic growth. Examination of the way in which respiratory structures are assembled on changing from anaerobiosis to aerobiosis and closer investigation of oxygen toxicity in strictly anaerobic and other micro-organisms may also provide much needed information.

CARBON DIOXIDE AS REGULATOR OF GROWTH AND METABOLISM

Unfortunately, to consider carbon dioxide as a regulator molecule, as we have done with oxygen, is enormously difficult, although probably just as important. It is hard because it has very seldom been investigated experimentally in this light. This discussion will be restricted as far as possible to cases where CO_2 has been shown to affect the growth or metabolism of micro-organisms.

Carbon dioxide requiring micro-organisms

For many years it has been known that certain bacteria yeasts and moulds require CO_2 (Rockwell & Highberger, 1927), and Rahn (1941) demonstrated a need for CO_2 among the protozoa. Walker (1933) showed that *Escherichia coli* can grow on complex media without CO_2 but needs the latter when growing on a simple salts medium and this observation has been extended to a variety of organisms grown in simple and complex media (Gladstone, Fildes & Richardson, 1935). Many pathogenic bacteria on primary isolation show the 'CO_2 effect', that is a dependence on CO_2 which is often lost by serial subculture in laboratory media (see, for example, Topley & Wilson, 1964).

Evidence that CO_2-fixation reactions occurred and were of utmost importance to microbial metabolism came from accurate analysis of the products of glycerol fermentation by propionic acid bacteria (Wood & Werkman, 1935). The use of isotopes of carbon made subsequent analysis of reactions involving CO_2 simpler and more precise: see, for example, Barker, Rubin & Beck (1940) and Slade *et al.* (1942) using heterotrophic bacterial systems and van Niel, Thomas, Ruben & Kamen (1942) in protozoa. The assimilation of CO_2 by heterotrophic organisms has been reviewed extensively by Wood & Stjernholm (1962), Wood & Utter (1965) and Lynen (1967) and will not be discussed in detail here.

The requirement for CO_2 observed in many organisms grown under certain conditions, as for example when TCA cycle intermediates are removed for biosynthetic purposes, has been one factor that has led to the discovery of a group of reactions designed to replenish these TCA cycle members—the so-called 'anaplerotic' reactions (Kornberg, 1965). Some anaplerotic reactions are CO_2-fixation steps, whose products are dicarboxylic acids and often *initiation* of growth rather than subsequent growth maintenance is CO_2 dependent. In this case small amounts of dicarboxylic acids may have a 'sparking effect'. Growth is subsequently maintained because the cell then produces sufficient CO_2 from endogenous metabolism to maintain anaplerotic reactions.

In some cases a CO_2 requirement exists even when the fixation product or a related product is present in the medium. For example, *Streptococcus anginosus* requires traces of oleate or elevated CO_2 for growth. The CO_2 is used to form aspartate but aspartate in the medium cannot get into the cell unless, it is suggested, oleate is present in the medium (Martin & Niven, 1960). In most organisms, however, the amount of CO_2 required can be related directly to the presence of the fixation product in the medium. *Streptococcus bovis* P-10 can grow well on a mixture of arginine, aspartate, glutamate, proline and lysine, but requires CO_2 for growth when arginine or ammonium acetate are nitrogen sources. Labelling experiments have confirmed that the CO_2 is fixed into glutamate and aspartate and that the amount of CO_2 fixed is decreased in the more complex medium (Prescott, Ragland & Hurley, 1965). In other organisms, particularly some of the pathogenic organisms, CO_2 fixation is required for pyrimidine synthesis and the requirement for CO_2 can be abolished with hypoxanthine (Mueller & Hinton, 1941; Steinman, Oyama & Schulze, 1954). An even more striking CO_2 requirement is seen in some rumen bacteria. *Bacteroides rumenicola* and *Succinomonas amylolytica* (both succinate producers) require high partial pressures of CO_2 for optimum growth and it has been suggested that they may need CO_2 to synthesize oxaloacetate or malate which are suitable electron acceptors for the highly reducing conditions found in the gut (Bryant, Small, Bouma & Chu, 1958; White, Bryant & Caldwell, 1962).

So many CO_2-fixing reactions are now known that it is difficult to establish which particular enzyme is the one for which the CO_2 is rate limiting. Some information may be gained by studies of CO_2-requiring mutants of nutritionally competent micro-organisms. Table 4 lists some of the CO_2-requiring mutants of *Escherichia coli* isolated by Charles & Roberts (1968). These strains all grow in 20 %, v/v, CO_2 in air and show interesting growth requirements and inhibitions when

Table 4. *Characteristic of CO_2-requiring mutants of* Escherichia coli
(*from Charles & Roberts*, 1968)

Mutant	Locus	Nutritional requirement in absence of CO_2	Growth on CO_2 inhibited by
A-17	*pur-E(ade* f)	Purines	—
S-19	*suc*	Succinate + 2-oxoglutarate or lysine + methionine	Aspartate
L-23	—	Isoleucine + valine or isoleucine + ketoisovalerate	—
H-13	—	Histidine	Aspartate, purines
M-12	*pyr*-F	Pyrimidines	—
B-22	*cap*	Arginine or citrulline ± a pyrimidine	Pyrimidines
45–1	Not done	Methionine or cysteine	Aspartate
'Unknown'	Not done	Unknown	Purines, histidine, tyrosine

grown in the absence of CO_2. At present the reason for inhibition of some mutants by compounds which replace the CO_2 fixation reaction in others is not known but probably involves interaction between the different carboxylating reactions at the level of regulation by feedback or repression.

Whilst few of these CO_2-requiring mutants have been examined at the enzymic level, the enzyme deficiency of a mutant requiring TCA cycle dicarboxylic acids has been investigated by Ashworth & Kornberg (1966). Both mutant and wild-type parent possessed two of the CO_2-fixing enzymes (phosphopyruvate carboxykinase and 'malate enzyme') and lacked two others (pyruvate carboxylase and phosphopyruvate carboxytransphosphorylase) but the mutant differed from the parent strain in that it lacked phosphopyruvate carboxylase. It follows that anaplerotic reactions in this organism and possibly in *Salmonella typhimurium* (Theodore & Englesberg, 1964) were almost completely mediated through phosphopyruvate carboxylase. With hindsight it would be most interesting if these mutants of both *Escherichia coli* and *S. typhimurium* were examined for ability to grow under elevated CO_2 tensions when a different CO_2-fixing enzyme might be of importance.

Whilst much remains to be done concerning the interactions and integration of CO_2-fixing reactions in micro-organisms, clearly studies with mutant strains of bacteria as nutritionally undemanding as *Escherichia coli* will throw light perhaps on the metabolism of organisms that show an obligatory requirement for CO_2.

The effects on CO_2 on growth and morphology

This section contains a small collection of apparently unconnected observations which however demonstrate some of the changes that CO_2 can mediate, but which are not yet explicable in biochemical terms.

High CO_2 brings about the development of sex organs in coelenterate *Hydra* (Ivanovics, 1937), and causes *Mucor*—a filamentous fungus—to adopt a yeast-like mode of growth (Loomis, 1961; Bartinicki-Garcia & Nickerson, 1962). CO_2 is essential for spore germination and sporeling growth in *Chaetomium globosum*, and its concentration influences the development of diatomaceous algae (Levshina, 1965). King (1966) investigated the effect of high concentrations of CO_2 on growth and metabolism of *Pseudomonas aeruginosa*: CO_2–air mixtures (50 %, v/v) increased both the lag and the generation time of this organism by a factor of 2–3, but had no effect on oxidation of substrates by washed cells. Although both malate dehydrogenase and isocitrate dehydrogenase were partially inhibited by 50 % CO_2–air mixtures, oxaloacetate decarboxylase was not affected and King does not think the growth inhibitory effect of CO_2 involves these enzymes.

CO_2, HCO_3^- and H_2CO_3 may not always act in the same way as regulators of growth and metabolism. Sporangium formation in *Blastocladiella emersonii* depends under certain conditions on HCO_3^- but not on CO_2 tension or pH (Griffin, 1963). In contrast Tremmel & Levedah (1966) show that the yield of bleached *Euglena* grown on succinate, and to a lesser extent acetate, is proportional to CO_2 partial pressure up to 10 %, v/v; in this case HCO_3^- is actually inhibitory.

Many more examples can be found in the literature, but those given should be sufficient to show the range of the possible effects of CO_2 on growth.

Carbon dioxide and enzyme activity

Several enzymes have been shown to be inhibited by carbon dioxide. For example, phosphopyruvate carboxylase is partially inhibited by 10 %, v/v, CO_2 in the gas phase and completely inhibited by 70 % CO_2 (Walker & Brown, 1957). Some mitochondrial enzymes in *Ricinus* are inhibited by CO_2 (Ranson, Walker & Clarke, 1960) as is mitochondrial oxidative phosphorylation in another plant, the cauliflower (Miller & Hsu, 1965). The amount of CO_2 causing inhibition may not necessarily be artificially high. Nyiri & Lengyel (1965) have shown that good aeration and agitation is necessary to prevent inhibition by the metabolically produced CO_2 of hyphal metabolism and penicillin production in a strain of *Penicillium*.

Autotrophy versus heterotrophy

It is not my intention to consider purely autotrophic micro-organisms with respect to CO_2. By definition they require CO_2 for growth and an extensive literature exists on microbial photosynthesis and on the CO_2-fixation reactions of the chemolithotrophs. More interesting in the present context are those organisms which are facultative with respect to CO_2. Gest (1963) has coined the word 'autoheterotrophic' to describe this type of cell. It may be photosynthetic such as *Chromatium*, a strictly anaerobic purple sulphur bacterium, and *Rhodospirillum rubrum*, a facultatively anaerobic purple non-sulphur bacterium, or chemosynthetic such as *Hydrogenomonas* or *Micrococcus denitrificans*.

These autoheterotrophic micro-organisms have quite different sets of enzymes according to whether they are behaving as autotrophs or heterotrophs. The enzymes necessary for the operation of the reductive pentose cycle, including the key CO_2-fixing enzyme ribulose-1,5-diphosphate carboxylase, are present when the organisms are grown with CO_2 as sole source of carbon. But when the organisms are grown, for example, on acetate, the pentose cycle enzymes are repressed and those of the glyoxylate cycle are induced (Kornberg, Collins & Bigley, 1960; Fuller, Smillie, Sisler & Kornberg, 1961; McFadden & Tu, 1967). A most interesting aspect of autoheterotrophy concerns the metabolism of formate or oxalate by *Pseudomonas oxaliticus*. When grown on formate this organism is an autotroph with a functional reductive pentose cycle, some of the formate being oxidized to provide the energy to drive the fixation of CO_2—itself derived from formate. In contrast, growth on oxalate represses the formation of these enzymes and the enzymes of the glycerate pathway for glyoxylate metabolism are induced since oxalate is converted to glyoxylate for further metabolism (Quayle, 1961).

Although CO_2 clearly has an important part to play in these transitions from autotrophic to heterotrophic metabolism, its role as a regulator has been virtually ignored. It is to be hoped that further experiments will clarify the picture.

Discussion

There seems very little to discuss: Carbon dioxide is necessary for the growth of certain bacteria, many of which are pathogenic to man and for microbial species grown under conditions that lead to diminution of the TCA cycle pool constituents. In these circumstances CO_2-fixation reactions make up the deficiency. However, there are mutants which require elevated partial pressures of CO_2 or the end products of the carboxylation reactions. In this case interesting cross-relations exist between apparently unrelated metabolic pathways. Carbon dioxide has a variety of unexplained effects on morphology, spore germination and other physiological sites and it can also be inhibitory. It must also play some part in the induction and repression of enzymes which occur when autoheterotrophs change from one form of metabolism to another. But a great deal more experiment will be needed before we can explain in molecular terms the role of CO_2 as a regulator of growth.

CONCLUSIONS

The more that we find out about the regulation of metabolism, the more we must admire the infinite cunning of living systems. Whilst all cellular regulatory mechanisms are by definition interrelated, it is obvious that a variety of spheres of influence can be discerned; these include the regulation of biosynthetic pathways, the regulation of catabolic pathways, the regulation of energy metabolism, the regulation of permeability and the regulation of genetic information. Oxygen and CO_2 can now be considered as a part of two further realms. Oxygen is included in the redox control area—a series of homeostatic reactions designed to preserved the constancy of the internal redox environment; and CO_2 as part of an anaplerotic regulatory area, a homeostatic mechanism designed to maintain the constancy of the internal biosynthetic precursor pool. A brief consideration of the information reviewed in this paper will help to justify such a conclusion. It has been reported that large changes in external E_h occur in facultative bacteria under different aeration conditions. Clearly the metabolic pool must be buffered against changes of this type. The E_0' of the NADH + H$^+$/NAD couple is -320mV.: if the external and internal E_h were the same, a change from -400 to $+300$ mV. environmental E_h would lead to a change from almost completely reduced to oxidized NAD; this in turn would result in concomitant drastic changes in the equilibria of all NAD-linked enzyme systems. The cell therefore, perhaps using external E_h as a control variable, sets in play regulatory mechanisms which tend to counteract gross changes in internal E_h. This it does by altering the levels of key oxidation reduction enzymes, and it seems logical to propose that important redox couples such as NAD/NADH + H$^+$ might themselves be the regulator molecules. In contrast, energy control is not necessarily affected by the redox environment, all sorts of short-term events might intervene preventing or stimulating energy utilization. In this case short term (positive or negative feedback) regulation mediated by energy-important molecules, such as the adenine nucleotides, seems desirable and indeed this is the weight of the present evidence. The case for regulation of CO_2 metabolism is less clear, but certainly anaplerotic reactions are concerned in the maintenance of a constant pool. From this pool are drawn precursors for monomer synthesis destined eventually for polymer biosynthesis, and it is perhaps here that we can distinguish between anaplerotic regulation and biosynthetic regulation, which acts a step later on monomer synthesis from such precursors.

Considering the fields of oxygen and CO_2 regulation in this way, opens up, I think, many new and fruitful avenues for research.

REFERENCES

ARIMA, K. & OKA, T. (1965), Cyanide resistance in *Achromobacter*. I. Induced formation of cytochrome a_2 and its role in cyanide-resistant respiration. *J. Bact.* **90**, 734.

ARNOLD, B. H. & STEELE, R. (1958). Oxygen supply and demand in aerobic fermentations. In *Biochemical Engineering*. Ed. R. Steele. New York: Macmillan.

ASHWORTH, J. M. & KORNBERG, H. L. (1966). The anaplerotic fixation of carbon dioxide by *Escherichia coli*. *Proc. R. Soc.* B, **165**, 179.

BARKER, H. A., RUBIN, S. & BECK, J. V. (1940). Radioactive carbon as an indicator of carbon dioxide reduction. IV. The synthesis of acetic acid from carbon dioxide by *Clostridium acidi-urici*. *Proc. natn. Acad. Sci. U.S.A.* **26**, 477.

BARTHOLOMEW, W. H. (1960). Scale up of submerged fermentation. *Adv. appl. Microbiol.* **2**, 289.

BARTHOLOMEW, W. H., KAROW, E. O., SFAT, M. R. & WILHEML, R. H. (1950). Oxygen transfer and agitation in submerged fermentations. *Ind. Eng. Chem.* **42**, 1801.

BARTNICKI-GARCIA, S. & NICKERSON, W. J. (1962). Induction of yeast-like development in *Mucor* by carbon dioxide. *J. Bact.* **84**, 829.

BAUCHOP, T. & ELSDEN, S. R. (1960). The growth of micro-organisms in relation to their energy supply. *J. gen. Microbiol.* **23**, 457.

BETZ, A. & CHANCE, B. (1965). Phase relationship of glycolytic intermediates in yeast cells with oscillatory metabolic control. *Arch. Biochem. Biophys.* **109**, 585.

BISHOP, D. H. L., PANDYA, K. P. & KING, H. K. (1962). Ubiquinone and vitamin K in bacteria. *Biochem. J.* **83**, 606.

BLUM, J. J. & BEGIN-HEICK, N. (1967). Metabolic changes during phosphate deprivation in *Euglena* in air and in oxygen. *Biochem. J.* **105**, 821.

BOND, G. (1961). The oxygen relation of nitrogen fixation in root nodules. *Z. allg. Mikrobiol.* **1**, 93.

BRYANT, M. P., SMALL, N., BOUMA, C. & CHU, H. (1958). *Bacteroides rumenicola*, nov. sp. and *Succinomonas amylolytica*, nov. gen.—species of succinic acid producing anaerobic bacteria of the bovine rumen. *J. Bact.* **76**, 5.

CALDERBANK, P. H. (1967). In *Biochemical and Biological Engineering Science*. Ed. N. Blakeborough. London: Academic Press.

CALDWELL, J. (1965). Effects of high partial pressures of oxygen on fungi and bacteria. *Nature, Lond.* **206**, 321.

CHAIX, P. & LABBE, P. (1965). A propos de l'interprétation du spectre d'absorption de cellules de levures récoltées à la fin ou après la phase exponentielle de leur croissance anaérobie. *Colloques Internationaux du C.N.R.S. Marseilles* 1963, no. 24. Paris: Editions du Centre National de la recherches scientifiques.

CHANCE, B. (1965). Reaction of oxygen with the respiratory chain in cells and tissues. *J. gen. Physiol.* **49**, 163.

CHANG, J. P. & LASCELLES, J. (1963). Nitrate reductase in cell free extracts of a haemin requiring strain of *Staphylococcus aureus*. *Biochem. J.* **89**, 503.

CHARLES, H. P. & ROBERTS, G. A. (1968). Carbon dioxide as a growth factor for mutants of *Escherichia coli*. *J. gen. Microbiol.* **51**, 211.

CLARK-WALKER, G. D., RITTENBERG, B. & LASCELLES, J. (1967). Cytochrome synthesis and its regulation in *Spirillum itersonii*. *J. Bact.* **94**, 1648.

COHEN-BAZIRE, G. & KUNISAWA, R. (1963). The fine structure of *Rhodospirillum rubrum*. *J. cell Biol.* **16**, 401.

COLE, H. A., WIMPENNY, J. W. T. & HUGHES, D. E. (1967). The ATP pool in *Escherichia coli* I. Measurement of the pool using a modified luciferase assay. *Biochim. biophys. Acta*, **143**, 445.

COLE, J. A. (1967). *The function and control of anaerobic respiratory pigments in facultative bacteria*. D. Phil. Thesis, University of Oxford.

COLLINS, F. M. & LASCELLES, J. (1962). The effect of growth conditions on oxidative and dehydrogenase activity in *Staphylococcus aureus. J. gen. Microbiol.* **29**, 531.

DALTON, H. & POSTGATE, J. R. (1967). Inhibition of growth of *Azotobacter* by oxygen. *J. gen. Microbiol.* **48**, v.

DAVIES, H. C. & DAVIES, R. E. (1965). Biochemical aspects of oxygen poisoning. In *Handbook of Physiology: Respiration*. Ed. W. O. Fenn and H. Rahn. Washington D.C.: Amer. Physiol. Soc.

DOLIN, M. I. (1961). Cytochrome-independent electron transport enzymes of bacteria. In *The Bacteria*, vol. II. Ed. I. C. Gunsalus and R. Y. Stanier. New York: Academic Press.

DOWNEY, R. J. (1966). Nitrate reductase and respiratory adaptation in *Bacillus stearothermophilus. J. Bact.* **91**, 634.

DREW, G. & GIESBRECHT, P. (1963). Zur morphogenese der Bakterien 'Chromatophoren' (Thylakoide) und zur synthese des Bacteriochlorophylls bei *Rhodopseudomonas spheroides* und *Rhodospirillum rubrum. Zbl. Bakt.* (1. Abt. Orig.), **190**, 508.

ENGLESBERG, E., GIBOR, A. & LEVY, J. B. (1954). Adaptive control of terminal respiration in *Pasteurella pestis. J. Bact.* **68**, 146.

ENGLESBERG, E. & LEVY, J. B. (1955). Induced synthesis of tricarboxylic acid cycle enzymes as correlated with the oxidation of acetate and glucose by *Pasteurella pestis. J. Bact.* **69**, 418.

ESTABROOK, P. K. & MAITRA, R. W. (1962). A fluorimetric method for the quantitative microanalysis of adenine and pyridine nucleotides. *Anal. Biochem.* **3**, 369.

FINN, R. K. (1967). In *Biochemical and Biological Engineering Science*. Ed. N. Blakeborough. London: Academic Press.

FORGET, P. & PICHINOTY, F. (1967). Le cycle tricarboxylique chez *Aerobacter aerogenes. Ann. Inst. Pasteur*, **112**, 261.

FOSTER, J. W. (1949). *Chemical activities of the fungi*. New York: Academic Press.

FRERMAN, F. E. & WHITE, D. C. (1967). Membrane lipid changes during formation of a functional electron transport system in *Staphylococcus aureus. J. Bact.* **94**, 1868.

FUJITA, T. (1966). Studies on soluble cytochromes in Enterobacteriaceae. I. Detection, purification and properties of cytochrome c_{552} in anaerobically grown cells. *J. Biochem.* **60**, 204.

FUJITA, T. & SATO, R. (1967). Studies on soluble cytochromes in Enterobacteriaceae. V. Nitrite-dependent gas evolution in cells containing cytochrome c_{552}. *J. Biochem.* **62**, 230.

FULLER, R. C., SMILLIE, R. M., SISLER, E. C. & KORNBERG, H. L. (1961). Carbon metabolism in *Chromatium. J. biol. Chem.* **236**, 2140.

GARRARD, W. & LASCELLES, J. (1968). Regulation of *Staphylococcus aureus* lactate dehydrogenase. *J. Bact.* **95**, 152.

GEST, H. (1963). Metabolic aspects of bacterial photosynthesis. In *Bacterial Photosynthesis*. Ed. H. Gest, A. San Pietro and L. P. Vernon. Yellow Springs, Ohio: The Antioch Press.

GHOSH, A. & CHANCE, B. (1964). DPNH oscillations in a cell free extract of *S. carlsbergensis. Biochem. biophys. Res. Commun.* **16**, 182.

GLADSTONE, G. P., FILDES, P. & RICHARDSON, G. M. (1935). Carbon dioxide as an essential factor in the growth of bacteria. *Br. J. exp. Path.* **16**, 335.

GOLDFINE, H. (1965). The evolution of oxygen as a biosynthetic reagent. *J. gen. Physiol.* **49**, 253.

GOLDFINE, H. & BLOCH, K. (1963). Oxygen and biosynthetic reactions. In *Control Mechanisms in Respiration and Fermentation*. Ed. B. Wright. New York: Ronald Press.

GOTTLIEB, S. F. (1966). Bacterial nutritional approach to mechanisms of oxygen toxicity. *J. Bact.* **92**, 1021.

GOTTLIEB, S. F. & PAKMAN, L. M. (1968). Effect of high oxygen tensions on the growth of selected aerobic Gram-negative pathogenic bacteria. *J. Bact.* **95**, 1003.

GRAY, C. T., WIMPENNY, J. W. T., HUGHES, D. E. & RANLETT, M. (1963). A soluble *c* type cytochrome from anaerobically grown *Escherichia coli* and various Enterobacteriaceae. *Biochim. biophys. Acta*, **67**, 157.

GRAY, C. T., WIMPENNY, J. W. T. & MOSSMAN, M. R. (1966). Regulation of metabolism in facultative bacteria. II. Effects of aerobiosis, anaerobiosis and nutrition on the formation of Krebs cycle enzymes in *Escherichia coli*. *Biochim. biophys. Acta*, **117**, 33.

GRAY, C. T., WIMPENNY, J. W. T., HUGHES, D. E. & MOSSMAN, M. R. (1966). Regulation of metabolism in facultative bacteria. I. Structural and functional changes in *Escherichia coli* associated with shifts between the aerobic and anaerobic states. *Biochim. biophys. Acta*, **117**, 22.

GRIFFIN, D. H. (1963). The role of carbon dioxide in morphogenesis in *Blastocladiella emersonii*. *Plant Physiol.* **39** (Suppl. XIII).

GUNDERSEN, K. (1966). The growth and respiration of *Nitrocystis oceanus* at different partial pressures of oxygen. *J. gen. Microbiol.* **42**, 387.

GUNDERSEN, K., CARLUCCI, A. F. & BORSTRÖM, K. (1966). Growth of some chemoautotrophic bacteria at different oxygen tensions. *Experientia*, **22**, 229.

GUNSALUS, I. C. & SHUSTER, C. W. (1961). Energy-yielding metabolism in bacteria. In *The Bacteria*, vol. II. Ed. I. C. Gunsalus and R. Y. Stanier. New York: Academic Press.

GUERIN, B. & JACQUES, R. (1968). Photoinhibition de l'adaptation respiratoire chez *Saccharomyces cerevisiae*. II. Le spectre d'action. *Biochim. biophys. Acta*, **153**, 138.

HADJIPETROU, L. P., GRAY-YOUNG, T. & LILLY, M. D. (1966). Effect of ferricyanide on energy production by *Escherichia coli*. *J. gen. Microbiol.* **45**, 479.

HADJIPETROU, L. P. & STOUTHAMER, A. H. (1965). Energy production during nitrate respiration by *Aerobacter aerogenes*. *J. gen. Microbiol.* **38**, 29.

HARRISON, D. E. F. & PIRT, S. J. (1967). The influence of dissolved oxygen concentration on the respiration and glucose metabolism of *Klebsiella aerogenes* during growth. *J. gen. Microbiol.* **46**, 193.

HAUGAARD, N. (1946). Oxygen poisoning. XI. The relation between inactivation of enzymes by oxygen and essential sulphydryl groups. *J. biol. Chem.* **164**, 265.

HAUGAARD, N. (1955). Effect of high oxygen tensions upon enzymes. *Proc. Underwater Physiol. Symp. Natn. Acad. Sci.*, p. 8. National Research Council publication, no. 377.

HAUGAARD, N. (1968). Cellular mechanism of oxygen toxicity. *Physiol. Rev.* **48**, 311.

HEDEN, G. C. & MALMBORG, A. S. (1961). Aeration under pressure and the question of free radicals. *Sci. Report 1st. Super. Sanit.* **1**, 213.

HERNANDEZ, E. & JOHNSON, M. J. (1967a). Anaerobic growth yields of *Aerobacter aerogenes* and *Escherichia coli*. *J. Bact.* **94**, 991.

HERNANDEZ, E. & JOHNSON, M. J. (1967b). Energy supply and cell yield in aerobically grown micro-organisms. *J. Bact.* **94**, 996.

HEWITT, L. F. (1950). *Oxidation-Reduction Potentials in Bacteriology and Biochemistry*. Edinburgh: Livingstone, pp. 98–121.

HIGASHI, T. (1960). Physiological study of the oxygen and nitrate-respiration of *Pseudomonas aeruginosa*. *J. Biochem.* **47**, 326.

Hino, S. & Maeda, M. (1966). Effect of oxygen on the development of respiratory activity in *E. coli* K12. *J. gen. appl. Microbiol.* **12**, 247.

Hirose, Y., Sonoda, H., Kinoshita, K. & Okada, H. (1967). Studies on oxygen transfer in submerged fermentations. Part VI. The effect of aeration on glutamic acid fermentation. *Agric. Biol. Chem.* **13**, 1210.

Hirsch, C. A., Rasminsky, B., Davis, B. D. & Lin, E. C. C. (1963). A fumarate reductase in *Escherichia coli* distinct from succinate dehydrogenase. *J. biol. Chem.* **238**, 3770.

Hirsch, H. M. (1952). A comparative study of aconitase, fumarase and isocitrate dehydrogenase in normal and respiration deficient yeast. *Biochim. biophys. Acta*, **9**, 674.

Hommes, F. A. (1964). Oscillatory reductions of pyridine nucleotides during anaerobic glycolysis in brewer's yeast. *Arch. Biochem. Biophys.* **108**, 36.

Hurlbert, R. E. (1967). Effect of oxygen on viability and substrate utilization in *Chromatium*. *J. Bact.* **93**, 1346.

Ivanovics, G. (1937). Unter welchen Bedingungen werden bei der Nahrboden— Zuchtung der Milzbrandbazillen Kapseln gebildet. *Zbl. Bakt.* (1 Abt. Orig.) **138**, 449.

Jacobs, N. J. & Conti, S. F. (1965). Effect of haemin on the formation of the cytochrome system of anaerobically grown *Staphylococcus epidermidis*. *J. Bact.* **89**, 675.

Jacobs, N. J., Maclosky, E. R. & Conti, S. F. (1967). Effects of oxygen and haem on the development of a microbial respiratory system. *J. Bact.* **93**, 278.

Jacobs, N. J., Maclosky, E. R. & Jacobs, J. M. (1967). Role of oxygen and heme in heme synthesis and the development of hemoprotein activity in an anaerobically grown *Staphylococcus*. *Biochim. biophys. Acta*, **148**, 645.

Johnson, M. J. (1967). Aerobic microbial growth at low oxygen concentration. *J. Bact.* **94**, 101.

Kashket, E. R. & Brodie, A. F. (1960). Subcellular distribution of a biologically active naphthoquinone in *Mycobacterium phlei*. *Biochim. biophys. Acta*, **40**, 550.

Khmel, I. A. & Ierusalemski, N. D. (1967). Effect of aeration on *Azotobacter vinelandii* growth in continuous culture. *Mikrobiology, Moscow*, **36**, 530.

King, A. D. Jr. (1966). *The effect of carbon dioxide on microbial growth and metabolism*. Ph.D. Thesis, Washington State University, U.S.A.

Knight, B. C. J. G. & Fildes, P. (1930). Oxidation-reduction potentials in relation to bacterial growth. III. The positive limit of oxidation-reduction potential required for the germination of *Bacillus tetani* spores *in vitro*. *Biochem. J.* **24**, 1496.

Kornberg, H. L. (1965). The co-ordination of metabolic routes. In *Functions and Structure in Micro-organisms. Symp. Soc. gen. Microbiol.* **15**, 8. Ed. M. R. Pollock and M. H. Richmond. Cambridge University Press.

Kornberg, H. L., Collins, J. F. & Bigley, D. (1960). The influence of growth substrates on metabolic pathways in *Micrococcus denitrificans*. *Biochim. biophys. Acta*, **39**, 9.

Lascelles, J. (1964). *Tetrapyrrole biosynthesis and its regulation*. New York: W. A. Benjamin.

Lenhof, H. M., Nicholas, D. J. D. & Kaplan, N. O. (1956). Effects of oxygen, iron and molybdenum on routes of electron transport in *Pseudomonas fluorescens*. *J. biol. Chem.* **220**, 983.

Lester, R. L. & Crane, F. L. (1959). The natural occurrence of coenzyme Q and related compounds. *J. biol. Chem.* **234**, 2169–2175.

Levshina, N. A. (1965). Effect of CO_2 on the development of diatomaceous algae. *Byul. Moskov. Obschest. Ispytatelei. Prirody Otd. Biol.* **70**, 139.

LIN, E. C. C., LEVIN, A. P. & MAGASANIK (1960). The effect of aerobic metabolism on the inducible glycerol dehydrogenase of *Aerobacter aerogenes. J. Biol. Chem.* **235**, 1824.

LINDENMAYER, A. & SMITH, L. (1964). Cytochromes and other pigments of baker's yeast grown aerobically and anaerobically. *Biochim. biophys. Acta*, **93**, 445.

LINEWEAVER, H. (1933). Characteristics of oxidation by *Azotobacter. J. biol. Chem.* **99**, 575.

LINNANE, A. W. (1965). *Oxidases and Related Redox Systems*, vol. 2. Ed. T. E. King, H. S. Mason and M. Morrison. New York: Wiley.

LINNANE, A. W., VITOLS, E. & NOWLAND, P. G. (1962). Studies on the origin of yeast mitochondria. *J. cell Biol.* **13**, 345.

LONDON, J. (1968). Regulation and function of lactate oxidation in *Streptococcus faecium. J. Bact.* **95**, 1380.

LONDON, J. & KNIGHT, M. (1966). Concentrations of nucleotide coenzymes in micro-organisms. *J. gen. Microbiol.* **44**, 241.

LONGMUIR, I. S. (1954). Respiration rate of bacteria as a function of oxygen concentration. *Biochem. J.* **57**, 81.

LOOMIS, W. F. (1961). Cell differentiation. A problem in selective gene activation through self-produced micro-environmental differences of carbon dioxide tension. In *Biological Structure and Function*. Ed. T. W. Goodwin and O. Lindberg. New York: Academic Press.

LUKINS, H. B., THAM, S. H., WALLACE, P. G. & LINNANE, A. W. L. (1966). Correlation of membrane bound succinate dehydrogenase with the occurrence of mitochondrial profiles in *Saccharomyces cerevisiae. Biochem. biophys. Res. Commun.* **23**, 363.

LYNEN, F. (1967). Role of biotin dependent carboxylations in biosynthetic reactions. *Biochem. J.* **102**, 381.

MACKLER, B., COLLIPP, P. J., DUNCAN, H. M., RAO, N. A. & HUENEKENS, F. M. (1962). The genetic control of the cytochrome system in yeasts. *J. biol. Chem.* **237**, 2968.

MAITRA, P. K. & ESTABROOK, R. W. (1964). A fluorometric method for the enzymic determination of glycolytic intermediates. *Analyt. Biochem.* **7**, 472.

MARTIN, W. R. & NIVEN, C. F. (1960). Mode of carbon dioxide fixation by the minute streptococci. *J. Bact.* **79**, 295.

MCFADDEN, B. A. & TU, C.-C. L. (1967). Regulation of autotrophic and heterotrophic carbon dioxide fixation in *Hydrogenomonas facilis. J. Bact.* **93**, 886.

MCPHEDRAN, P., SOMMER, B. & LIN, E. C. C. (1961). Control of ethanol dehydrogenase levels in *Aerobacter aerogenes. J. Bact.* **81**, 852.

MILLER, G. W. & HSU, W. J. (1965). Effects of carbon dioxide-bicarbonate mixtures on oxidative phosphorylation by cauliflower mitochondria. *Biochem. J.* **97**, 615.

MOSS, F. (1952). The influence of oxygen tension on respiration and cytochrome a_2 formation of *Escherichia coli. Aust. J. exp. Biol. med. Sci.* **30**, 531.

MOSS, F. (1956). Adaptation of the cytochromes of *Aerobacter aerogenes* in response to environmental oxygen tension. *Aust. J. exp. Biol. med. Sci.* **34**, 395.

MUELLER, J. H. & HINTON, J. (1941). A protein-free medium for primary isolation of the *Gonococcus* and *Meningococcus. Proc. Soc. exp. Biol. Med.* **48**, 330.

NING KWAN, C., APIRION, D. & SCHLESSINGER, D. (1968). Anaerobiosis induced changes in an isoleucyl transfer ribonucleic acid and the 50s ribosomes of *Escherichia coli. Biochemistry*, **7**, 427.

NYGAARD, A. P. (1961). D(−)-Lactic cytochrome *c* reductase, a flavoprotein from yeast. *J. biol. Chem.* **236**, 920.

NYIRI, L. & LENGYEL, Z. L. (1965). Studies on automatically aerated biosynthetic processes. I. The effect of agitation and carbon dioxide on penicillin formation in automatically aerated liquid cultures. *Biotech. Bioeng.* **7**, 343.

O'HARA, J., GRAY, C. T., PUIG, J. & PICHINOTY, F. (1967). Defects in formate hydrogen lyase in nitrate-negative mutants of *Escherichia coli*. *Biochem. biophys. Res. Commun.* **28**, 951.

OKA, K. & ARIMA, K. (1965). Cyanide resistance in *Achromobacter*. II. Mechanism of cyanide resistance. *J. Bact.* **90**, 744.

PASCAL, M-C. (1966). Regulation de la biosynthése et du fonctionnement des D- et L- lactate deshydrogenase chez *Aerobacter aerogenes*. *Bull. Soc. Frç. Physiol. Veg.* **12**, 123.

PICHINOTY, F. (1962). Inhibition par l'oxygène de la biosynthèse et de l'activité de l'hydrogènase et de l'hydrogènelyase chez les bactéries anaérobies facultatives. *Biochim. biophys. Acta*, **64**, 111.

PICHINOTY, F. (1965a). L'effet oxygène et la biosynthèse des enzymes d'oxydoreduction bacteriens. *Colloques Internationaux du C.N.R.S.* Marseilles 1963, no. 24. Paris: Editions du Centre National de la recherche scientifique.

PICHINOTY, F. (1965b). L'inhibition par l'oxygène de la denitrification bactérienne. *Annls Inst. Pasteur, Paris*, **109**, 248.

PICHINOTY, F. (1966). Propriétés, régulation et fonctions physiologiques des nitrate-reductases bactériennes. A. et B. *Bull. Soc. Frç. Physiol. Veg.* **12**, 97.

PICHINOTY, F. & BIGLIARDI-ROUVIER, J. (1962). Etude et mise au point d'une methode permettant de mésurer l'activité des tetrathionate-réductases d'origine bactériénne. Inhibition par l'oxygène de la biosynthèse et de l'activité de l'enzyme d'*Escherichia intermedia*. *Antonie van Leeuwenhoek*, **28**, 134.

PICHINOTY, F. & D'ORNANO, L. (1961). Inhibition de la réduction assimilatrice du nitrate et du nitrite par des pressions partielle élevées d'oxygène chez les bactéries, les champignons et les levures. *Z. Allg. Mikrobiol.* **1**, 376.

PIRT, S. J. (1957). The oxygen requirement of growing cultures of an *Aerobacter* species, determined by means of the continuous culture technique. *J. gen. Microbiol.* **16**, 59.

POLAKIS, E. S., BARTLEY, W. & MEEK, G. A. (1964). Changes in the structure and enzyme activity of *Saccharomyces cerevisiae* in response to changes in the environment. *Biochem. J.* **90**, 369.

POLLOCK, M. R. & KNOX, R. (1943). Bacterial reduction of tetrathionate. (A report to the Medical Research Council.) *Biochem. J.* **37**, 476.

PRESCOTT, J. M., RAGLAND, R. S. & HURLEY, R. J. (1965). Utilization of carbon dioxide and acetate in amino acid synthesis by *Streptococcus bovis*. *Proc. Soc. exp. Biol. Med.* **119**, 1097.

QUAYLE, J. R. (1961). Metabolism of C_1 compounds in autotrophic and heterotrophic microorganisms. *Annl. Rev. Microbiol.* **15**, 119.

RAHN, O. (1941). Protozoa need carbon dioxide for growth. *Growth*, **5**, 197.

RANSON, S. L., WALKER, D. A. & CLARKE, I. D. (1960). Effects of carbon dioxide on mitochondrial enzymes from *Ricinus*. *Biochem. J.* **76**, 216.

ROCKWELL, G. E. & HIGHBERGER, J. H. (1927). The necessity of carbon dioxide for the growth of bacteria, yeasts and molds. *J. infect. Dis.* **40**, 438.

ROODYN, D. B. (1966). In *Regulation of Metabolic Processes in Mitochondria*. Ed. J. M. Tager, S. Papa, E. Quagliariello and E. C. Slater. Amsterdam: Elsevier.

ROODYN, D. B. & WILKIE, D. (1968). *The Biogenesis of Mitochondria*. London: Methuen.

ROXBURGH, J. M., SPENCER, J. F. T. & SALANS, H. R. (1954). Factors affecting the production of ustilagic acid by *Ustilago zeae*. *J. Agric. Fd. Chem.* **2**, 1121.

SACKS, L. E. & BARKER, H. A. (1949). The influence of oxygen on nitrate and nitrite reduction. *J. Bact.* **58**, 11.

SANDERS, A. P., HALL, I. H. & WOODHALL, B. (1965). Succinate: protective agent against hyperbaric oxygen toxicity. *Science*, **150**, 1830.

SCHAEFFER, P. (1952). Recherches sur le métabolisme bactérien des cytochromes et des porphyrines. I. Disparition partielle des cytochromes par culture anaérobie chez certaines bactéries aérobies facultatives. *Biochim. biophys. Acta*, **9**, 261.

SCHON, G. (1965). Untersuchungen über den Nutz effeckt von *Nitrobacter winogradskiya. Buch. Arch. Mikrobiol.* **50**, 111.

SHERMAN, F. (1965). Mécanismes de regulation des activités cellulaires chez les microorganismes. *Colloque Internationaux du C.N.R.S.* Marseilles 1963, no. 24, 465. Paris: Editions du Centre National de la recherche scientifique.

SLADE, H. D., WOOD, H. G., NIER, A. O., HEMINGWAY, A. & WERKMAN, C. H. (1942). Assimilation of heavy carbon dioxide by heterotrophic bacteria. *J. biol. Chem.* **143**, 133.

SLONIMSKI, P. P. (1953). *La formation des enzymes respiratoires chez la levure.* Paris: Masson.

SLONIMSKI, P. P. (1956). Adaptation respiratoire: développement du système hémoprotéique induit par l'oxygène. *Proc. 3rd Intern. Congr. Biochem.* Brussels, p. 242.

SLONIMSKI, P. P. & HIRSCH, H. M. (1952). Rôle de l'oxygène dans le déterminisme de la constitution enzymatique de la levure. *C. r. hebd. Séanc. Acad. Sci., Paris*, **235**, 914.

SOMLO, M. & FUKUHARA, H. (1965). On the necessity of molecular oxygen for the synthesis of respiratory enzymes in yeast. *Biochem. biophys. Res. Commun.* **19**, 587.

STEINMAN, H. G., OYAMA, V. I. & SCHULZE (1954). Carbon dioxide, co-carboxylase, citrovorum factor and coenzyme A as essential growth factors for a saprophytic treponeme (S–69). *J. biol. Chem.* **211**, 327.

STUART, B., GERSCHMAN, R. & STANNARD, J. N. (1962). Effect of high oxygen tensions on potassium retention and colony formation of baker's yeast. *J. gen. Physiol.* **45**, 1019.

SUGIMURA, T., OKABE, K. & RUDNEY, H. (1964). Ubiquinone in respiration-deficient mutants of *Saccharomyces cerevisiae. Biochim. biophys. Acta*, **82**, 350.

SUGIMURA, T. & RUDNEY, H. (1960). The adaptive formation of ubiquinone 30 (coenzyme Q_6) in yeast. *Biochim. biophys. Acta*, **37**, 560.

TERUI, G., KONNO, N. & SASE, M. (1960). Analysis of the behaviour of some industrial microbes towards oxygen. 1. Effect of oxygen concentration upon the rates of oxygen adaptation and metabolism of yeasts. *Technol. reports, Osaka Univ.* **10**, 527.

THEODORE, T. S. & ENGLESBERG, E. (1964). Mutant of *Salmonella typhimurium* deficient in the carbon dioxide-fixing enzyme phosphoenol pyruvic carboxylase. *J. Bact.* **88**, 946.

TOPLEY & WILSON'S *Principles of Bacteriology* (5th ed. 1964). Ed. G. S. Wilson and A. A. Miles. London: Edward Arnold.

TREMMEL, R. D. & LEVEDAH, B. H. (1966). The effects of carbon dioxide on growth of bleached *Euglena. J. cell. Physiol.* **67**, 361.

TUSTANOFF, E. R. & BARTLEY, W. (1964). Development of respiration in yeast grown anaerobically on different carbon sources. *Biochem. J.* **91**, 595.

VAN NIEL, C. B., THOMAS, J. O., RUBEN, S. & KAMEN, M. D. (1942). Radioactive carbon as an indicator of carbon dioxide utilization. IX. The assimilation of carbon dioxide by protozoa. *Proc. natn. Acad. Sci. Wash.* **28**, 157.

VERHOEFEN, W. & TAKEDA, Y. (1956). The participation of cytochrome *c* in nitrate reduction. In *Inorganic Nitrogen Metabolism.* Ed. W. D. McElroy and B. Glass. Baltimore: Johns Hopkins Press.

WALKER, D. A. & BROWN, J. M. A. (1957). Effects of carbon dioxide concentration on phosphoenol pyruvate carboxylase activity. *Biochem. J.* **67**, 79.

WALKER, H. H. (1933). Carbon dioxide as a factor affecting lag in bacterial growth. *Science*, **76**, 602.

WALLACE, P. G. & LINNANE, A. W. (1964). Oxygen induced synthesis of yeast mitochondria. *Nature, Lond.* **201**, 1191.

WEBB, F. C. (1964). *Biochemical Engineering*. London: Van Nostrand.

WHITE, D. C. (1962). Cytochrome and catalase patterns during growth of *Haemophilus parainfluenzae*. *J. Bact.* **83**, 851.

WHITE, D. C. (1963). Factors affecting the affinity for oxygen of cytochrome oxidases in *Haemophilus parainfluenzae*. *J. biol. Chem.* **238**, 3757.

WHITE, D. C. (1964). Differential synthesis of five primary electron transport dehydrogenases in *Hemophilus parainfluenzae*. *J. Bact.* **239**, 2055.

WHITE, D. C. (1965). Synthesis of 2-demethyl vitamin K_2 and the cytochrome system of *Haemophilus*. *J. Bact.* **89**, 299.

WHITE, D. C., BRYANT, M. P. & CALDWELL, D. R. (1962). Cytochrome linked fermentation in *Bacteroides ruminicola*. *J. Bact.* **84**, 822.

WIMPENNY, J. W. T. (1967). Regulation of enzyme profiles in *Escherichia coli*: The effects of oxygen, nitrate or nitrite. *Biochem. J.* **106**, 34 P.

WIMPENNY, J. W. T., RANLETT, M. R. & GRAY, C. T. (1963). Repression and derepression of cytochrome c biosynthesis in *Escherichia coli*. *Biochim. biophys. Acta*, **73**, 170.

WINZLER, R. J. (1941). The respiration of baker's yeast at low oxygen tensions *J. cell. comp. Physiol.* **17**, 263.

WISEMAN, G. M., VIOLAGO, F. C., ROBERTS, E. & PENN, I. (1966). The effect of hyperbaric oxygen on aerobic bacteria. I. *In vitro* studies. *Can. J. Microbiol.* **12**, 521.

WOOD, H. G. & STJERNHOLM, R. L. (1962). Assimilation of carbon dioxide by heterotrophic organisms. In *The Bacteria*, vol. III. Ed. I. C. Gunsalus and R. Y. Stanier. New York: Academic Press.

WOOD, H. G. & UTTER, M. (1965). The role of carbon dioxide fixation in metabolism. *Essays in Biochemistry*, **1**, 1.

WOOD, H. G. & WERKMAN, C. H. (1935). The utilisation of CO_2 by the propionic acid bacteria in the dissimilation of glycerol. *J. Bact.* **30**, 332.

WYN-JONES, R. G. (1967). Ubiquinone deficiency in an auxotroph of *Escherichia coli* requiring 4-hydroxybenzoic acid. *Biochem. J.* **103**, 714.

WYN-JONES, R. G. & LASCELLES, J. (1967). The relationship of 4-hydroxybenzoic acid to lysine and methionine formation in *Escherichia coli*. *Biochem. J.* **103**, 709.

ZOBELL, C. E. & HITTLE, L. L. (1967). Some effects of hyperbaric oxygenation on bacteria at increased hydrostatic pressure. *Can. J. Microbiol.* **13**, 1311.

MICROBIAL GROWTH AND PRODUCT FORMATION

S. J. PIRT

*Department of Microbiology, Queen Elizabeth College,
University of London, Campden Hill Road, London, W.8*

INTRODUCTION

The purposes for which microbes are grown in culture may be classified into five categories: (1) to maintain the organism in a viable state without loss of function; (2) to identify the organism; (3) to study the role of the organism in nature; (4) to study the structure and function of the organism; (5) to obtain microbial products. For each of these purposes the optimum growth conditions may be different, which implies that growth can be directed in various ways. The main theme of this article is consideration of purpose (5) and how to direct growth so as to maximize the formation of a particular product and to minimize the extraneous activities of the organism.

The culture media used for microbial syntheses are frequently conventional rather than based on a quantitative assessment of the growth requirements and selection of the growth-limiting factor. Also it is common to accept the drift in conditions such as pH value, dissolved oxygen concentration, nutrient concentration and growth rate brought about by growth of the organisms. Drift in the culture conditions, which complicates the interpretation of results, often can be prevented only by the use of instruments to control the process. Even the more advanced processes such as penicillin production are little understood and have evolved by trial and error. It is remarkable that in the old and better understood processes such as production of ethanol or lactic acid the kinetics which provide the basis for control have not been established (Holzberg, Finn & Steinkraus, 1967).

To a great extent industrial microbiological synthesis is still based on pure empiricism despite the great increase in knowledge about cell mechanisms. A case in point is the production of citric acid from sugar by the fungus *Aspergillus niger*, which must depend upon a disturbance of the regulation of the citric acid cycle so that citric acid is over-produced. It has been empirically established that the process requires a deficiency of the ions of certain trace metals (probably iron and manganese) at a very low pH (about 2·0) and that only certain strains of the mould are efficient producers. Under optimum conditions, which are quite critical, the yield of citric acid relative to hexose consumed can be as high as 90%, w/w. Most citric acid is produced by a surface-culture method, but the process can be carried out successfully in submerged cultures where the culture is much more homogeneous. In either case the duration of the process is about 9 days. The submerged culture conditions require the most careful control, so that the more primitive surface-culture process has remained attractive as a production method. In order to give this process a sound scientific basis, considerably more needs to be known about the regulation of the citric acid cycle and the effects of growth conditions on it.

One of the important requirements for development of microbial biosynthetic processes is improvement in culture techniques. This is an engineering problem concerned with culture-vessel design, instrumentation and development of continuous flow techniques. The other important requirement is enhanced knowledge of growth physiology. Both technically and physiologically the chemostat culture method is a key to development in this field. The advantage of the chemostat is that it opens up a whole range of different growth environments and allows them to be maintained indefinitely. The difference between the range of growth conditions possible in the chemostat and in batch culture is shown by a consideration of the relation between substrate concentration and growth rate. This relation for most purposes is adequately represented by the expression,

$$\mu = \mu_m \frac{S}{K_s + S},\tag{1}$$

where μ = specific growth rate (hr.$^{-1}$), μ_m = maximum specific growth rate, S = concentration of growth-limiting substrate (it is assumed that all other substrates are in excess so that they represent a constant factor), K_s is the 'saturation constant'. It is found empirically that the practicable range of growth rates in the chemostat is about 0·05–0·95 times the maximum value. Substituting for μ in (1) gives $0·05 K_s$ to $19 K_s$ for the corresponding range of growth-limiting substrate concentrations. In contrast, in batch cultures, since the initial substrate concentration normally has to be of the order $500 K_s$, the organism is saturated with substrate throughout almost the entire growth phase (over 95% of the growth will occur with $\mu > 0·95 \mu_m$). The ability to obtain steady-state conditions in which the organism is not saturated with substrate is a unique feature of the chemostat. One might prolong substrate-limited growth in batch culture by continuous feeding of substrate, but this is not a practicable method of obtaining steady-state conditions.

Classification of microbial products

A classification of microbial products has been given in Table 1. The products in classes 1–6 all have some obvious role in the cell. Class 7, the 'secondary metabolites', contains compounds such as antibiotics and virulence factors which could play an ecological role; that is, influence the organism's survival in nature. In culture, without competing species, secondary metabolite synthesis is not indispensable and consequently selection against it is possible. Such selection of less productive strains is referred to as 'strain degeneration'. The bioderivatives or transformation products in class 8 are abnormal metabolites formed in response to the presence of certain compounds in the culture medium. Bioconversion, that is, transformation of a chemical by a cell—may, like secondary metabolite synthesis, have evolved to give the organism a selective advantage in the natural habitat; steroid bioconversion might be a detoxification mechanism. If detoxification is, indeed, the role of bioconversion, then there should be a selective pressure in favour of it, even in pure culture; however, information on the stability of bioconversion systems in cells is lacking. Viruses form a category of cell products with production problems which are conveniently unified with those of other cell products. Some microbial products such as vaccines, beer and food material may be classified as mixtures of two or more of the classes of products that are given in Table 1.

Some responses to environmental influences correlated with the different classes of products are indicated in Table 1. The pH value and dissolved oxygen concentration are important in determining the end products of energy metabolism. The pH effect could reflect the importance of H^+ in NAD^+ reductions. The importance of dissolved oxygen tension is well illustrated by *Klebsiella aerogenes* where three types of reaction (the third is complete oxidation of carbon source to CO_2 and H_2O) may occur depending on the dissolved oxygen tension (for further examples see Wimpenny, this Symposium). Accumulation of energy storage compounds such as glycogen requires an excess of the energy source, while high yields of RNA require the organism to be grown at the maximum rate. The addition of precursors of the product sometimes stimulates its production; an example is synthesis of benzylpenicillin where addition of the side-chain precursor, phenylacetic acid, stimulates the rate of penicillin production about 10-fold; in contrast,

Table 1. *Classification of microbial products*

Class of product	Process or product examples	Organisms	Some special conditions
1. End products of energy metabolism			
(a) Aerobic	Ethanol⟶ acetic acid Sulphide⟶ sulphate Ammonia⟶ nitrite	*Acetobacter* *Thiobacillus* *Nitrosomonas*	— — —
(b) Partially aerobic	Hexose⟶ 2, 3-butanediol	*Klebsiella aerogenes*	Acid pH, oxygen tension, < 1 mm. Hg
(c) Anaerobic	Hexose⟶ ethanol Hexose⟶ glycerol + acetate	*Saccharomyces cerevisiae* *S. cerevisiae*	Acid pH Alkaline pH
2. Energy storage compounds	Glucose⟶ glycogen Acetate ⟶ poly β-hydroxybutyrate	*Escherichia, Saccharomyces* *Bacillus megaterium*	Glucose in excess Acetate in excess
3. Enzymes			
(a) Extracellular	Ammonia + sugar⟶ amylases	*Aspergillus*	Use constitutive strain or supply inducer
(b) Intracellular	Ammonia + sugar⟶ β-galactosidase	*Escherichia*	Use constitutive strain or supply inducer
4. Nucleic acids	Ammonia + sugar + phosphate + magnesium + potassium⟶ ribosomal RNA	Bacteria, fungi	Maximum specific growth rate

Class of product	Process or product examples	Organisms	Some special conditions
5. Structural components of cells	Carbon, nitrogen and phosphate sources⟶ cell walls	Bacteria, fungi	—
6. Intermediary metabolites	Vitamin B_{12}	Propionibacterium	Cobalt ions + benzimidazole as precursors
	Citric acid	Aspergillus niger	pH 2, iron and manganese deficiencies
	Glutamic acid	Corynebacterium glutamicus	Biotin deficiency
7. Secondary metabolites*	Benzylpenicillin	Penicillium chrysogenum	Phenylacetate as precursor
	Cephalosporin C	Cephalosporium sp.	Methionine as stimulant
	Tetanus toxin Fraction I antigen (virulence factor)	Clostridium tetani Pasteurella pestis	Temperature $> 30°$
8. Bioderivatives†	Hydroxylation of steroids D-Sorbitol⟶ l-sorbose	Rhizopus Acetobacter	Two stages: (1) growth of organisms, (2) bioconversion†
9. Viruses	T-phage	Escherichia coli	Two stages: (1) growth of organisms, (2) virus production

* A 'secondary metabolite' (Bu'Lock, 1965) is defined as a product which is not essential for growth.
† 'Bioderivative' is a term coined to replace the somewhat ambiguous term 'transformation product'. 'Bioconversion' is used instead of 'transformation' (of a compound).

the other precursors of the penicillin molecule, L-valine and L-cysteine, have no such effect. Nutrient deficiencies are sometimes important as a means of stimulating production. For example, the production of glutamic acid by *Corynebacterium glutamicus* is stimulated by biotin deficiency. The use of auxotrophic mutants to stimulate excretion of intermediary metabolites will certainly increase in importance as stressed by Demain (1966). A survey of auxotrophic bacteria as a means of producing amino acids from hydrocarbons has been reported by Ishii, Otsuka & Shiio (1967).

A new control principle was introduced by the discovery that methionine stimulates the production of cephalosporin C by *Cephalosporium* species. Cephalosporin C is an analogue of penicillin but, unlike penicillin, the production of cephalosporin C is stimulated if methionine or its analogue, norleucine, is supplied during the growth phase of the fermentation (Demain, 1966). The stimulation is much reduced if methionine is added after the end of the growth phase. Demain has linked these observations with the known repressor effects of methionine and norleucine on synthesis of enzymes which convert cysteine to methionine or degrade cysteine to pyruvate. Since cysteine is a precursor of cephalosporin C, the repression of enzymes which convert cysteine to other products would spare cysteine and thus make more available for antibiotic synthesis.

Both the methionine effect on cephalosporin C synthesis and the controlled feeding of growth factors to auxotrophs illustrate the importance of metabolic regulation of metabolism to achieve overproduction of a metabolite. Temperature, besides its general effect on chemical reaction rate, can exert highly selective effects on metabolic pathways such as repression of particular protein syntheses. Classical examples of such effects are provided by *Pasteurella pestis*; maximum yield of certain of its antigens are obtained at 37°C. and they are not produced at all at 28°C., at which temperature the growth yield is at a maximum (Pirt, Thackeray & Harris-Smith, 1961). High temperatures may also induce phenotypic auxotrophy. The growth factor requirements of *P. pestis* are increased by raising the growth temperature from 28 to 37°C. (Hills & Spurr, 1952). The observation by Ware (1951) that at 44°C. *Escherichia coli* becomes auxotrophic is of interest as a possible means of studying the role of the growth factors in metabolism and possibly of inducing accumulation of intermediary metabolites.

Different states of the cell's biosynthetic apparatus

In the application of biological synthesis to obtain single compounds such as penicillin or glutamic acid the organism is only required to reproduce that part of its enzyme system which is concerned with the synthesis of the desired product. The required enzyme system may be present in one of the following four forms: (1) in growing organisms, (2) in non-growing (that is, 'resting') organisms, (3) in dead organisms (that is, not able to reproduce), (4) in a cell-free enzyme system. In virtually all applications of biosynthesis either growing or non-growing organisms are used. There is a mounting interest in cell-free enzymes, but their use in the foreseeable future seems likely to be confined to the simple processes not requiring energy such as hydrolysis of poly-saccharides or proteins, oxidations (for example, ethanol to acetic acid, glucose to gluconic acid) and formation of sugar degradation products by the Embden–Meyerhof pathway. The extent to which dead organisms will retain their biosynthetic functions must depend upon the nature of the lesion which causes death. The use of lipid solvents such as toluene to disrupt plasma membranes is one way of killing or 'fixing' cells so as to preserve enzyme systems but systematic studies of methods of 'fixing' whole organisms so as to preserve their biosynthetic systems are lacking.

Synthesis by growing organisms

If growth of the wild-type of *Penicillium chrysogenum* is glucose-limited the organism converts the nutrients (ammonia, sugar, inorganic salts, oxygen) almost exclusively into more organism and the end products of energy metabolism, carbon dioxide and water. The secondary metabolites, such as penicillin and any excreted intermediary metabolites, account for only a minute proportion of the total carbon substrate utilized. This virtual 100% efficiency in conversion of substrate carbon into cell structural material, which is characteristic of the wild-type organism, reflects the precise regulation of the rates of synthesis of every component in the cell so that none is overproduced. In contrast, the aim of microbiological synthesis usually is to maximize the formation of one product and minimize the formation of the others. To achieve overproduction of any particular substance the normal regulation of the metabolic pathways must be upset by either gene mutation or an extreme environmental condition or a combination of both.

A model of the growing cell

A model of the growing cell is required to provide a guide for thought on the nature of growth and how the growing organism may differ from the non-growing organism. Inevitably one must begin with greatly simplified systems, but even the simplest model may point out principles of control and lead to formulation of better models. The Dean & Hinshelwood (1966) model will be discussed here since it illustrates the basic approach to kinetic models and because it leads to some important predictions of cell behaviour which are borne out by experiment.

In the Dean & Hinshelwood model the cell is considered to consist of an array of structural units which include nucleic acids, enzymes, lipids and polysaccharides. For growth to occur the structure must be primed with the intermediary metabolites (amino acids, nucleotides, fatty acids, sugars and their precursors) and supplied with exogenous substrates which are the raw materials for cell synthesis. Then, under the appropriate physico-chemical conditions, the structural units, by interdependent action, reproduce themselves. Obviously the relative proportions of the structural units may vary under transient conditions but under steady-state conditions the proportions of the structural units will remain constant; the latter state is equivalent to the 'balanced growth' of Campbell (1957). In the simplest form of the model (Fig. 1a) the organism is represented by two structural components X_1 and X_2 whose syntheses are interdependent; in the diagram, X_1 —○ X_2 means that the synthesis of X_1 is dependent upon the action of X_2. One such dependence occurs between protein synthesis and ribosomal RNA; another would arise if X_2 produced an amino acid which is incorporated into the structure of X_1. The basic concept of the cell as a closed loop of interdependent processes was introduced by Hinshelwood (1952). He expressed the dependencies mathematically in this form

$$\frac{dX_1}{dt} = \alpha_1 X_2; \quad \frac{dX_2}{dt} = \alpha_2 X_1, \tag{2}$$

where the αs are constants dependent on the concentrations of intermediary metabolites and the physical conditions. In actual fact the cell must consist of a branched but closed network of such dependencies. The more complex model represented in Fig. 1b shows two additional features. One of these is alternative dependencies for synthesis; synthesis of X_1 is dependent upon the function of either X_1 or Y_1. Such a dual dependency may be represented as

$$\frac{dX_1}{dt} = \alpha_1 X_2 + \beta_1 Y_1. \tag{3}$$

Another feature in Fig. 1 b, represented as $Z_1 \bigcirc\!\!-\!\!+ X_5$, is the degradation of a structural component or a 'reverse reaction' in the terminology of Dean & Hinshelwood (1967); Z_1 is considered to be an enzyme which degrades component X_5. Turnover of protein and nucleic acid clearly requires the presence of such degradative processes. In the mathematical model the production of X_5 may be represented as

$$\frac{dX_5}{dt} = \alpha_5 X_1 - \gamma_1 Z_1. \tag{4}$$

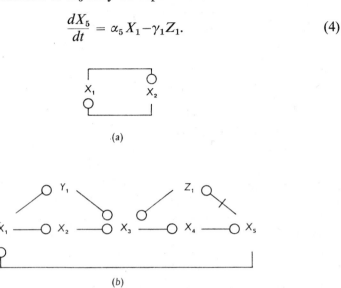

(a)

(b)

Fig. 1. Hinshelwood models of self-reproducing systems. The Xs, Y and Z are structural components which are assumed to have a supply of the requisite intermediary metabolites. The connections indicate dependencies for synthesis: for example, synthesis of X_1 depends upon the function of X_2; the connection with a bar indicates a dependency for degradation (X_5 is degraded by Z_1).

From analysis of the kinetics of this model Dean & Hinshelwood (1967) deduced that the proportions of the structural components must spontaneously vary to cope with changes in the chemical and physical environment which determine the values of the α, β and γ coefficients. In balanced growth, the specific rates of synthesis of each component $[(1/X_1) . [dX_1/dt]$, etc.) will equal the specific growth rate (μ) of the organism, then from the solution of the equations in (2), $X_1/X_2 = \alpha_1/\mu$. Continuous variation in the proportions of cell structural components, with environmental change, which is predicted by the models, is in agreement with chemostat studies where sufficient time can be allowed for balanced growth to be achieved; for example, continuous variation in 'constitutive' enzyme content of yeast with change in growth rate was observed by Tempest & Herbert (1965).

Another interesting consequence of the model follows if there are alternative dependencies. For instance, if in Fig. 1 b, Y_1 and X_2 duplicate each other's functions, the model kinetics show that one of these components could spontaneously disappear with an appropriate change in the environment. Such an effect would be analogous to repression of an enzyme, although in contrast to repressor action the purely dynamic response could be slow. The degradative processes which involve relations such as (4) are of significance in autolysis and viral infection of the cell. The existence of the degradative processes is made apparent by the onset of autolysis which often follows immediately after sudden exhaustion of a substrate in cultures undergoing steady-state growth; the latter phenomenon was observed in *Bacillus subtilis* cultures by Monod (1942) and in *Penicillium* cultures by Pirt & Righelato (1967).

The continuous change in enzyme concentration in cells with environmental change which is predicted by the Dean & Hinshelwood theory is analogous to 'proportional control' and in contrast to the 'on-off' control provided by a gene repressor. The continuous regulation which is predicted by the kinetic principles could be the original primitive means of metabolic regulation. It does not preclude gene repressor effects and feed-back inhibition which could have evolved later as a refinement to speed up the basic kinetic control mechanism. Gene repressor effects should be more obvious in short-term culture. Longer term culture extending over many generations in a constant environment may be essential to test the significance of the underlying kinetic control of autosynthesis predicted by the Hinshelwood model. Little is known about the time required for the cell to return to a steady-state condition after a change in the environment. The duration of the adjustment is dependent upon the nature and magnitude of the change required; Mateles, Ryu & Yasuda (1965) found with *Escherichia coli* that about three doubling times (2·5 hr.) was required for a change in growth rate from 0·4 to 0·8 hr.$^{-1}$ after the corresponding 'shift up' in the dilution rate. A much longer time, about 30 doubling times (100 hr.), was required to develop the maximum rate of production of 2,3-butanediol in a chemostat culture of *Klebsiella aerogenes* (Pirt, 1964) after inoculation with a batch culture.

Single cell and mycelial modes of growth and effect of organism age

The question of how much microbial productivity depends upon the age of the organism is often raised. In order to discuss this question it is necessary to consider the definition of age, and the definition depends upon the mode of growth of the organism. The three modes of growth considered here and represented in Fig. 2 are: (*a*) cell enlargement followed by binary fission, (*b*) budding, and (*c*) apical growth of hyphae. It is customary to measure the age of bacteria which multiply by binary fission from the time when division is complete. At the moment just before division is complete the age of the mother cell is given by *g*, the doubling time. By this definition the mean cell age in a

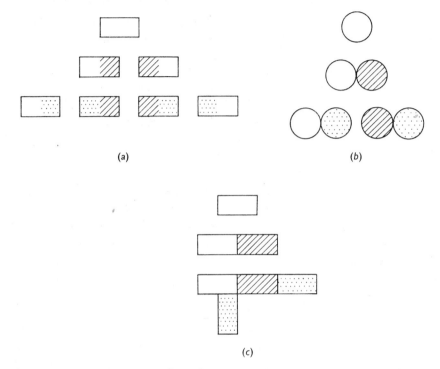

(a)

(b)

(c)

Fig. 2. Diagrammatic representations of age distribution with different modes of growth. The new material in each doubling is shown by different shading; (*a*) intercalatory growth and binary fission, (*b*) budding, (*c*) mycelial habit with branching.

growing culture is less than *g*. In the budding habit of growth typical of yeasts, division is considered to lead to one new cell and one mother cell and the latter may be several generations old (Beran, Malek, Streiblova & Leiblova, 1967). In fungal hyphae new cell material is laid down at the tips of the hyphae and as a result the age of a hypha must vary continuously along its length. Since fungi are either non-septate or if septate have septa with communication pores through which material can be transported by 'protoplasmic streaming', it seems preferable to consider the whole network of hyphae as one organism, not as new and old parts. Aiba & Hara (1965) have proposed a method of determining the 'mean age' of a mycelial organism in a growing culture. They show that in a small time-interval $\triangle\ t$ an element of organism $\mu\ x\ \triangle\ t$ will be formed where μ is the specific growth rate and x is the biomass present at time t. This element of the biomass will have an age $\triangle t^1$ where $\triangle t^1 \leqslant t$. The mean age is obtained by summing the ages of all small elements of the biomass and dividing by the sum of the amounts of elements having each particular age. Thus Aiba & Hara show that in an exponentially growing batch culture or in a chemostat, with increase in the number of generations the mean organism age approaches $1/\mu$, that is $g/\ln 2$. On this basis, therefore, the mean age is greater than g, whereas in cells dividing by binary fission the mean age is represented as less than g. Until this conflict is resolved it seems more convenient to relate the productivity of growing organisms to the growth rate rather than to the variable concept of age.

Synthesis by non-growing organisms

Non-growing organisms are obtained by two basic means: (1) by withdrawal of an essential nutrient, that is, starvation; an example of this type is the so-called 'resting bacterial suspension' (Stephenson, 1950), (2) by inhibition of growth by either a chemical inhibitor such as chloramphenicol or a physical condition such as adverse temperature. It often seems to be implicitly assumed that when growth ceases the organism becomes a stable structure; however, the contrary is indicated by studies on the effects of starvation (Dawes & Ribbons, 1964). It would be interesting to compare the sequence of events following starvation with those which follow the cessation of growth caused by an antibiotic. Biochemical studies on the effects of growth inhibitors have been more concerned with the primary site of action rather than with the *sequence* of events following growth inhibition. Studies on starvation of the mould *Penicillium chrysogenum* by depriving it only of the energy source (Righelato, Trinci, Pirt & Peat, 1968) showed that structural change occurred in the cell immediately starvation commenced. The sequence of events was turn over of DNA, protein and RNA followed by formation of conidia. In microbial synthesis the problems are to determine how breakdown of the vegetative cell affects the desired synthetic activity and whether the decay of activity can be prevented.

The effects of cessation of growth on penicillin production by *Penicillium chrysogenum* were investigated by Pirt & Righelato (1967). In this study the mould was grown in a chemostat with glucose limitation of growth and the effect of different growth rates on metabolic activities was determined. Metabolic quotients of energy metabolism

(q_{O_2}, q_{CO_2}, $q_{glucose}$) were all linear functions of the growth rate of the form,

$$q_{O_2} = \frac{\mu}{Y_0} + m_0, \qquad (5)$$

where μ = specific growth rate, Y_0 is the growth yield (g. of biomass formed per g. of oxygen), m_0 is the maintenance coefficient for oxygen; analogous expressions apply to glucose utilization and to CO_2 production. In contrast, the $q_{pen.}$ (units penicillin produced/g. biomass ×

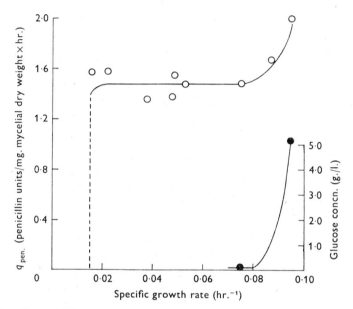

Fig. 3. Specific penicillin production rate ($q_{pen.}$) as a function of specific growth rate in a glucose-limited chemostat culture of *Penicillium chrysogenum*. Upper curve, $q_{pen.}$; lower curve, glucose concentration.

hr.) was found to be independent of the growth rate over a wide range (Fig. 3). Processes such as CO_2 production are sometimes termed 'growth associated' because they are linked to the growth rate, and, conversely, because it shows a certain independence of the growth rate, penicillin production is said to be 'non-growth associated'. This latter term is somewhat misleading because penicillin synthesis is not maintained when growth ceases.

The existence of the maintenance coefficient (m) implies that the uptake of glucose (as energy source) must exceed the maintenance ration (m × biomass) before growth can occur. When the penicillium was fed with the maintenance ration of glucose the dry weight of mould remained constant indefinitely (> 200 hr.); however, substantial numbers of conidia appeared at 45 hr. (Righelato *et al.* 1968). Although the maintenance ration effectively prevented the autolysis of the mould, it did not prevent decay of the penicillin synthetic system. Further it was observed that the rate of decay of penicillin synthetic activity in the 'maintained' organism was inversely related to the steady-state growth rate before growth was stopped. This effect of growth rate provides further evidence for the quantitative changes in cell mechanism associated with growth rate as predicted by the Dean & Hinshelwood model. The glucose supply rate necessary to prevent the decay of penicillin synthetic activity was $2 \cdot 5 \times$ maintenance ration, which corresponded to a growth rate of $0 \cdot 014$ hr.$^{-1}$. In summary this study showed that: (1) the rate of decay of an enzyme activity in the resting organism can be a function of the previous growth rate, (2) the prevention of decay of penicillin synthetic activity required a glucose supply of $2 \cdot 5$ m, (3) the optimal glucose supply for conidia formation was 1 to $2 \times m$, (4) the maintenance ration of glucose was required to prevent autolysis.

These studies on *Penicillium* show that the growth rate cannot be decreased indefinitely, but that at a finite value ($0 \cdot 014$ hr.$^{-1}$) the mould begins to break down and to set in motion the process of conidiation. Tempest, Herbert & Phipps (1967) studied the effect of decreasing growth rate in a bacterial culture and found that the minimum growth rate (allowing for the death of cells, which was considerable) was about $0 \cdot 009$ hr.$^{-1}$ with glycerol-limited growth; unlike the mould culture results, this corresponds to a glycerol feed rate considerably less than the maintenance ration.

Some kinetics of important culture systems

The kinetics of growth in various culture systems have been developed elsewhere by Herbert, Elsworth & Telling (1956), Herbert (1961), Powell & Lowe (1964) and Herbert (1964). Here the kinetics of some of the more important culture systems are considered from the point of view of product formation rather than growth.

Batch culture

In a batch culture in which a culture vessel is charged with culture medium and inoculated with the organism, the amount of product formed in an infinitely small time interval dt is given by

$$dp = q_p \, x \, dt,$$

where x = biomass concentration (g./l.); q_p = specific rate of product formation. If q_p is constant the product concentration at time t is given by

$$p = q_p \int_0^t x\,dt + p_0, \tag{6}$$

where p_0 = product concentration when $t = 0$. Thus q_p can be evaluated either by substituting a function of t for x and integrating or by evaluating the integral graphically. If growth is exponential one can substitute $x = x_0 e^{\mu t}$ and integrate. This method may be applied to compare specific penicillin production rates in batch cultures.

The chemostat and penicillin production

It has been shown that the consumption of energy source by bacteria and fungi in growing cultures may be represented as the sum of two terms; one is the consumption of energy source for growth and the other is the consumption of energy source for maintenance (Pirt, 1965; Righelato *et al.* 1968). Thus if s is the concentration of the energy source in the medium the substrate balance is given by

$$-ds = \frac{\mu x}{Y_G}\,dt + mx\,dt, \tag{7}$$

where Y_G is the 'true growth yield' and m is the 'maintenance coefficient'. If dp is the amount of end product of energy metabolism, $dp = -\alpha\,ds$ where α is a constant given by the stoichiometry of the energy-yielding reaction. Substitution for ds in equation (7) gives for the rate of product formation

$$\frac{dp}{dt} = \left(\frac{\mu}{Y_G} + m\right)\alpha x. \tag{8}$$

Hence the relation between the specific rate of product formation (q_p) and the growth rate is

$$q_p = \left(\frac{\mu}{Y_G} + m\right)\alpha. \tag{9}$$

If the process is carried out in a chemostat the product balance is given by

rate of product accumulation in culture
  = rate of product formation − rate of washout of product,

that is
$$\frac{dp}{dt} = \left(\frac{\mu}{Y_G} + m\right)\alpha x - Dp. \tag{10}$$

In the steady-state $dp/dt = 0$, so that

$$\bar{p} = \left(\frac{\mu}{Y_G} + m\right)\frac{\alpha}{D}\bar{x}. \tag{11}$$

(a bar over a symbol denotes that it is a steady-state value). In deriving equation (11) it was supposed that the energy source was used exclusively for energy production. When the same substrate is used as a source of both carbon and energy, Y_G must be based on the amount of substrate used in energy production (Pirt, 1967). As yet there has been little systematic effort to establish the kinetics of the formation of end products of energy metabolism in growing cultures although they include such important processes as the production of ethanol, lactic acid, acetone and butanol, acetic acid. It has apparently only just been recognized in brewing, as a result of the introduction of continuous-flow techniques, that alcoholic fermentation follows zero-order and not first-order kinetics (Shore & Royston, 1968). Luedeking & Piret (1959) studied the kinetics of the production of lactic acid by a lactobacillus and showed empirically that it followed relation (8). Holzberg *et al.* (1967) reported that the same kinetics did not apply to ethanol production by a yeast; however, the data reported by these workers does not seem adequate to afford a test of the theory.

Penicillin production is the only example of secondary metabolism for which a relation between growth rate and product formation has been definitely established for a wide range of growth rates (Fig. 3). Between growth rates varying from about 15 to 90 % of the maximum, the specific penicillin production rate ($q_{\text{pen.}}$) was constant. The penicillin production rate is given by

$$\frac{dp}{dt} = q_{\text{pen.}}\, x - Dp,$$

where D = dilution rate and p = penicillin concentration. In the steady-state when $dp/dt = 0$

$$\bar{p} = q_{\text{pen.}}\, x/D. \tag{12}$$

There was a critical growth rate below which the penicillin synthetic rate could not be maintained. At the other extreme of the growth rate, when the organism became saturated with glucose, penicillin production was stimulated, which suggests that under steady-state conditions with glucose-limited growth there is some repression of penicillin synthesis. Wright & Calam (1968) have provided further evidence for this view by their comparison of penicillin production in batch culture and in chemostat culture. This work shows that in batch culture penicillin production is in a transient state in which the $q_{pen.}$ rises to a maximum many times greater than the steady-state value in the chemostat. This result shows that the potential $q_{pen.}$ of the culture is several times greater than the steady-state value with glucose-limited growth. The problem of how to realize the full productivity in a continuous steady-state may be solved by a study of the regulation of the metabolic pathway leading to penicillin. The pathway represented in Fig. 4 and the

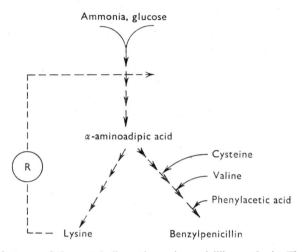

Fig. 4. Key features of the metabolic pathway in penicillin synthesis. The broken line labelled ® indicates the supposed site of feed-back inhibition or repression of enzyme synthesis.

evidence for it are discussed by Demain (1966). Penicillin and lysine compete for the common intermediate α-aminoadipic acid, and lysine inhibits production of penicillin probably by either feed-back inhibition or enzyme repression at a stage before aminoadipic acid. This being the mechanism of regulation, one may draw the conclusion that if one makes a lysine auxotroph of the mould in which lysine synthesis is blocked at a stage after aminoadipic acid, then supplying lysine at a low growth-limiting level could relieve the inhibition or repression and stimulate penicillin synthesis several-fold. With such a lysine auxotroph, the lysine could be the growth-limiting factor needed in a chemostat process to release the full potential for penicillin synthesis.

Chemostat with feed-back

In this variant of the chemostat, shown diagrammatically in Fig. 5a, some device such as centrifugation or sedimentation is used to separate a fraction of the effluent organisms and return them to the culture vessel as a concentrated suspension. The chemostat with feed-back possesses certain special features which may be advantageous for both organism production and formation of cell products. The following symbols are used for the parameters:

x_1(g./l.)   concentration of biomass in the culture vessel.
x_2(g./l.)   concentration of biomass in the effluent from the system.
s(g./l.)   concentration of growth-limiting substrate in culture vessel.
s_R(g./l.)   concentration of growth-limiting substrate in fresh medium.
F(l./hr.)   rate of addition of fresh medium.
V(l.)   culture volume.
D(hr.$^{-1}$) over-all dilution rate $= F/V$.
a fraction of flow out of culture vessel which is fed back.
b fraction of biomass leaving the culture vessel which is fed back.

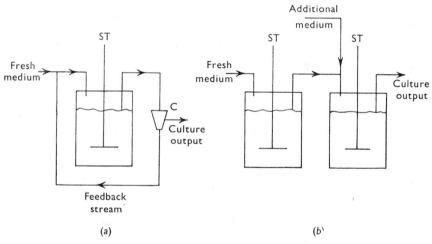

Fig. 5. Diagrammatic representations of (a) chemostat with feed-back, (b) chain of two chemostats. 'C' represents a separator for concentrating the organisms for feed-back; ST represents the stirrer to maintain homogeneity.

The outflow (F_s) from the culture vessel will be given by the sum of the inflow (F) and the amount of the outflow fed back (aF_s), that is $F_s = F + aF_s$; hence $F_s = F/(1-a)$. The concentration of organisms in the feed-back stream is given by $(b/a) x_1$. Considering unit volume the organism balance equation for the culture vessel is

$$\frac{dx_1}{dt} = \mu x_1 + \frac{b}{1-a} Dx_1 - \frac{Dx_1}{1-a}, \tag{13}$$

that is, net accumulation = growth + feed-back – output. Similarly, the substrate balance is given by

$$\frac{ds}{dt} = Ds_R + \frac{aDs}{1-a} - \frac{Ds}{1-a} - \frac{\mu x_1}{Y}. \tag{14}$$

Also in the steady-state

$$F\bar{x}_2 = \frac{F}{1-a}\bar{x}_1 - \frac{bF}{1-a}\bar{x}_1 \tag{15}$$

Solving equations (13), (14) and (15) for the steady-state values when $dx/dt = 0$ and $ds/dt = 0$ gives

$$\mu = \alpha D; \quad \bar{x}_1 = \frac{Y}{\alpha}(s_R - \bar{s}); \quad \bar{x}_2 = Y(s_R - \bar{s}); \quad \bar{s} = \frac{K_s \alpha D}{\mu_m - \alpha D}, \tag{16}$$

where $\alpha = (1-b)/(1-a)$. Herbert (1964) discussed the advantages of this system for organism production and showed that if $b/a > 1$ the rate of output of organisms can be increased above the maximum possible in the simple chemostat. The rate of product formation in the culture vessel is given by

$$\frac{dp}{dt} = q_p x_1 + \frac{aDp}{1-a} - \frac{Dp}{1-a}, \tag{17}$$

where it is assumed that the product is exocellular. In the steady state when $dp/dt = 0$, the product concentration both in the culture vessel and in the effluent is given by

$$\bar{p} = q_p \bar{x}_1/D. \tag{18}$$

It follows from relation (18) that if q_p (specific rate of product formation) is independent of growth rate the effect of introducing feed-back in the chemostat will be to increase the product concentration in proportion to the increase in organism concentration; the proportionality factor will be practically $(1-a)/(1-b)$. The product instead of being exocellular may remain attached to the organism as is the case with energy storage compounds like glycogen or the antibiotic griseofulvin. In this case also it can be shown that the product concentration in the effluent is given by relation (18), although the concentration in the culture vessel is higher owing to the concentration of organism in the feed-back stream. The advantages of feed-back, therefore, if the q_p is independent of growth rate, are that both output rate (Dp), and the product yield (p/s_R) are increased.

Chain of chemostats

The chain of chemostats (Fig. 5*b*) is of interest for product formation because it offers new possibilities for control over the process. The advantage is that in the second and subsequent stages stable steady-states can be obtained with growth rates from zero to the maximum, that is, with the concentration of growth-limiting substrate variable from the maintenance level to the saturation level. In contrast, in the single-stage chemostat the organism cannot have zero-growth rate, and stable steady-states are not practicable if the organism is saturated with substrate unless some form of 'turbidostat' control (Herbert, 1958) is used. Powell & Lowe (1964) have pointed out that a chain of several chemostats can be made to approximate closely to a 'plug flow' or 'pipe-flow' system which ideally occurs when the culture flows along a pipe without any mixing.

The two-stage system is required if the optimum conditions for growth are different from those for product formation. The growth kinetics of the two-stage system, which were developed by Herbert (1964), contain the assumption that the organism can adjust its metabolic rates immediately, without lag, on passing from the first to the second stage. This assumption can be valid only for relatively small changes in the environment. A further use of the two-stage system is that it may be used to exploit the residual productivity of the organism after growth has ceased.

Tests of the theory of the two-stage process applied to penicillin production were discussed by Pirt & Righelato (1967) and Pirt (1968). The conditions which give maximum penicillin concentration in the minimum time may be achieved with a first stage in which the growth rate is the minimum compatible with maintenance of the $q_{pen.}$. In the second stage the organism should be in the maintained state, that is, fed the maintenance ration of glucose. Under these circumstances in the second stage, on account of the decay of the $q_{pen.}$, the mould's contribution to the $q_{pen.}$ will be dependent on its residence time. In the ideal chemostat there is a characteristic distribution of the residence times of elements of the culture dependent on the fact that any element of the material in the vessel irrespective of its time of entry has an equal chance of leaving the vessel in a given time interval. Consequently if $q_{pen.}(t)$ is a function which represents the decay rate in the maintained state the penicillin concentration in the second stage is given by

$$p_2 = p_1 + x_2 \int_0^T q_{pen.}(t) e^{-D_2 t} \, dt, \tag{19}$$

where p_1 = concentration of penicillin in the ingoing medium, D_2 = dilution rate in the second stage, T = time for $q_{pen.}$ to fall to zero. The experimental data available for the output from the second stage agree well with the theoretical values predicted by equation (19) (Pirt & Righelato, 1967).

CONCLUSION

Modern microbial culture methods largely adhere to the traditional prescientific approach to fermentation processes, that is, an arbitrary set of nutritional and other conditions are set up initially and the culture is left to take its course. Improved control can be achieved by instrumental control over important medium conditions such as pH value and dissolved oxygen concentration and by use of the chemostat culture method.

The necessity for a purely empirical approach to development of processes for cell growth and product formation is being questioned by microbiologists. An alternative is to develop a mathematical model which represents the basic mechanism of the process. The work of Maxon & Chen (1966) typifies this approach. The main advantage of mathematical models for biological systems at the moment is that they stimulate thought about the fundamental control of the processes. The key to obtaining massive production of particular products must be a knowledge of metabolic regulation by end-product inhibition, enzyme repression and induction, and other means. Once the cell mechanism is understood it should be possible to interfere with the cell regulation so as to cause overproduction of the desired product. So far in industrial processes this has been achieved by accident rather than design. Biochemical studies have mostly focused on qualitative aspects of metabolic control rather than the fully quantitative approach essential to optimize production. The use of microbes for synthetic purposes calls for a modified concept of the organism. The wild-type of organism possesses a host of characters which probably are essential to the organism's survival in the natural habitat but are unnecessary in pure culture where the most favourable conditions for growth can be artificially contrived. The aim of microbial culture for efficient product formation should be to ensure that the autosynthetic processes of the organism are limited to reproduction of only that enzyme system (the 'minimal system') which performs the required function, such as conversion of sugar to ethanol. Such induced degeneration of the organism is the antithesis of the normal aim of the biologist, which is concerned with preserving as many as possible of the organism's characters.

REFERENCES

AIBA, S. & HARA, M. (1965). Studies on continuous fermentation. Part I. The concept of the mean cumulative age of microbes. *J. gen. appl. Microbiol.* **11**, 25.

BERAN, K., MALEK, I., STREIBLOVA, E. & LEIBLOVA, J. (1967). The distribution of the relative age of cells in yeast populations. In *Microbial Physiology and Continuous Culture*. Ed. E. O. Powell *et al.* London: H.M.S.O.

BU'LOCK, J. D. (1965). *The Biosynthesis of Natural Products*. London: McGraw-Hill.

CAMPBELL, A. (1957). Synchronization of cell division. *Bact. Rev.* **21**, 263.

DAWES, E. A. & RIBBONS, D. W. (1964). Some aspects of the endogenous metabolism of bacteria. *Bact. Rev.* **28**, 126.

DEAN, A. C. R. & HINSHELWOOD, C. (1966). *Growth Function and Regulation in Bacterial Cells*. Oxford: Clarendon Press.

DEAN, A. C. R. & HINSHELWOOD, C. (1967). Kinetics of cell growth in conditions of phage infection, substrate imbalance and synchronization. *Nature, Lond.* **214**, 1081.

DEMAIN, A. L. (1966). Industrial fermentations and their relation to regulatory mechanisms. *Adv. Appl. Microbiol.* **8**, 1.

HERBERT, D. (1958). Some principles of continuous culture. In *Recent Progress in Microbiology. VII Int. Congr. for Microbiology*, p. 381. Ed. G. Tunevall. Stockholm: Almqvist and Wiksell.

HERBERT, D. (1961). A theoretical analysis of continuous culture systems. In *Continuous Culture*, monograph no. 12, 21. London: Soc. Chem. Ind.

HERBERT, D. (1964). Multi-stage continuous culture. In *Continuous Cultivation of Microorganisms*. Ed. by I. Malek *et al.* Prague: Czechoslovak Academy of Sciences.

HERBERT, D., ELSWORTH, R. & TELLING, R. C. (1956). The continuous culture of bacteria: a theoretical and experimental study. *J. gen. Microbiol.* **14**, 601.

HILLS, G. M. & SPURR, E. D. (1952). The effect of temperature on the nutritional requirements of *Pasteurella pestis*. *J. gen. Microbiol.* **6**, 64.

HINSHELWOOD, C. N. (1952). On the chemical kinetics of autosynthetic systems. *J. Chem. Soc.* p. 745.

HOLZBERG, I., FINN, R. K. & STEINKRAUS, K. H. (1967). A kinetic study of the alcoholic fermentation of grape juice. *Biotech. & Bioeng.* **9**, 413.

ISHII, R., OTSUKA, S. & SHIIO, I. (1967). Microbial production of amino acids from hydrocarbons. II. Isolation of good hydrocarbon utilizers and amino acid production by their auxotrophs. *J. gen. appl. Microbiol.* **13**, 217.

LUEDEKING, R. & PIRET, E. L. (1959). Transient and steady states in continuous fermentation: theory and experiment. *J. Biochem. Microb. Tech. Eng.* **1**, 431.

MATELES, R. I., RYU, D. Y. & YASUDA, T. (1965). Measurement of unsteady state growth rates of micro-organisms. *Nature, Lond.* **208**, 263.

MAXON, W. D. & CHEN, J. W. (1966). Kinetics of fermentation product formation. *J. Ferm. Technol.* **44**, 255.

MONOD, J. (1942). *Recherches sur la croissance des cultures bactériennes*, 2nd ed. Paris: Hermann.

PIRT, S. J. (1964). Microbial synthesis in industry and its relation to microbial physiology. *Chem. & Ind.* p. 1772.

PIRT, S. J. (1965). The maintenance energy of bacteria in growing cultures. *Proc. Roy. Soc.* B, **163**, 224.

PIRT, S. J. (1967). Steady-state conditions for synthesis of exocellular products. In *Microbial Physiology and Continuous Culture*. Ed. E. O. Powell *et al.* London: H.M.S.O.

PIRT, S. J. (1968). Application of continuous-flow methods to research and development in the penicillin fermentation. *Chem. & Ind.* p. 601.

PIRT, S. J. & RIGHELATO, R. C. (1967). Effect of growth rate on the synthesis of penicillin by *Penicillium chrysogenum* in batch and chemostat cultures. *Appl. Microbiol.* **15**, 1284.

PIRT, S. J., THACKERAY, E. J. & HARRIS-SMITH, R. (1961). The influence of environment on antigen production by *Pasteurella pestis* studied by means of the continuous flow culture technique. *J. gen. Microbiol.* **25**, 119.

POWELL, O. & LOWE, J. R. (1964). Theory of multi-stage continuous cultures. In *Continuous Cultivation of Micro-organisms.* Ed. I. Malek *et al.* Prague: Czechoslovak Academy of Sciences.

RIGHELATO, R. C., TRINCI, A. P. J., PIRT, S. J. & PEAT, A. (1968). The influence of maintenance energy and growth rate on the metabolic activity, morphology and conidiation of *Penicillium chrysogenum.* *J. gen. Microbiol.* **50**, 399.

SHORE, D. T. & ROYSTON, M. G. (1968). Chemical engineering of the continuous brewing process. *Chem. Engineer*, CE 99.

STEPHENSON, M. (1950). *Bacterial Metabolism*, 3rd ed. London: Longmans Green.

TEMPEST, D. W. & HERBERT, D. (1965). Effect of dilution rate and growth-limiting substrate on the metabolic activity of *Torula utilis* cultures. *J. gen. Microbiol.* **41**, 143.

TEMPEST, D. W., HERBERT, D. & PHIPPS, P. J. (1967). Studies on the growth of *Aerobacter aerogenes* at low dilution rates in a chemostat. In *Microbial Physiology and Continuous Culture.* Ed. E. O. Powell *et al.* London: H.M.S.O.

WARE, G. C. (1951). Nutritional requirements of *Bacterium coli* at 44°. *J. gen. Microbiol.* **5**, 880.

WRIGHT, D. G. & CALAM, C. T. (1968). Importance of the introductory phase in penicillin production using continuous flow culture. *Chem. & Ind.* p. 1274.

GROWTH DYNAMICS AND SYNCHRONIZATION OF CELLS

B. C. GOODWIN

School of Biological Sciences, University of Sussex

CELLULAR CONTROL CIRCUITS

Molecular biology has provided us with some precise concepts about the control of macromolecular synthesis and activity in cells. These concepts derive largely from work done with bacteria, especially the models established for the control of gene activity. Whereas the mechanism of feed-back inhibition carries over unchanged from prokaryotes to eukaryotes, only the more formal aspects of the gene control model can be applied, at present, to both levels of cellular organization.

These molecular models of control and others derived from them form the foundation for much current thinking about the physiological behaviour of cells. The basic unit of behaviour is a control circuit, defined as a set of causally connected variables whose time variations are both correlated and stable; stable in the sense that for a given cellular environment the mean values of the variables will eventually become constant. For example, the arginine biosynthetic enzymes form part of the control circuit for the regulation of arginine production in *Escherichia coli* K 12. Other variables involved in this circuit are gene activities and messenger RNA (m-RNA) for the eight enzymes involved, arginine concentration, and, possibly, aporepressor level (if this varies with the state of the control circuit, and so varies in some correlated manner with the arginine concentration). Not included in the control circuit are general non-specific factors involved in macromolecular synthesis such as RNA polymerase, the activated bases for m-RNA synthesis, transfer RNA (t-RNA)—amino acid species for protein synthesis, etc. Under certain physiological conditions these non-specific factors could become rate-limiting for arginine production, but they do not correlate necessarily in time with variations in arginine level. They act as parameters of the control circuit, not as variables (Goodwin, 1963).

Another example of a control circuit in bacteria is that for the regulation of ribosomal RNA (r-RNA) synthesis. According to the results of Maaløe & Kjeldgaard (1966) and others, the relevant variables are the protein synthetic capacity of the cell, determined largely by ribosome concentration, t-RNA concentration, and the ribosomal RNA genes.

The negative feedback signal is believed to be the t-RNA concentration. The details of this regulatory process have not yet been elucidated, but its formal properties are clear and conform to the basic model for gene control in prokaryotes.

The dynamics of cellular control circuits

The dynamic behaviour of control circuits such as those described above has been studied in asynchronous bacterial cultures by observing responses of the arginine enzymes to variations in exogenous arginine concentrations, for example; or by shift-up, shift-down experiments in the case of the r-RNA control circuit. Such investigations have established the response characteristics of the circuits, averaged over time and over the growth cycle in the cells. Only recently has the study of the dynamic behaviour of these circuits during the cycle of cell growth and division begun. To carry out such studies it is of course necessary to use cultures synchronized with respect to cell division, or to use techniques which allow one to calculate single cell responses. A major objective in this work is the description and analysis of the dynamic behaviour of the control circuits during the cell cycle.

This cycle is by definition periodic. The set of processes leading to cell division is repeated by each daughter cell. The initiation of cell division is a periodic process in the sense that at some point in the cell cycle a state is reached which initiates division, and this same state is reached again in the daughter cells after a relatively constant time interval, given a constant environment. The variables I am referring to as defining cell states are concentrations of molecular species within cells, concentrations being assumed to be the correct variables for describing the dynamics of cellular control processes (Goodwin, 1969a).

It is evident that DNA replication must be initiated by some cell state which occurs once per cell doubling time in normal growth. It is known that DNA replication and cell division can be dissociated from one another. For example, Rörsch & van der Kamp (1961) showed that cells of *Escherichia coli* B induced to form filaments by irradiation continued to grow normally with respect to DNA and RNA synthesis, respiratory rate, etc. This shows an uncoupling of division from DNA replication.

Uncoupling between growth and DNA replication also occurs: a thymine-requiring strain of *Escherichia coli*, deprived of thymine, will continue to grow for 2 to 3 doublings of cell mass without replicating its DNA. This increase in cell mass occurs without cell division, so that growth and division are also uncoupled. When thymine is given to these cells, they synthesize DNA and divide to form viable cells (Donachie & Hobbs, 1967), and normal growth resumes.

These observations allow us to draw the obvious but important conclusion that the physiological state of a bacterium normally changes in a periodic manner, and that different processes are initiated at different times in the periodic cycle. Furthermore, different processes can be interrupted while others continue, showing that sets of variables can be dissociated or uncoupled from one another. Those variables which cannot be uncoupled without interrupting the operation of some process then constitute the variables of a control circuit. The set of processes leading from one DNA initiation to the next and specifically involved in this initiation form a causally connected set which constitutes a control circuit for the regulation of DNA replication. This has been called the replicon (Jacob, Ryter & Cuzin, 1966). A similar circuit must exist for cell division, involving cross-wall synthesis.

The use of the term control circuit in the sense thus defined is a shorthand way of referring to strongly coupled variables in cells with specific regulatory properties. The distinction between which variables are to be regarded as part of one circuit and which are not is dependent upon the strength of the coupling, and is thus somewhat arbitrary. However, the language is a convenient one in many respects.

The fact that these control circuits normally operate in a periodic dynamic mode means that at some level of its organization the bacterial cell generates an oscillation or a set of oscillations. Furthermore, since certain observable periodicities such as cell separation can be eliminated, while others such as DNA replication and septation continue, it is evident that some circuits can be shut off or modified in their operation, while others continue to operate normally. Another example is the pyrimidine control circuit involving aspartate transcarbamylase (ATCase). In synchronized cultures of *Bacillus subtilis* the synthesis of ATCase is periodic if pyrimidines are not present in the growth mechanism (Masters & Pardee, 1965), but this circuit can be shut off by the addition of uracil to the culture and the periodicity vanishes. The cessation of activity in this circuit in no way interferes with the growth and division cycle of the cell. Any dynamic element which the pyrimidine control circuit may have contributed to the over-all physiological organization of the cell cycle is evidently inessential to its continued operation.

Explanations of periodicities in cellular control circuits

This brings us to an observation which is somewhat puzzling at first sight. When the activities of a number of enzymes, such as ornithine transcarbamylase (OTCase), ATCase and histidase, were measured throughout the cell cycle in synchronized cultures of *Bacillus subtilis* or *Escherichia coli* (Kuempel, Masters & Pardee, 1965; Masters & Pardee, 1965), it was found that their synthesis was actually periodic, not constant, with a periodicity equal to that of cell doubling. The enzymes are made during a restricted and characteristic time interval in the cell cycle. If the bacteria are given arginine or uracil, the respective enzyme (OTCase, ATCase) is not made, and the periodic intracellular production of arginine or CTP, assumed to occur in relation to the periodicity of enzyme production, will stop. It should be emphasized that there has as yet been no direct proof that intracellular end-product pools such as arginine or CTP vary periodically in relation to OTCase or ATCase activity in synchronized cultures, although there is circumstantial evidence for this (Masters & Donachie, 1966).

It is thus evident that whatever dynamic contribution may be made to the organization of events during the cell cycle by the periodic synthesis of enzymes such as OTCase or ATCase, it is not an essential one. The growth and division cycle continues normally in the absence of these periodicities. Why, then, do they occur? There are a number of possible answers, varying in their sophistication, and possibly all partly correct.

It could be argued that it is in the nature of negative feed-back control circuits of the type being considered to show oscillatory behaviour, so that their dynamics are normally periodic (Goodwin, 1963). We must then explain why the period of the control circuit oscillation is the same as that of the cell growth cycle. A possible mechanism for this was given by Goodwin (1966), where it was shown that one of the consequences of the proposed explanation is a correlation between the linear order of genes on the bacterial chromosome and the time order of synthesis of corresponding enzymes during the cell cycle, as has been reported by Masters & Pardee (1965).

Another possible explanation follows the line of reasoning presented above, where it was argued that cellular physiology must involve periodicities in the growth cycle in order that specific events such as initiation of DNA replication and cell division should occur in a particular temporal order. It seems reasonable to propose that in a similar manner there is a phase in the cell cycle when RNA synthesis is maximal, and a phase when protein synthesis is maximal. Kogoma & Nishi (1965) have shown that proteolytic enzyme activity varies rhythmically during the cell cycle in *Escherichia coli* with the maximum coinciding with the period of cell division. Intracellular amino acid pools may be expected to increase in size during this period, with the consequence that biosynthetic enzymes such as those for arginine production would be repressed. After division, with reduced proteolytic activity, these enzymes would be derepressed. A periodicity would then be introduced into the operation of certain control circuits which was the result of a periodicity in another circuit, that regulating proteolytic enzyme activity. This latter periodicity might also be a forced oscillation, not an autonomous, self-generating one, arising from some other periodicity. At some point, however, there must be an autonomous oscillation, or more probably a set of autonomous oscillators, which provide the basic periodic dynamics around which the cell cycle is organized. The basis of this reasoning is the observation that in a system where there are multiple interactions, as there are between control circuits in growing and dividing cells, autonomous oscillations in a few circuits can drive the whole system into periodic behaviour. Under different growth conditions, different control circuits may be operating in autonomously periodic modes, and it may be difficult to sort out which are doing the driving and which are being driven. However, we can deduce that there is always some set of circuits with self-sustaining periodicities, in order to generate the time structure necessary for the organization of events in the growth and division cycle.

Periodicities in the control circuit operation may not only be very probable in a system organized around a basic cycle of processes as is the cell cycle, they may have adaptive significance in terms of the reliability of performance of the control circuits. It can be estimated that the numbers of m-RNA molecules of any particular type required for maintaining a protein population at a particular mean level in bacteria is, on average, 10 or less (Goodwin, 1963). Assuming only one copy of a particular operon in a bacterial cell (on average there will be more), a messenger transcription rate of 1 m-RNA synthesized in 1 min. and a specific messenger decay rate of 0.4 min.$^{-1}$ (that is a half life of about 2 min.), the net rate of m-RNA synthesis is given by

$$\frac{dx}{dt} = 1 - 0.4x, \tag{1}$$

where x = number of m-RNA molecules per cell and t = time (min.). The first term on the right-hand side in equation (1) represents the rate of synthesis of m-RNA, and the second term the rate of decay in molecules per minute. Solving (1) gives $x = 25 (1 - e^{0.4t})$. If a gene transcribes at the above rate for 20 min. and protein is made at the rate of one protein molecule every 5 sec. on m-RNA, then the number of protein molecules (y) produced as a result of the 20 min. of gene activity is

$$y = 12 \int_0^{20} 25 \, (1 - e^{0 \cdot 4t}) \, dt + 12 \int_{20}^{\infty} 25 \, e^{-0 \cdot 4t} \, dt. \qquad (2)$$

The first integral represents the protein produced during the period of gene activity; the second integral represents the protein produced afterwards (x at time, 20 min. is taken as 25). The value for y turns out to be about 3600. It has been estimated that the number of protein molecules in a bacterium is of the order of 10^6 (Guild, 1956) and with about 1000 different species of protein required by a bacterium growing in minimal medium (generation time about 60 min.), this gives about 1000 molecules of protein per species. Thus it is evident that ≮ two operons have the capacity to direct the synthesis of enough molecules of an average protein species for a new bacterium in 20 min. of full activity. If, on the other hand, a gene transcribes continuously throughout a 60 min. cell cycle, its transcription rate must be kept at about one-third of its full potential, thus making on average 1 m-RNA molecule every 3 min. in order to produce no more than the required number of protein molecules for a new cell. If alkaline phosphatase is at all typical of the control characteristics of feed-back repression, then the transition from basal rate of synthesis to full gene activity occurs over a small range of co-repressor level (Torriani, 1960). To hold repression at an intermediate level between basal and 100 % derepression requires accurate control of corepressor level. Given the inevitable 'noisiness' of control circuit operation where small numbers of molecules are involved and where there is a high sensitivity of repression to corepressor concentration, it is possible that greater accuracy of control is achieved by timing the relative duration of full-on to full-off periods than by trying to hold gene activity at an intermediate value. Such on-off frequency control would be observed as a periodicity of enzyme synthesis in a synchronized population of cells.

This argument suggests, then, that the inevitable periodicities of state underlying the cyclic process of cell growth and division have been exploited by bacteria, and possibly also by the eukaryotes, to achieve greater accuracy and reliability of gene control by means of a fairly discontinuous on-off periodicity in message transcription. Regulation of the relative durations of the on to off period provides continuous regulation of the amplitude and mean of the concentrations of enzymes, endproducts, and other cell variables. The dynamic behaviour of cellular control circuits would then become smoothly periodic rather than noisily constant, with evident advantages in respect of the reliable production of two similar daughter cells.

The model being advanced here for the cell cycle is thus a set of control circuits operating with oscillatory dynamics, periodic cell division being the normal consequence of their cooperative interaction. Potentially any circuit could be uncoupled from the rest and made to run either at a different frequency or to stop oscillating altogether, if the appropriate modifications in circuit operation can be made. I have carried out studies on the control circuit for β-galactosidase synthesis, involving the *lac* operon, in synchronized cultures of *Escherichia coli* B, which show how such uncoupling and frequency changes relative to the over-all growth and division cycle can be brought about by perturbations (Goodwin, 1969 b). These investigations showed that certain relations must hold between the frequency of cell division and the state of the *lac* control circuit for entrainment to occur so that there is an observable periodicity in the control circuit which matches that of the cell cycle. By changing either the condition of the circuit (by altering inducer or repressor levels, or both), or by changing the doubling time of the culture, control circuit oscillations which are entrained to the cell cycle can be made to appear or disappear. What these studies were not able to establish with any certainty was whether or not the *lac* control circuit could be in a state of undamped oscillation at a frequency different from that of cell division, although the evidence suggests this.

Support for this possibility is provided by the very interesting observations reported by Boddy, Clarke, Houldsworth & Lilly (1967) on sustained but irregular oscillations in amidase activity in chemostat cultures of *Pseudomonas aeruginosa*. The mean frequency of these oscillations was greater than that of cell division, and the authors interpreted the phenomenon as one arising from negative feed-back control of amidase biosynthesis by catabolite repression.

The results reported by Knorre (1968) on oscillations in the rate of synthesis of β-galactosidase in batch cultures of *Escherichia coli* on transfer from glucose to lactose as carbon source provide direct evidence for the occurrence of periodicities in the *lac* control circuit. The observed damping of these oscillations could be due either to dynamic damping in each cell, or to a loss of synchrony in control circuit operation throughout the cells of the culture after the step function perturbation on transfer of cells from glucose to lactose. The frequency of the oscillation was about the same as that observed in chemostat cultures of *Escherichia coli* B after a similar perturbation (Goodwin, 1969 b).

RESPONSE OF CELLS TO ENVIRONMENTAL
PERIODICITIES

If the cell is indeed an intrinsically rhythmic or oscillatory system with all or many of its operative constituents undergoing periodic variations during the growth and division cycle, then it should be possible to synchronize a population of cells by introducing any one of many different periodicities into the environment, providing, of course, the frequency of the environmental periodicity is close to that of the cell cycle so that they can become entrained. I have shown that it is possible to synchronize *Escherichia coli* B/1, a non-filamentous B strain, in a chemostat by introducing a periodicity of phosphate into the cell's environment, the phosphate periodicity being set equal to the doubling time of the culture as determined by the flow rate (Goodwin, 1968*a*). This demonstrates that, under the particular growth conditions used, there is a periodic variation in the rate of phosphate utilization in each cell during the growth cycle. If there were no such intracellular periodicity, there could be no synchronization of cell division: it requires an internal oscillation with a particular phase relationship to the cell division process to match up with an external oscillation and so result in division synchrony.

However, this observation does not prove that phosphate utilization rate is periodic in a cell growing in a constant environment. This may be the case, and it has been shown to be true for other organisms (e.g. *Chlorella*: Schmidt, 1966), but it cannot be concluded from the result reported above. It is likely that the periodicity in extra-cellular phosphate causes more oscillatory activity in the cell during the growth cycle than would normally occur. The 'oscillatory energy' of the environment will be transmitted to the cell, which resonates in response to it. A study of the way the culture responds to varying strength or intensity of the environmental periodicity should reveal something about the nature of the entrainment process causing synchrony.

This investigation was carried out by running the continuous culture at a fixed dilution rate, giving a generation time of 90 min., and varying the amount of phosphate added at each pulse. The culture medium was M9 buffered with Tris-HCl at pH 7·4 in place of phosphate, with 0·2 % glycerol and 0·02 % lactose as carbon source. The conditions of culture and methods of enzyme assay and cell counting were essentially the same as those described previously (Goodwin 1969*a*, *b*) with the exception that the size of the phosphate pulse added once per generation time was varied.

The results of these experiments are shown in Fig. 1. Four variables were measured: turbidity (OD_{540}), cell number, alkaline phosphatase (APase), and β-galactosidase (β-Gase), enzyme values being expressed as specific activities. Of these variables, APase ought to be most sensitive to the environmental phosphate periodicity. Figure 1 a shows the results of an experiment in which the phosphate delivered during the pulse, marked by the arrow on the abscissa, produced an initial concentration of about $5\cdot2 \times 10^{-4}$M KH_2PO_4 in the culture. Some 40–50 min. after the pulse, APase synthesis recommenced, so the phosphate concentration had evidently dropped below 10^{-5}M, the concentration estimated by Torriani (1960) to cause derepression of the APase gene in *Escherichia coli*. The minimum phosphate (P) concentration in the cells during the cycle is not known, but it is clear from the results that none of the other three variables is significantly affected by this range of P variation. As the amount of P added during the pulse is progressively reduced, making the maximum KH_2PO_4 concentrations 4·5 (Fig. 1 b), 3·6 (Fig. 1 c), 2·7 (Fig. 1 d), and 1·8 (Fig. 1 e) $\times 10^{-4}$M, β-galactosidase, turbidity, and cell number all begin to show cyclic variations in response to the environmental P periodicity.

As P limitation becomes more severe, two complementary processes presumably occur which ultimately give rise to the highly synchronized condition of Fig. 1 e, where cell division has become strongly entrained to the P periodicity. The first of these is a progressive alignment of cellular periodicities with the environmental periodicity. For example, if there are phases of more active protein and nucleic acid synthesis during the cell cycle, they will tend to become aligned with the period of relatively higher phosphate concentration, while periods of more active degradation, such as revealed by the observations of Kogoma and Nishi (1965), will tend to become phased with the period of phosphate starvation. At the same time, the environmental periodicity will tend to exaggerate internal cellular periodicities and to force oscillations which do not normally occur at all. The periodicity in APase activity is itself one of these.

The periodicity in β-galactosidase is undoubtedly due in part to a cyclic variation in the size of the intracellular pool of catabolite repressor (Magasanik, 1961), since catabolite repression is known to increase in certain strains of bacteria during phosphate starvation (McFall & Magasanik, 1962). However, previous studies (Goodwin, 1969 b) have shown that this is only a partial explanation of the dynamic behaviour of β-Gase in cells synchronized by this method. Under certain conditions of lactose induction, there is no observable oscillation in β-Gase at a

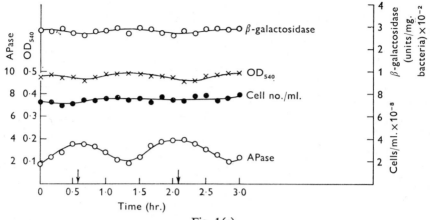

Fig. 1(a).

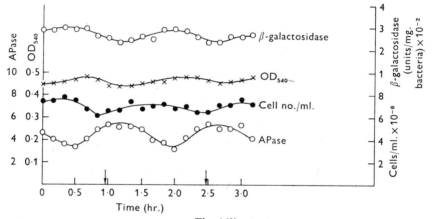

Fig. 1(b).

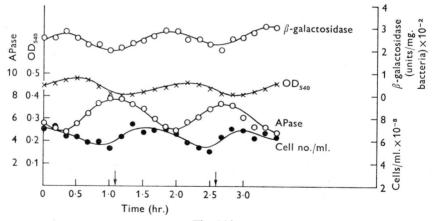

Fig. 1(c).

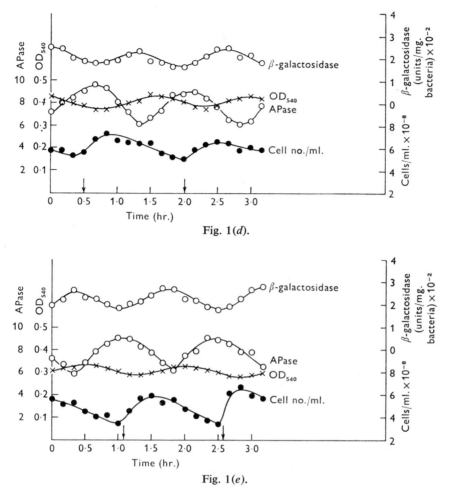

Fig. 1(d).

Fig. 1(e).

Fig. 1, a–e. Variations in turbidity (OD 540), cell number per ml., and specific activities of β-galactosidase and alkaline phosphatase (APase, arbitrary units) in continuous cultures of *Escherichia coli* B growing on M9 Tris minimal salts, pH 7·4, with 0·2 % glycerol and 0·02 % lactose as carbon source. Phosphate (KH_2PO_4) added at arrows on abscissae to give initial concentrations of about $5·2 \times 10^{-4}$ M (a), $4·5 \times 10^{-4}$ M (b), $3·6 \times 10^{-4}$ M (c), $2·7 \times 10^{-4}$ M (d), and $1·8 \times 10^{-4}$ M (e).

frequenc y equal to that of cell division in the culture. The results reported in Fig. 1 were obtained with 0·02 % lactose in the culture medium, a concentration known to give an oscillation in β-Gase which is stably entrained to the cell cycle. Under these conditions there is a characteristic phase relation between the β-Gase oscillation and cell division, the oscillation rising after the P-pulse and falling during the period of P star vation. What the experimental results of Fig. 1 a–e show is that the *lac* control circuit does not respond to the environmental P cycle until it is strong enough to begin to affect over-all cell growth, as measured by turbidity (Fig. 1*b*). The amplitude of the *β*-Gase periodicity then increases progressively as the degree of P-limitation is increased, the amplitude increase correlating with greater degrees of cell division synchrony.

It would thus appear that division synchrony arises in this procedure from the two complementary processes described above: the entrainment of intracellular periodicities to the environmental P periodicity, different intracellular oscillations having different phase relations to the P cycle and to cell division; and the generation of exaggerated and possibly abnormal cellular periodicities which contribute to organizing the cell growth and division cycle into a temporal pattern which is determined by the nature of the environmental periodicity. If a macronutrient other than P, such as K^+ or Mg^{2+}, were used as the growth-limiting factor and supplied periodically to the culture in the same manner as P, it seems likely that cell division synchrony would again result, but that the particular temporal patterns of intracellular events during the cell cycle would be different.

It is of interest to observe that there is no evidence in the results presented in Fig. 1 for any threshold in the response of the cell culture to increasing intensity of the environmental P periodicity. All variables respond more or less continuously, although there are clearly different ranges of sensitivity for the different cell variables observed. This suggests that the degree of cell division synchrony achieved by this means depends simply upon the degree to which the dynamic behaviour of control circuits gets temporally ordered in relation to the P periodicity, and how many control circuits are so ordered. As the environmental periodicity increases in intensity, its 'oscillatory energy' gets transmitted to more and more control circuits, which then resonate at a frequency determined by that of the environmental periodicity and by their own dynamic characteristics. When the environmental frequency matches closely that of cell division, as in the experiments described above, synchrony will gradually emerge as the intensity of the periodic stimulus increases. There does not appear to be any threshold point at which the cells suddenly become locked to the environmental cycle.

It is not possible, of course, to extrapolate directly from the behaviour of a mass culture to that of single cells. Single bacteria could respond with a sharp phase transition from an un-entrained to an entrained state. If this were the case, then the smooth, continuous response of the mass culture could be due to a fairly continuous range of threshold response points for entrainment in individual cells; i.e. the cell population would have to be quite heterogeneous with respect to its response characteristics. From the results obtained, it is not possible to discount such an interpretation.

CONCLUSIONS

The model presented in this paper for the dynamic organization of growing and dividing bacteria carries over in its formal properties also to eukaryotic organization. A similar functional partitioning of cells into partially autonomous control circuits, revealed experimentally by dissociation techniques such as heat shock, change of medium, shift from aerobic to anaerobic conditions, allows one to reconstruct the cell cycle in terms of the dynamic behaviour of cooperatively interacting control circuits. The molecular nature of these circuits in the eukaryotes is not nearly so well understood as are those of the prokaryotes, but the underlying dynamic organization of the system cannot be very different, since the over-all process is subject to the same basic considerations. The cell must have a time structure in order that specific molecular species, such as DNA, mitotic spindle protein, centriole constituents, etc., be synthesized at specific times in relation to one another and to the observable events of mitosis and cytokinesis. In order to carry out an ordered, repeatable process such as growth and division reliably, cellular activities must be organized dynamically in time. Since the particular process I am considering is itself periodic, the underlying dynamics must be periodic. The detailed molecular nature of the control circuits in eukaryotic cells and the particular temporal strategy involved in the organization of the processes constituting the cell cycle in these higher forms is a subject for current and future research.

ACKNOWLEDGEMENTS

The experimental work reported in this paper was largely carried out by Miss Jacqueline Hall, whose excellent technical assistance I would like to acknowledge. The research was supported by a grant gratefully received from the Medical Research Council.

REFERENCES

BODDY, A., CLARKE, P. H., HOULDSWORTH, M. A. & LILLY, M. D. (1967). Regulation of amidase synthesis by *Pseudomonas aeruginosa* 8602 in continuous culture. *J. gen. Microbiol.* **48**, 137.

DONACHIE, W. D. & HOBBS, D. G. (1967). Recovery from 'thymineless death' in *Escherichia coli* 15T⁻. *Biochem. biophys. Res. Commun.* **29**, 172.

GOODWIN, B. C. (1963). *Temporal Organization in Cells*. London: Academic Press.

GOODWIN, B. C. (1966). An entrainment model for timed enzyme synthesis in bacteria. *Nature, Lond.* **209**, 479.

GOODWIN, B. C. (1969*a*). Synchronization of *E. coli* B in chemostat by periodic phosphate feeding. *Europ. J. Biochem.* (in the press).

GOODWIN, B. C. (1969*b*). Control dynamics of β-galactosidase in relation to the bacterial cell cycle. *Europ. J. Biochem.* (in the press).

GUILD, W. R. (1956). *J. cell. comp. Physiol.* **47** (Suppl. 1), Discussion, p. 71.

JACOB, F., RYTER, A. & CUZIN, F. (1966). On the association between DNA and membrane in bacteria. *Proc. R. Soc.* B, **164**, 267.

KNORRE, W. A. (1968). Oscillations of the rate of synthesis of β-galactosidase in Escherichia coli ML30 and ML308. *Biochem. biophys. Res. Commun.* **31**, 812.

KOGOMA, T. & NISHI, A. (1965). Rhythmic variations in proteolytic activity during the cell cycle of *Escherichia coli*. *J. gen. appl. Microbiol.* **11**, 321.

KUEMPEL, P. L., MASTERS, M. & PARDEE, A. B. (1965). Bursts of enzyme synthesis in bacterial duplication cycle. *Biochem. biophys. Res. Commun.* **18**, 858.

MAALØE, O. & KJELDGAARD, N. O. (1966). *Control of Macromolecular Synthesis*. New York: W. A. Benjamin.

MAGASANIK, B. (1961). Catabolite repression. *Cold Spring Harb. Symp. quant. Biol.* **24**, 249.

MASTERS, M. & PARDEE, A. B. (1965). Sequence of enzyme synthesis and gene replication during the cell cycle of *Bacillus subtilis*. *Proc. natn. Acad. Sci. U.S.A.* **54**, 64.

MASTERS, M. & DONACHIE, W. D. (1966). Repression and the control of cyclic enzyme synthesis. *Nature, Lond.* **209**, 476.

MCFALL, E. & MAGASANIK, B. (1962). Effects of thymine and of phosphate deprivation on enzyme synthesis in *Escherichia coli*. *Biochim. biophys. Acta*, **55**, 900.

RÖRSCH, A. & VAN DER KAMP, CORNELIA (1961). The effect of ^{32}P decay on radiation-sensitive mutants of *Escherichia coli*. *Biochim. biophys. Acta*, **46**, 401.

SCHMIDT, R. P. (1966). Intracellular control of enzyme synthesis and activity during synchronous growth of *Chlorella pyrenoidosa*. In *Cell Synchrony*, p. 189. Ed. I. L. Cameron. New York: Academic Press.

TORRIANI, A. (1960). Influence of inorganic phosphate in the formation of phosphatases in *Escherichia coli*. *Biochim. biophys. Acta*, **38**, 460.

GROWTH AND DIVISION OF INDIVIDUAL BACTERIA

A. G. MARR, P. R. PAINTER, AND E. H. NILSON

*Department of Bacteriology, University of California,
Davis, California 95616, U.S.A.*

INTRODUCTION

The growth of a microbial population is obviously a summation of the growth and division of the individual micro-organisms which comprise that population; thus, a knowledge of the kinetics of growth and division of individual micro-organisms is fundamental to understanding the behaviour of populations.

Because of their small size and the attendant difficulties in direct measurement, much less is known about the growth of individual bacteria than about growth of bacterial populations. The ease of measuring the growth of bacterial populations by turbidimetry or by counting has resulted in a large body of significant physiological information and is reflected in many of the contributions to this Symposium. The results of such measurements lend themselves to analysis by analogy with chemical kinetics (Monod, 1942; Hinshelwood, 1946). Although such analysis is useful in predicting the behaviour of populations in balanced exponential growth, it cannot take into account the underlying processes of the growth and division of the individual bacteria.

The basic features of the growth of individual micro-organisms are revealed by observing the growth of a bacterium in micro-culture (Ward, 1895; Kelly & Rahn, 1932; Errington, Powell & Thompson, 1965; Schaechter, Williamson, Hood & Koch, 1962). The size of the bacterium will increase with time, and at some point in time it will divide and will be replaced by two bacteria (sisters). At any given age (interval of time from time of birth to time of observation) the sisters, in general, will have different sizes. Ultimately, both sisters will divide; but, in general, the ages of sisters at division (interdivision times) will be different. In this paper we will ignore the rare event of loss of viability; i.e. the failure of a cell to divide before reaching an arbitrarily old age.

One major difficulty encountered in the course of microscopic measurement of interdivision times and growth rates is the design of a culture chamber in which bacteria can be observed while they grow for several generations in a constant environment. The conventional experiment is to observe bacteria growing on moist agar blocks. Somewhat better control of the environment is afforded by a chamber in which the bacteria under observation are partitioned from liquid medium by a cellophane membrane, through which metabolites diffuse (Powell, 1956). Another difficulty is the small number of bacteria which may be observed at one time. Microscopic measurement of bacteria growing in microculture is less satisfactory as a basis for determining the kinetics of growth than for measurement of interdivision times. The random error of measurement by microscopy obscures the detailed kinetics of growth, and pooling of measurements on different bacteria is difficult because of the variability in birth size.

Both the distribution of interdivision times and the kinetics of growth of individual bacteria may be estimated accurately on large samples, avoiding both the difficulty of controlling the environment in microcultures and the large error of measurement by optical microscopy. Such estimates depend upon two experimental techniques which, together with appropriate mathematical analysis (Painter & Marr, 1968), permit accurate estimates of the growth rate and distribution of interdivision times of individual bacteria with a speed and convenience approaching that of the measurement of the growth of populations. One technique is that of selection with a minimum of physiological disturbance of a population consisting of newly formed bacteria. This technique, developed by Helmstetter & Cummings (1963, 1964), is based on the nonselective adhesion of cells of *Escherichia coli* strain B/r to a membrane filter and the selective elution of newly formed cells resulting from the division of an attached cell to form one attached and one free daughter cell.

The second technique is that of high-speed electronic counting and measurement of volume of bacteria based on the conductometric principle developed by Coulter in conjunction with a pulse-shaping amplifier and an automatic pulse height analyser (Harvey & Marr, 1966). Our equipment permits a measurement in less than 1 min. of the volumes of 10^5 bacteria with a precision of $\pm 1\%$ and with an accuracy superior to electron microscopy. Accurate measurements require that the culture medium be passed through a filter with 0·22 μ average pore diameter shortly before use. Also, the measurements must be corrected for the contribution of both electronic and particulate noise.

We shall be concerned with large populations, all members of which have almost identical, invariant, local environments. For *Escherichia coli* this state is attained by cultivation in well-stirred liquid medium with the density maintained at less than 10^6 bacteria per ml. by periodic transfer to fresh medium. After a sufficient period of growth the distribution of any intensive random variable (size, age, or life-length) is independent of the time at which the sample is chosen. We designate this condition as a steady-state even though the number of cells in the population obviously is not constant. Constancy of the distribution of intensive random variables implies exponential increase at the same specific rate of all extensive variables of the population (Painter & Marr, 1968). Thus, exponential increase in number is a necessary but not a sufficient condition to establish that a population is in a steady-state. The most convenient confirmation of a steady-state is constancy of the distribution of size of extant cells.

If from a population in the steady-state one selects a sample of newly formed cells without altering the local environment, the new population will not be in a steady-state. Over a considerable period of time the number will remain essentially constant and the mean size will increase rapidly. The degree of synchrony in the first generation and the decay of synchrony in subsequent generations reflects the kinetics of growth and the regulation of division of the individual bacteria in that population. We shall focus attention on three models for the control of cell division by comparing experimental results for synchronous growth of *Escherichia coli* with numerical predictions based on each of the models.

KINETICS OF GROWTH OF BACTERIA

Estimation of the kinetics of growth of individual bacteria depends upon measurements (or knowledge) of the distribution of volumes of cells in three types of random samples drawn from a population in the steady-state of exponential growth. The samples are extant cells, dividing cells, and newly formed cells, and the frequency functions of the distributions are termed $\lambda(x)$, $\phi(x)$, and $\psi(x)$, respectively. It is easily shown (Painter, & Marr, 1968; Harvey, Marr & Painter, 1967) that the average rate of growth, $V(x)$, of cells of given size, x, is given by the equation of Collins & Richmond (1962),

$$V(x) = \frac{k}{\lambda(x)} \int_0^x [2\psi(\theta) - \phi(\theta) - \lambda(\theta)]\, d\theta \tag{1}$$

for a population in a steady-state with a constant specific growth rate, k.[1]

---

[1] The symbol θ is a dummy variable of integration. The dummy variable with which the integration is performed takes on the physical implications of the limits of that integration.

Obviously, the functions $\psi(x)$ and $\phi(x)$ are not independent. For binary fission of a mother into daughters of equal size

$$\psi(x) = \phi 2(x), \tag{2}$$

and, in general, $$\psi(x) = \int_x^\infty \phi(\theta) \, K^*(\theta, x) \, d\theta,$$

in which $K^*(m, x)$ is the frequency function for the conditional distribution of size of daughter, x, produced by division of a mother of size m. The function $K^*(m, x)$ is difficult to estimate from existing experimental data, but the frequency function $K(x/m)$ of the distribution of the ratio of size of daughter (x) to size of mother (m) has been estimated (Marr, Harvey & Trentini, 1966; Harvey et al. 1967). If $K(x/m)$ is independent of m, we have

$$\psi(x) = \int_0^x [\phi(\theta)/\theta] \, K(x/\theta) \, d\theta, \tag{3}$$

which was deduced by Powell (1964). For cells of *Escherichia coli* growing in a glucose minimal medium the coefficient of variation of $K(x/m)$ is quite small, only 0·04 (Marr et al. 1966); thus, the simple relationship in equation (2) is an adequate approximation.

In order to compute from equation (1) the average growth rate of each size class it is necessary to estimate k, $\lambda(x)$, and either $\phi(x)$ or $\psi(x)$. Figure 1 shows the estimation of $\lambda(x)$ and $\psi(x)$ from experimental data. The values of $\lambda(x)$ were estimated from conductometric measurement of the volumes of the cells in a sample of a culture of *E. coli* strain B/r/1 growing exponentially ($k = 0·554$ hr.$^{-1}$) at 30°C. in a glucose minimal medium prepared according to the recipe of Helmstetter & Cummings (1964). The values of $\psi(x)$ were estimated from measurement of the cell volumes in a sample of the effluent stream from a culture growing attached to a membrane filter and percolated with glucose minimal medium at 30°C. for 2 hr.

The values of $\phi(x)$ in Fig. 1 were computed from $\psi(x)$ using equation (2). The function $\phi(x)$ may be estimated more directly from measurement of the distribution of volumes of bacteria before and shortly after a shift to a medium lacking a carbon source or before and immediately after subjecting the culture to mild shear by ejection of a sample from a hypodermic syringe (K. F. Hedden, Ph.D. Thesis, University of California, Davis, California, U.S.A.). These estimates of $\phi(x)$, the mean of which should be biased low, are in good agreement with the values of $\phi(x)$, obtained as in Fig. 1, the mean of which should be biased high.

The average growth rates, $V(x)$, were computed with a digital com-

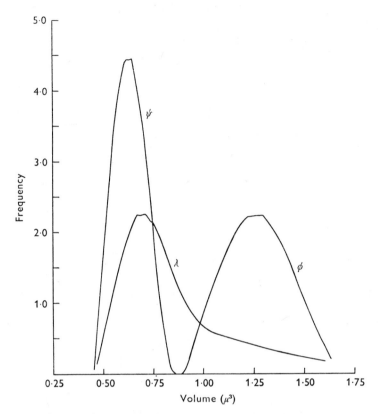

Fig. 1. Frequency functions of the distribution of volume of newly formed cells (ψ), extant cells (λ), and dividing cells (ϕ) of *Escherichia coli* strain B/r/1 growing in glucose minimal medium at 30°C. The function $\lambda(x)$ was estimated from conductometric measurement of the distribution of volumes of bacteria in an untreated sample drawn from a population in the steady state of exponential growth (specific growth rate, $k = 0.554$ hr.$^{-1}$). The function $\psi(x)$ was estimated from conductometric measurement of the distribution of volumes of bacteria in an untreated sample of the effluent stream from a membrane culture 120 minutes after loading the membrane (Millipore GS, 0.22 μ pore size, 142 mm. diameter with an effective surface of 111 cm.²) with 1 l. of culture containing 5×10^6 bacteria/ml. and irrigating with fresh medium at 10 ml./min. Both $\lambda(x)$ and $\psi(x)$ were corrected for noise by subtracting the function, $A \exp \{B(0.36\text{-}x)\}$, in which x is the volume in μ^3 and $B = 6.4175$ μ^{-3}; A was chosen to reduce the corrected function to zero at 0.36 μ^3. In addition, the function $\psi(x)$ was trimmed by subtracting from it the function, $C \exp \{-[(0.85\text{-}x)/0.085]^2/2\}$, with C chosen to reduce the trimmed function $\psi(x)$ to zero at 0.85 μ^3. The function $\phi(x)$ was computed from equation (2).

puter according to equation (1) and are presented in Fig. 2. Smaller cells grow at a low rate; as the cells increase in volume, the growth rate increases to a maximum and then declines. The results in Fig. 2 are in general agreement with the growth rates estimated for *Escherichia coli* strain ML 30 (Harvey *et al.* 1967).

Analysis of the steady-state does not establish whether all cells in the population of given size grow at the same rate or whether the growth rate of those cells are distributed in some manner. Size and age of individuals from a given population are strongly correlated, but the fact of variability in size at birth implies that the ages of individual cells of

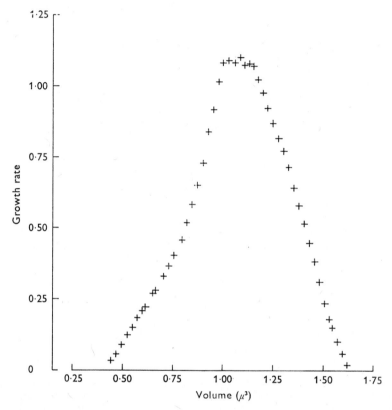

Fig. 2. Growth rate as a function of volume of *Escherichia coli* strain B/r/1 in glucose minimal medium at 30° C. The growth rates in μ^3/hr. were computed from the data in Fig. 1 according to equation (1).

given size must be distributed. If age rather than size or both age and size govern the rate of growth, the growth rates of cells of given size will be distributed. This possibility was tested by determining the growth rate as a function of size for individual cells from a synchronous culture, in which, before the first division, all cells have (nearly) the same age.

The equation of Collins & Richmond is applicable only to a population in a steady-state with respect to the distribution of age or size; the equation does not apply to a synchronous culture. A more general relationship has been developed (Eakman, Fredrickson & Tsuchiya, 1966; Painter & Marr, 1968) which is applicable to a synchronous culture growing in a constant environment. This may be stated by the differential equation

$$\frac{d}{dt} \log \lambda(x,t) = \mu(t) \{[2\psi(x,t) - \phi(x,t)]/\lambda(x,t) - 1\} - \frac{\partial}{\partial x} V(x,t), \quad (4)$$

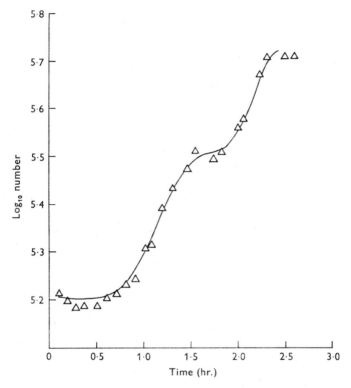

Fig. 3. Synchronous growth of *Escherichia coli* strain B/r/1 in glucose minimal medium. The effluent from the membrane culture described in the legend of Fig. 1 was collected for 3 min. and incubated at 30°C.

in which λ, ϕ, and ψ and the instantaneous specific growth rate, $\mu(t)$, are functions of time. If one assumes that the functions $\phi(x, t)$ and $\psi(x, t)$ do not differ appreciably from the steady-state functions $\phi(x)$ and $\psi(x)$, one obtains the approximation

$$V(x,t) \approx \frac{2\int_0^x \{\lambda(\theta,t) - [1 + \mu(t)\Delta t]\,\lambda(\theta, t + \Delta t) + \mu(t)\Delta t\,[2\psi(\theta) - \phi(\theta)]\}\,d\theta}{\lambda(x,t) + [1 + \mu(t)\Delta t]\,\lambda(x, t + \Delta t)},$$

$$(5)$$

which permits calculation of the growth rate, $V(x, t)$, for cells of size x in a synchronous culture at time t.

A synchronous culture of *Escherichia coli* strain B/r/1 was made by incubating the effluent from a culture growing on a membrane filter. The synchronous culture was sampled periodically, and the number and distribution of volumes of the cells were measured. Figure 3 shows the number of cells as a function of time.

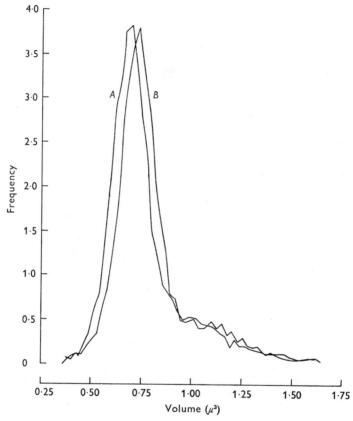

Fig. 4. Distribution of volumes of the cells of *Escherichia coli* strain B/r/1 in samples drawn from the synchronous culture depicted in Fig. 3. A, 16·4 min.; B, 21·5 min.

From pairs of samples taken a few minutes apart during the first generations, values were obtained for $\lambda(x, t)$ and $\lambda(x, t + \triangle t)$ such as those shown in Fig. 4. From this and similar pairs of samples we computed with a digital computer the growth rates, $V(x, t)$, according to equation (5). The results of these calculations, presented in Fig. 5, show that size has a major influence and age only a minor influence on the rate of growth. The growth rates of cells of all ages vary with size in the same manner as in the steady-state (solid line), but the maximum growth rate is somewhat higher for older cells. To a good approximation the rate of growth is a function of size only.

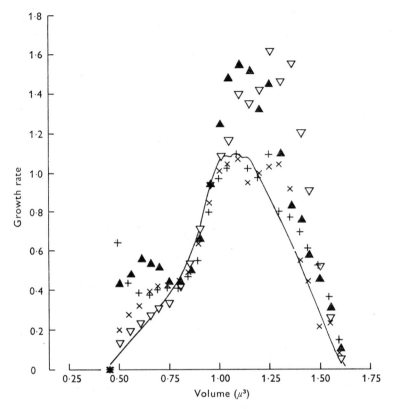

Fig. 5. Growth rate as a function of volume for cells of various ages: x, 11·5–21·5 min.; $+$, 36·1–42·3 min.; ▲, 42·3–48·2 min.; ▽, 137·5–149·6 min. The growth rates (μ^3/hr.) were computed according to equation (5) from pairs values of $\lambda(x,t)$ and $\lambda(x,t+\triangle t)$, such as the pair shown in Fig. 4 and the values of $\psi(x)$ and $\phi(x)$ shown in Fig. 1. The instantaneous specific growth rate, $\mu(t)$, was estimated from the data in Fig. 3 by the approximation $\mu(t) \approx [\log N(t+\triangle t) - \log N(t)] / \triangle t$.

DISTRIBUTION OF INTERDIVISION TIMES

The interdivision time of a bacterium is the interval of time from birth (by the division of its mother) to its division. The distribution of interdivision times is fundamental to the statistics of age of bacterial populations. For example, the distribution of ages of bacteria in a population in the steady-state of exponential growth is uniquely determined by the distribution of interdivision times (Lotka, 1913, 1922). The distribution of interdivision times has been estimated from observations made of bacteria growing in microculture (Ward, 1895; Kelly & Rahn, 1932; Errington *et al.* 1965; Schaechter *et al.* 1962), but the accuracy of such estimation is limited both by the difficulty of obtaining a population in a steady-state in microculture and by the small size of the sample in a given experiment and the bias which results from pooling of results from differential experiments.

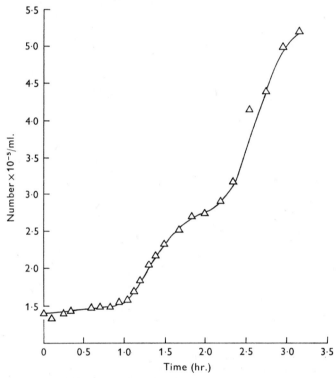

Fig. 6. Synchronous growth of *Escherichia coli* strain B/r/1 in glucose minimal medium at 30° C. Conditions of this experiment were the same as for the experiment depicted in Fig. 3.

In general, the distribution of interdivision times estimated from any set of data will depend on the choice of sample. Most investigators have chosen a sample consisting of all cells formed during a specified interval of time. This is equivalent to a sample of cells newly formed by division during an arbitrarily short interval of time. The resulting distribution of interdivision times has been called the τ-distribution and is denoted by $F(\tau)$. The derivative of $F(\tau)$, denoted by $f(\tau)$, is the frequency function of interdivision times.

The bacteria in the effluent stream from a culture growing on a membrane is a sample to which the τ-distribution applies, and an analysis of the subsequent synchronous growth of such a sample gives a reliable estimate of distribution of interdivision times. The synchronous growth of a sample of newly formed cells of *Escherichia coli* B/r/1 collected

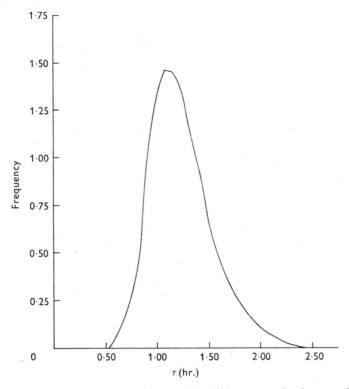

Fig. 7. Frequency function of interdivision time, f(τ), computed using equation (6) and the data in Fig. 6. The parameters of f(τ) are as follows: mean $= 1\cdot234$ hr.; mode $= 1\cdot12$ hr.; standard deviation $= 0\cdot295$ hr.; coefficient of variation $= 0\cdot239$; $g_1 = 0\cdot671$; and $g_2 = 0\cdot434$. The parameters g_1 and g_2 are measures of skewness and kurtosis, respectively (Fisher, 1946).

in the effluent from a membrane culture is shown in Fig. 6. The slope of the curve during the first cycle of division is evidently related to the frequency of interdivision times. A precise formulation has been given by Harris (1959), who has shown that, if the interdivision times of mothers and their daughters are independent, then the growth rate of a synchronous culture is given by

$$N'(t) = N_0 f(t) + 2 \int_0^t N'(t-\theta) f(\theta) \, d\theta. \tag{6}$$

Here N_0 is the number of newly formed bacteria collected at zero time. The integral represents the part of the division rate contributed by cells formed by divisions after zero time. If interdivision times are not independent, the last term of equation (6) is an approximation, the error of which becomes significant only for long inter-division times.

The values of $f(\tau)$ calculated by use of equation (6) from the data presented in Fig. 6 are shown in Fig. 7.

HYPOTHESES OF RAHN AND KENDALL
Independence of interdivision times

There have been several attempts to explain the characteristic features of the τ-distribution in mechanistic terms. Both Rahn (1932) and Kendall (1948) proposed stochastic models in which the completion of a fixed number of events (perhaps the duplication of genes) triggers division. Rahn assumed that these events can occur in any order, and Kendall assumed that they must occur in sequence. Rahn's assumptions lead to a prediction that $f(\tau)$ is Yule's function, and Kendall's lead to a prediction of a Pearson type III function (incomplete gamma function). Both functions are unimodal and positively skew, and closely resemble the experimentally derived function in Fig. 7. It should be possible to modify Rahn's or Kendall's hypotheses in many ways (for example, the steps might not proceed at the same rate) to fit the experimental data as closely as one desires. Therefore, any reasonable model that successfully predicts $f(\tau)$ should merely be considered as one of many plausible models.

Both of these models embody the view that division is controlled by a clock, the statistical fluctuations of which account directly for the shape of $f(\tau)$. Since it is assumed that the clock is reset to zero in each newly formed cell and that the clocks do not affect one another, interdivision times must be independent in such models.

On the point of independence of interdivision times such models break down. Experimental results show a strong positive correlation between interdivision times of sisters (Schaechter *et al.* 1962; Powell & Errington, 1963; Kubitschek, 1962) and a negative correlation between the interdivision time of a cell and the average interdivision time of several of its ancestors (Kubitschek, 1966). Therefore, division cannot be controlled solely by the type of clocks proposed in the models mentioned above.

Analysis of the loss of synchrony provides a graphic example of the inadequacy of models that assume independence of interdivision times.

Figure 8 contrasts the growth of a synchronous culture, reported by Cummings (1965), with the growth predicted by equation (6). It is clear that synchrony of division in the second generation is greater than predicted by a model which assumes independence of interdivision times.

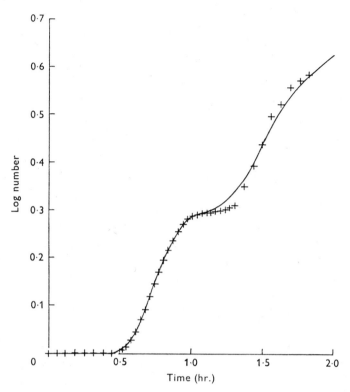

Fig. 8. Comparison of the synchronous growth with a prediction based on independence of interdivision times. The experimental data are taken from a paper by Cummings (1965) describing the synchronous growth of *Escherichia coli* strain B/r in glucose minimal medium at 37°C.; the plotted points were interpolated from a smooth curve drawn through the original data points. The prediction (solid line) was based on the assumption that f(τ) in an incomplete gamma function with mean = 0·79 hr. and standard deviation = 0·145 hr. These parameters were obtained by fitting the values computed with a digital computer to the data points for the first generation of synchronous growth.

HYPOTHESIS OF KOCH AND SCHAECHTER

Division occurs on attainment of critical size

From the preceding section it is clear that hypotheses for the control of cell division which imply independence of interdivision times must be rejected because of the observed correlations of interdivision times and because of the failure of numerical models derived from such hypotheses to predict the behaviour of synchronous cultures beyond the first generation (Fig. 8).

Primarily on the basis of the observation that the coefficient of variation of size (length) of dividing cells of *Escherichia coli* is less than the coefficient of variation of interdivision time (Schaechter *et al.* 1962), Koch & Schaechter (1962) proposed that division is controlled by *size* (implying by this term the amounts of various constituents of the cell as well as the volume) rather than *age*. Koch & Schaechter's hypothesis for control of division is based on two fundamental assumptions which may be phrased generally as follows: (i) the rate of growth in size of an individual bacterium is a deterministic function of its size; and (ii) division occurs on growth to a critical size which is independent of the size at birth. The validity of the first assumption has been demonstrated in a previous section ('Kinetics of Growth of Bacteria'). The second assumption will be examined in this section.

Specifically, Koch & Schaechter's model assumed that the rate of growth of a bacterium is proportional to its size, an assumption shown to be incorrect (Fig. 5). Also their model did not take into account the inequality in size of sisters and did not treat fully the distribution of size at division. Nevertheless, the model did predict a positive correlation in interdivision times of mothers and daughters. Powell (1964) retained the specific assumption of exponential growth but introduced more explicit formulation of the distributions of size of newly formed and of dividing cells.

A more general model of the hypothesis of Koch & Schaechter has been developed, permitting the calculation of $f(\tau)$ (Harvey *et al.* 1967) and of the growth of synchronous cultures (Painter & Marr, 1968).

The function $f(\tau)$ may be computed from the equation

$$f(\tau) = \int_0^\infty \psi^*(t) \, \phi_b^*(t+\tau) \, dt. \tag{7}$$

In equation (7) the variable t is given by

$$t = \int_{x_0}^x 1/V(\theta) \, d\theta,$$

where x_0 is the size of the smallest cell of interest. The functions $\psi^*(t)$ and $\phi_b^*(t)$ are transformations of functions of size

$$\psi^*(t) = \psi(x) \, V(x)$$

and $$\phi_b^*(t) = k\phi(x) \exp\left[-\int_0^x k\phi(\theta)/[\lambda(\theta) \, V(\theta)] \, d\theta\right]\Big/\lambda(x).$$

Since τ cannot be negative, the functions $\psi^*(t)$ and $\phi_b^*(t)$ must be non-overlapping (Powell, 1964). This requirement was met by arbitrarily trimming the tail of the measured ψ-distribution (see legend to Fig. 1) on the assumption that the largest cells in the effluent from a

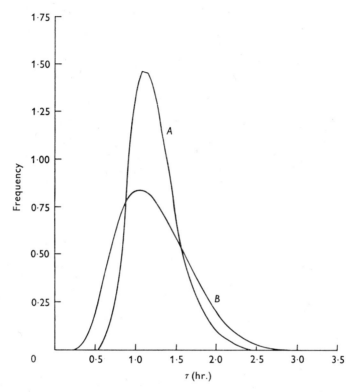

Fig. 9. Comparison of f(τ) estimated from experimental measurement (*A*) with f(τ) predicted from the Koch–Schaechter hypothesis (*B*). The parameters of the predicted f(τ) are as follows: mean$=1\cdot227$ hr.; mode$=1\cdot054$ hr.; standard deviation$=0\cdot430$ hr.; coefficient of variation$=0\cdot351$; $g_1=0\cdot497$; and $g_2=-0\cdot035$. The values represented by curve *A* are the same as those in Fig. 7.

membrane culture are not newly formed but are eluted by some other (random) process. From the values of $\psi(x)$, $\lambda(x)$ and $\phi(x)$ shown in Fig. 1, $f(\tau)$ was computed according to equation (7), and the results of this calculation are shown in Fig. 9. The coefficient of variation of $f(\tau)$ computed from equation (7) is somewhat larger, but the shape of the calculated function is in general agreement with direct measurement.

The functional relationships of the Collins–Richmond equation pertain only to populations in a steady state; a more general formulation (Painter & Marr, 1968) is

$$N'(t) \int_0^x \lambda(\theta,t)\, d\theta + N(t) \frac{\partial}{\partial t} \int_0^x \lambda(\theta,t)\, d\theta$$

$$= N'(t) \int_0^x 2\psi(\theta,t)\, d\theta - N'(t) \int_0^x \phi(\theta,t)\, d\theta - N(t)\, \lambda(x,t)\, V(x,t)$$

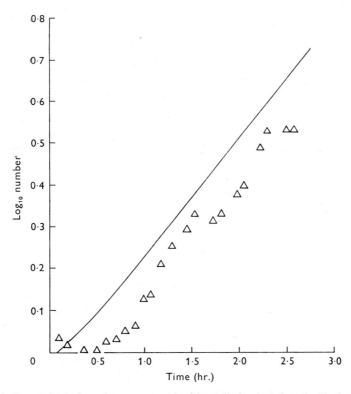

Fig. 10. Comparison of synchronous growth with prediction based on the Koch–Schaechter hypothesis. The experimental points are the data points of Fig. 3 rescaled to give unit number at zero time. The solid line is the prediction from equations (8) and (9) and the values of functions shown in Fig. 1 except that the frequency function of size at zero time, $\lambda(x, 0)$, were not trimmed as was $\psi(x)$ shown in Fig. 1.

from which one obtains the approximation

$$\lambda(x+\Delta x, t+\Delta t) \approx \lambda(x,t)\, \exp \left\{ \mu(t)\, \Delta t\, ([2\psi(x,t) - \phi(x,t)]/\lambda(x,t) - 1) \right.$$

$$\left. - \frac{\partial}{\partial x} V(x,t)\, \Delta t \right\}. \quad (8)$$

Equation (8) permits digital computation of the distribution of size of extant cells at any time. The instantaneous specific growth rate, $\mu(t)$, is given by

$$\mu(t) = \int_0^\infty \left[V(x,t)\,\lambda(x,t)\,\phi_b(x) \Big/ \int_x^\infty \phi_b(\theta)\,d\theta \right] dx. \tag{9}$$

The function ϕ_b is given by

$$\phi_b(x) = k\phi(x) \exp\left[-\int_0^x k\phi(\theta)/[\lambda(\theta)\,V(\theta)\,d\theta]/[\lambda(x)\,V(x)], \right.$$

the frequency function of size at division of a sample of newly formed cells, which function, according to the second fundamental assumption of the Koch–Schaechter hypothesis, applies to each newly formed cell.

From equations (8) and (9) and the values of $\psi(x)$ and $\phi(x)$ shown in Fig. 1 the expected number of bacteria in a synchronous culture was computed. The results of the calculations are compared with experimental values in Fig. 10. In these calculations the frequency function of size of newly formed cells, $\lambda(x, 0)$, was not trimmed to exclude large bacteria because all of the bacteria present, whether newly formed or randomly eluted, must be taken into account. The calculations are in striking disagreement with experimental results. Evidently the large cells in the effluent from a membrane culture do not divide as soon as predicted from the Koch–Schaechter hypothesis.

Although the second fundamental assumption of the hypothesis of Koch & Schaechter appears to be incorrect, the hypothesis has been fruitful in stimulating inquiry into physiological processes controlling division. The proposal that size, in some manner, regulates division remains attractive, but the regulation must be indirect.

HYPOTHESIS OF COOPER AND HELMSTETTER

Division is controlled by DNA replication

Recent experiments with synchronous cultures have located the position of the DNA replication cycle with respect to the division cycle in *Escherichia coli* strain B/r. Clark & Maaløe (1967) have shown that [3H]thymidine incorporation increases abruptly near the midpoint of the division cycle in cells growing in glucose minimal medium. Helmstetter (1967) and Helmstetter & Cooper (1968) have shown that over a wide range of growth rates in different media at 37° C. the end of a round of replication occurs 20–25 min. before division. This observation, together with the constant length of a round of replication (Cairns, 1963; Maaløe & Kjeldgaard, 1966), led Cooper & Helmstetter (1968) to propose a model for the division cycle of *E. coli* strain B/r. They assumed that for growth at given temperature the time C required for replication to proceed from one end of the genome to the other is constant and that the time D between the end of a round of genome replication and division is constant. Thus, division occurs $C + D$ units of time after initiation. (For further discussion see Pritchard, this Symposium.)

One shortcoming of this model is its failure to account for any distribution in the times C and D. Another is that it does not specify the control of initiation of rounds of replication. This control is crucial in this model, for interinitiation time (time between initiation of successive rounds of replication) determines interdivision time. In the following we will generalize the model to account for distribution and will add one assumption to account for initiation. The resulting model will predict the growth of synchronous cultures.

Virtually nothing is known about the control of initiation of rounds of replication but there must be some distribution in the times that elapse between initiations of successive rounds of replication (interinitiation times). Otherwise, synchrony of rounds of replication would persist indefinitely. The frequency function of interinitiation times (defined on a sample of newly initiated genomes) will be denoted by $f_i(\tau)$. We assume that interinitiation times are independent. Thus, by analogy with the growth of a microbial population (Painter & Marr, 1968), it is possible to calculate the rate at which new points of replication are added by the renewal equation

$$G'(t) = 2 \int_0^{\infty} G'(t - \theta) f_i(\theta) \, d\theta. \tag{10}$$

Here $G(t)$ is the number of complete or partially complete genomes at time t and is equal to one-half of the number of conserved units of DNA (Lark & Bird, 1965). By analysis of a careful study of the decay of synchrony of replication cycles it should be possible to estimate $f_i(\tau)$ and to see if interinitiation times are independent. Unfortunately, such data on synchronous cultures are not available.

The time (D) that elapses between the end of a round of replication and the subsequent division should also be distributed. It is more convenient to consider instead the time ($C + D$) that elapses between initiation of the replication of a genome and the division that follows the end of the round of replication. (The difference between these times is C, the constant length of a round of replication.) The frequency function of times from initiation to division will be denoted by $f_d(\tau)$ (again defined for a sample of newly initiated genomes). Now if $N(t)$ is the number of cells at time t, then the division rate is given by the integral equation

$$N'(t) = \int_0^\infty G'(t-\theta) f_d(\theta)\, d\theta. \tag{11}$$

In the steady-state the numbers of both cells and genomes must increase at the same specific rate, k, and if N_0 and G_0 are the numbers of cells and of genomes, respectively, at zero time then we have

$$N(t) = N_0 e^{kt}, \quad \text{and} \quad G(t) = G_0 e^{kt}.$$

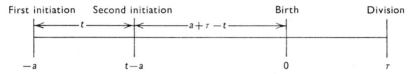

Fig. 11. Time scale for the events leading to birth at zero time and division at time τ (see text).

A useful relationship between G_0 and N_0 can be derived by considering the growth rate of the population at zero time, kN_0. From equation (11), we see that

$$kN_0 = \int_0^\infty kG_0 e^{-k\theta} f_d(\theta)\, d\theta,$$

which gives the relationship

$$\int_0^\infty \frac{G_0}{N_0} e^{-k\theta} f_d(\theta)\, d\theta = 1, \tag{12}$$

in which G_0/N_0 is the number of complete or partial genomes per cell (or one-half the number of conserved units of DNA per cell) in the steady-state. The function

$$(G_0/N_0)\, e^{-ka} f_d(a) \tag{13}$$

has an important interpretation: it is the frequency function of initiation in the past at any time $-a$ of rounds of replication that lead to division at the present (zero time).

Now we are in a position to derive an expression for $f(\tau)$. A cell newly formed at zero time results from the initiation of a round of replication at time $-a$, and expression (13) is the frequency function of this initiation. The original initiation will lead to a second initiation, which, in turn, will lead to a division at time τ of the cell newly formed at zero time. If the second initiation occurs t units of time after the first (at time $t-a$), then the subsequent division must occur $a+\tau-t$ units of time after the second initiation. The time scale for this sequence is shown in Fig. 11.

Thus, for a cell with its first initiation fixed at time $-a$ the frequency at which division occurs at any time τ is easily seen to be

$$\int_0^{a+\tau} f_i(t) f_d(a+\tau-t) \, dt.$$

Now $f(\tau)$ is obtained by integrating the above frequency over all probability elements given by expression (13); thus,

$$f(\tau) = \int_0^\infty (G_0/N_0) \, e^{-ka} f_d(a) \int_0^{a+\tau} f_i(t) f_d(a+\tau-t) \, dt \, da. \qquad (14)$$

The variance of $f(\tau)$ is found most easily by applying the convolution theorem to $f(\tau)$. This yields the result

$$Lf = Lf_i Lf_d \int_0^\infty e^{sa}(G_0/N_0) \, e^{-ka} f_d(a) \, da,$$

where L is the Laplace transform operator. The variance is

$$L''f|_{s=0} - [L'f|_{s=0}]^2,$$

which is computed to be

$$\text{var} \, [f] = \text{var} \, [f_i] + \text{var} \, [f_d] + \text{var} \, [(G_0/N_0) \, e^{-ka} f_d(a)]. \qquad (15)$$

We have introduced the approximation

$$\text{var} \, [(G_0/N_0) \, e^{-ka} f_d(a)] \approx \text{var} \, [f_d]$$

into our calculations as a convenient means of selecting the variances of f_i and f_d that give a specified variance of f by equation (15).

The predicted growth of a synchronous culture can be computed from equations (10) and (11). The first step is to compute the rate of initiation at time $-a$ for a sample of cells that divide at zero time as a result of this initiation. From expression (13) this rate is

$$G'(-a) = G_0 \, e^{-ka} f_d(a).$$

Now equation (10) can be used to compute successive values of $G'(t)$ for the synchronous culture, and from these values equation (11) gives $N'(t)$ which can then be compared with experimental measurements.

The calculations of synchronous growth were made assuming that f_i and f_d are incomplete gamma functions. The mean of $f_i(\tau)$ was specified as 0·79 hr. in accordance with the approximation,

$$E[f_i(\tau)] \approx E[f(\tau)],$$

and the value for the mean of $f(\tau)$ obtained previously (see legend for Fig. 8). The mean of $f_d(\tau)$ was specified as 1·26 hr.; this selection was based on the data of Helmstetter (1967), which demonstrates that for *Escherichia coli* growing in glucose minimal medium

$$E[f_d(\tau)] \approx 1·6 \, E[f_i(\tau)].$$

The value of the standard deviation of $f_i(\tau)$ was selected by fitting the predicted values of number of bacteria to the experimental values during the second generation. Both sets of values were displayed simultaneously on an oscilloscope attached to a data terminal of the digital computer. A new choice for the parameter could be entered

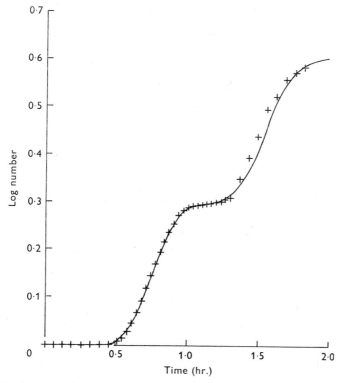

Fig. 12. Comparison of synchronous growth with the prediction based on the model developed from the Cooper–Helmstetter hypothesis. The experimental data are those depicted in Fig. 8. The prediction (solid line) was computed from equations (10) and (11) from the values of $f_i(\tau)$ and $f_d(\tau)$ shown in Fig. 13.

through the data terminal. The value of the standard deviation of $f_d(\tau)$ was computed from equation (15) such that the standard deviation of $f(\tau)$ would be 0·145 hr., a value which was previously determined (see legend to Fig. 8).

The results of the calculation are shown in Fig. 12, which should be compared with Fig. 8. The fit during the first generation is equally good in both figures, but the fit during the second generation is greatly improved in Fig. 12. The experimental values are displaced approximately 2 min. from the prediction in the second generation, reflecting a slightly shorter mean interdivision time in the second generation than in the first. This discrepancy is commonly observed (D. J. Clark, personal communication). It may be due to a slightly earlier separation of the progeny of a division of a mother cell growing attached to the membrane than in liquid culture.

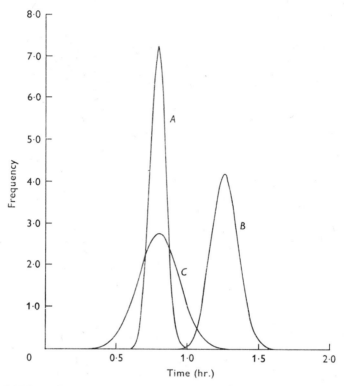

Fig. 13. Values of the functions used in computing the prediction shown in Fig. 12. *A*, $f_i(\tau)$; *B*, $f_d(\tau)$; *C*, $f(\tau)$. The functions f_i and f_d are incomplete gamma functions. The mean and standard deviation of f_i are 0·76 and 0·055 hr., respectively. The mean and standard deviation of f_d are 1·26 and 0·095 hr., respectively. The function $f(\tau)$ was computed according to equation (14); the parameters of $f(\tau)$ are as follows: mean=0·798 hr.; mode=0·790 hr.; standard deviation=0·145 hr.; coefficient of variation=0·181 hr.; $g_1=0·0085$; and $g_2=0·0130$ (for definition of g_1 and g_2, see Fig. 7).

The value of G_o/N_o, which is primarily determined by the relationship of the mean of $f_d(\tau)$ to mean of $f_i(\tau)$, was computed according to equation (12) for the values of the parameters given in the legend of Fig. 13, and was found to be 3·01, which is equivalent to 6·02 conserved units of DNA per cell. This may be compared with an experimental value of 7 to 8 conserved units per cell for a different strain of *Escherichia coli* (Lark & Bird, 1965).

The functions used in the calculations in Fig. 12 are shown in Fig. 13. Although the fit in Fig. 12 appears to be adequate during the first generation, the skewness of $f(\tau)$ is much less than that of the experimentally determined function in Fig. 7. Since the function f_d does not contribute to skewness of $f(\tau)$, the function f_i must be more skew. Possibly the function f_i would have been better approximated by a negative exponential function. A negative exponential distribution for initiation would have a straightforward kinetic interpretation: a one-step reaction to trigger initiation of DNA replication.

Finally, although this model was developed specifically to test the hypothesis of Cooper & Helmstetter, it is, in fact, more general. The model permits a partitioning of the variance of interdivision time into two parts. One part, represented by the function f_i, contributes to the decay of synchrony; the other part, represented by f_d, does not. This partitioning accounts for the observed positive correlations in inter-division times of sisters. Furthermore, the events which govern a particular division extend over more than one generation (Fig. 11). This feature can lead to ancestral control of division of the type reported by Kubitschek (1966).

The most ambiguous part of the model is the control of initiation. Our formal proposal for control by timing from the preceding initiation is mathematically convenient. However, it is likely that size (in the sense of Koch & Schaechter) mechanistically governs, at least in part, the initiation of new rounds of replication. Presently we are unable to develop a formal treatment of this possibility.

ACKNOWLEDGEMENT

We gratefully acknowledge the assistance of Mr James Vance in programming and digital processing of data, and the assistance of Mr Richard Kleker in some of the experiments. The experimental work was supported financially by a grant (GB 7027) from the National Science Foundation (U.S.A.). Data processing was supported financially by the U.S. Department of Defence (USA DA MD 49-193-66-G 9217). The expense of digital computation was borne by the National Institutes of Health (AI 05526-05).

REFERENCES

CAIRNS, J. (1963). The bacterial chromosome and its manner of replication as seen by autoradiography. *J. molec. Biol.* 6, 208.

CLARK, D. & MAALØE, O. (1967). DNA replication and the division cycle in *Escherichia coli. J. molec. Biol.* 23, 99.

COLLINS, J. & RICHMOND, M. (1962). Rate of growth of *Bacillus cereus* between divisions. *J. gen. Microbiol.* 28, 15.

COOPER, S. & HELMSTETTER, C. (1968). Chromosome replication and the division cycle of *Escherichia coli* B/r. *J. molec. Biol.* **31**, 519.

CUMMINGS, D. J. (1965). Macromolecular synthesis during synchronous growth of *Escherichia coli* B/r. *Biochim. biophys. Acta*, **85**, 341.

EAKMAN, J., FREDERICKSON, A. & TSUCHIYA, H. (1966). Statistics and dynamics of microbial cell populations. *Chem. Engng Prog. Monogr. Ser.* **62**, 37.

ERRINGTON, F., POWELL, E. & THOMPSON, N. (1965). Growth characteristics of some Gram-negative bacteria. *J. gen. Microbiol.* **39**, 109.

FISHER, R. A. (1946). *Statistical Methods for Research Workers*, 10th ed., pp. 70–76. London: Oliver and Boyd.

HARRIS, T. (1959). In *The Kinetics of Cellular Proliferation*. Ed. F. Stohlman, pp. 368–81. New York: Grune and Stratton.

HARVEY, R. J. & MARR, A. G. (1966). Measurement of size distribution of bacterial cells. *J. Bact.* **92**, 805.

HARVEY, R. J., MARR, A. G. & PAINTER, P. R. (1967). Kinetics of growth of individual cells of *Escherichia coli* and *Azotobacter agilis*. *J. Bact.* **93**, 605.

HELMSTETTER, C. E. (1967). Rate of DNA synthesis during the division cycle of *Escherichia coli* B/r. *J. molec. Biol.* **24**, 417.

HELMSTETTER, C. E. & COOPER, S. (1968). DNA synthesis during the division cycle of rapidly growing *Escherichia coli* B/r. *J. molec. Biol.* **31**, 507.

HELMSTETTER, C. E. & CUMMINGS, D. J. (1963). Bacterial synchronization by selection of cells at division. *Proc. natn. Acad. Sci. U.S.A.* **50**, 767.

HELMSTETTER, C. E. & CUMMINGS, D. J. (1964). An improved method for the selection of bacterial cells at division. *Biochim. biophys. Acta*, **82**, 608.

HINSHELWOOD, C. (1946). *The Chemical Kinetics of the Bacterial Cell*. Oxford: Clarendon Press.

KELLY, C. & RAHN, O. (1932). The growth rate of individual bacterial cells. *J. Bact.* **23**, 147.

KENDALL, D. G. (1948). On the role of variable generation time in the development of a stochastic birth process. *Biometrika*, **35**, 316.

KOCH, A. & SCHAECHTER, M. (1962). A model for the statistics of the cell division processes. *J. gen. Microbiol.* **29**, 435.

KUBITSCHEK, H. (1962). Normal distribution of cell generation rate. *Expl Cell Res.* **26**, 439.

KUBITSCHEK, H. (1966). Generation times: ancestral dependence and dependence upon cell size. *Expl Cell Res.* **43**, 30.

LARK, K. G. & BIRD, R. E. (1965). Segregation of the conserved units of DNA in *Escherichia coli*. *Proc. natn. Acad. Sci. U.S.A.* **54**, 1444.

LOTKA, A. (1913). Vital statistics. A natural population norm. *J. Wash. Acad. Sci.* **3**, 289.

LOTKA, A. (1922). The stability of the normal age distribution. *Proc. natn. Acad. Sci. U.S.A.* **8**, 339.

MAALØE, O. & KJELDGAARD, N. O. (1966). *Control of Macromolecular Synthesis*. New York: Benjamin.

MARR, A. G., HARVEY, R. J. & TRENTINI, W. C. (1966). Growth and division of *Escherichia coli*. *J. Bact.* **91**, 2388.

MONOD, J. (1942). *Recherches sur la Croissance des Cultures Bactériennes*. Paris: Masson.

PAINTER, P. R. & MARR, A. G. (1968). Mathematics of microbial populations. *A. Rev. Microbiol.* **22**, 519.

POWELL, E. (1956). An improved culture chamber for the study of living bacteria. *Jl. R. microsc. Soc.* **75**, 235.

POWELL, E. (1964). A note on Koch and Schaechter's hypothesis about growth and fission of bacteria. *J. gen. Microbiol.* **37**, 231.

POWELL, E. & ERRINGTON, F. (1963). Generation time of individual bacteria: Some corroborative measurements. *J. gen. Microbiol.* **31**, 315.

RAHN, O. (1932). A chemical explanation of the variability of the growth rate. *J. gen. Physiol.* **15**, 257.

SCHAECHTER, M., WILLIAMSON, J. P., HOOD, J. R. & KOCH, A. L. (1962). Growth cell and nuclear divisions in some bacteria. *J. gen. Microbiol.* **29**, 421.

WARD, H. (1895). On the biology of *Bacillus ramosus* (Fraenkel). A schizomycete of the River Thames. *Proc. R. Soc.* **58**, 265.

CONTROL OF DNA SYNTHESIS
IN BACTERIA

R. H. PRITCHARD, P. T. BARTH AND J. COLLINS

Department of Genetics, University of Leicester

Wide differences in the growth rate of a bacterial culture can be achieved by altering the composition of the growth medium or the environmental temperature. The rate of synthesis of all macromolecular components of the cells composing the culture must therefore be correspondingly adjustable and coordinated, one with another. In the case of DNA, where only one or a few molecules are present in each cell, the way in which its rate of synthesis might be coordinated with the growth rate is not obvious, and in this article we will consider how this relationship is maintained. We will be concerned primarily with the control of DNA synthesis, both chromosomal and episomal, in *Escherichia coli* but will also make use of data obtained from experiments with other bacteria. The specific hypothesis which we will discuss draws heavily on the extensive and painstaking observations of Maaløe and his collaborators on the size and composition of bacteria growing at different rates, and the changes which occur during a transition from one growth rate to another. Much of the relevant data has been reviewed in detail by Maaløe & Kjeldgaard (1966) and will not be described again here in detail.

The chromosome of *Escherichia coli* is a ring of double-stranded DNA. In the following discussion we will assume that a cycle of replication is initiated at a specific point on the chromosome defined here as the *vegetative origin* (*origin* for short) and that a replication fork traverses the chromosome in a unique direction starting at this point. It is possible that the position of the chromosome origin and the direction of replication may differ in different strains. This will be discussed more fully later. It is also possible that there may be more than one potential origin on a chromosome, but existing data strongly imply that successive acts of initiation in one cell and its progeny occur at the same site with high probability.

CHANGES IN RATE OF DNA SYNTHESIS

The rate of synthesis of DNA by any system will be determined (a) by the mean rate of replication of individual molecules (the rate of addition of nucleotides to a growing polynucleotide chain, or the rate of progression—*velocity*—of a replication fork along the molecule), and (b) by the mean density of replication forks (which will be a function of the frequency of insertion of replication forks into individual molecules: i.e. the frequency of *initiation*). The rate of DNA synthesis cannot permanently be altered by changing the replication velocity alone. Only if an increase in velocity is associated with an increase in the frequency of initiation (such that the mean density of replication forks remains constant) will there be a corresponding increase in the rate of DNA synthesis. The replication velocity could determine the rate of DNA synthesis only in the special case where there was a fixed time interval (or a variable interval determined by the replication velocity) between the completion of one round of replication and the start of another.

In *Escherichia coli* and related bacteria, there is considerable indirect evidence that changes in the rate of DNA synthesis are determined *solely* by changes in the frequency of initiation (Maaløe, 1961). This evidence has been discussed at length by Maaløe & Kjeldgaard (1966). We will not reiterate it in detail here but rather draw attention to the recent work of Helmstetter & Cooper (Helmstetter & Cooper, 1968; Cooper & Helmstetter, 1968) which suggests that in one strain of *E. coli* the replication velocity may be *constant* in batch cultures with doubling times varying between 22 and 53 min. If their data have been correctly interpreted, the rate of DNA synthesis must be determined solely by the frequency of initiation in this strain and under these conditions.

There is another argument (Pritchard, 1965), which leads to the same conclusion and is pertinent in that it suggests that the rate of DNA synthesis must be determined by the frequency of initiation generally. In cultural conditions permitting doubling times of about 40 min. or less there is no detectable fraction of the cell cycle in *Escherichia coli* in which no DNA synthesis is taking place—synthesis is continuous. Under these conditions the immediate DNA precursors must be present at all times and although the replication velocity might be determined by the concentration of these precursors, or of the polymerizing enzyme, the proper *sequence* of events could not be. There must be some mechanism which ensures that one act of initiation is not followed by another until some predetermined interval (which will be

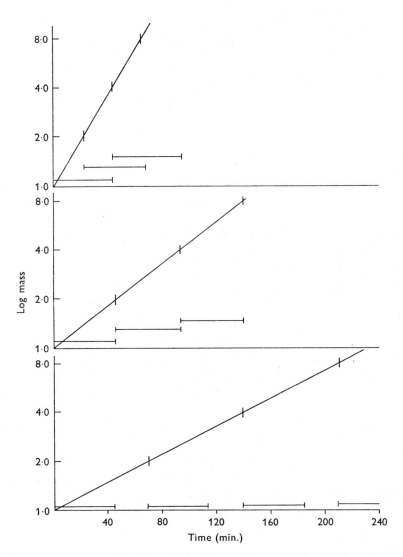

Fig. 1. Relationship of successive cycles of chromosome replication in cultures growing at different rates. The horizontal lines represent the period of replication of a chromosome. Gaps between adjacent lines (bottom graph) indicate a pause between successive cycles of replication. Overlapping lines (top graph) indicate dichotomous replication. The vertical bars on the curves mark off mass doublings.

fixed by the growth rate) has elapsed.[1] The logic which demands that under conditions of continuous DNA synthesis the rate must be determined by the frequency of initiation also leads to the same conclusion for systems (or under conditions) in which there is an interval between successive cycles of replication (Pritchard, 1966). There is evidence from at least one eukaryotic organism, *Aspergillus nidulans* (Rosenberger & Kessel, 1967; and personal communication), which indicates that increasing the growth rate leads to a reduction in the interval between successive cycles of DNA synthesis (i.e. to an increase in the frequency of initiation) without changing the rate of replication.

The properties of a system in which the replication velocity is constant and changes in the rate of DNA synthesis are determined by changing the frequency of initiation are illustrated diagrammatically in Fig. 1. At any growth rate the average time interval between successive acts of initiation must be equal to the time taken for the cytoplasmic mass to double if the cell composition is to remain constant. If this doubling time is greater than the time required for a replication fork to traverse the chromosome there will be a pause between successive rounds of replication. If the two times are equal, DNA synthesis will be continuous. If the doubling time is less than the replication time a new round of replication will commence before the previous round is completed, and each chromosome will, part of the time, have three replication forks instead of one. The existence of this so-called dichotomous replication at relatively fast growth rates is now well documented in *Bacillus subtilis* (Oishi, Yoshikawa & Sueoka, 1964) and *Escherichia coli* (Pritchard & Lark, 1964; Helmstetter & Cooper, 1968). Indeed, it should be pointed out that it was the ingenious demonstration of dichotomous replication in *B. subtilis*, growing fast under normal physiological conditions, by Sueoka and his collaborators which drew attention to the possibility that rate of DNA synthesis might be independent of the replication velocity.

CONTROL OF INITIATION

The question posed by the foregoing discussion is how the correlation between the frequency of initiation and the growth rate is achieved. An analogous problem which has received much attention (e.g. see Koch &

---

[1] It could be argued that the number of polymerase molecules in a cell is equal to the number of replication forks and that the rate of replication is therefore determined by the 'concentration' of this enzyme, but this is another way of saying that the rate of DNA synthesis is determined by the frequency of initiation and is a specific mechanism which might ensure this. This possibility will be considered later.

Schaechter, 1962; Kubitschek, 1962; Ycas, Sugita & Bensam, 1965) is the way in which the frequency of cell division is determined. In the latter case one suggestion has been that the attainment of a critical cell mass or surface : volume ratio might induce septum formation (by

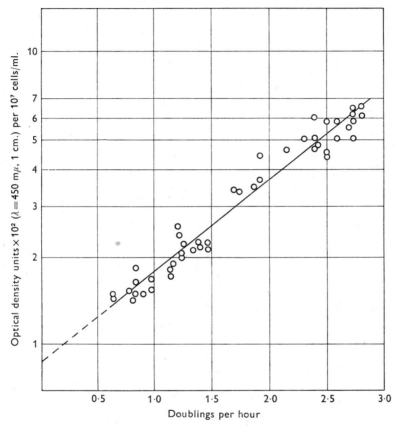

Fig. 2. Average cell mass as a function of growth rate in *Salmonella typhimurium*. Data of Schaechter *et al.* (1958) reproduced from control of macromolecular synthesis by Maaløe & Kjeldgaard (1966), by permission of the publisher, W. A. Benjamin, Inc., New York.

creating critical concentration gradients, for example) and it might be supposed that initiation of replication might be determined in a similar way. Interpretations of this sort seem unlikely in the case of *Escherichia coli* and related bacteria since it is well known that the mean cell size varies with the growth rate. Thus in *Salmonella typhimurium* (Schaechter, Maaløe & Kjeldgaard, 1958) and in *Aerobacter aerogenes* and *Bacillus megaterium* (Herbert, 1959) it has been shown that the mean cell mass is a continuous function of the growth rate (see Fig. 2). The largest cells in cultures growing at one rate may therefore be smaller than the smallest cells in cultures growing at a faster rate, and the cell mass at the time of initiation cannot be constant. Thus although the attainment of a critical cell mass or volume might be the determining factor at a given growth rate it seems unlikely that such a hypothesis

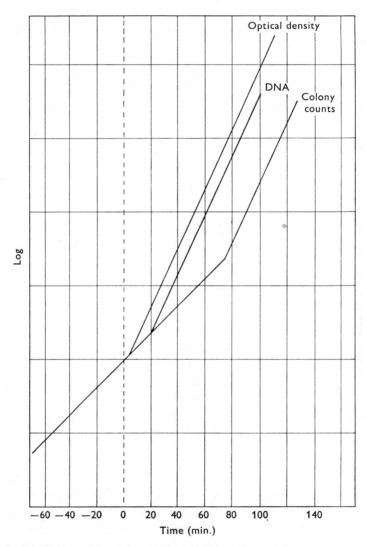

Fig. 3. Consequences of a 'shift up' in *Salmonella typhimurium*. Redrawn from Maaløe & Kjeldgaard (1966). The curves show how the rates of mass increase, DNA synthesis, and cell number change when a culture is shifted from a glucose synthetic medium to broth. The shift occurred at time O. The distance between horizontal lines corresponds to one doubling.

can provide a mechanism of general validity. Another possibility would be to assume that initiation is determined by some other discontinuous event in the cell cycle, such as septum formation. The behaviour of cultures of *S. typhimurium* transferred from one growth medium to another permitting a faster growth rate, analysed in detail by Kjeldgaard, Maaløe & Schaechter (1958) and Kjeldgaard (1961), seems to rule this out. In 'shift-up' experiments of this sort it was found that the rate of increase in mass changed to the new rate very rapidly after the shift to a new medium, the rate of DNA synthesis adjusted to the new rate more slowly, and cell division more slowly still (see Fig. 3). This sequence of events shows that initiation cannot be determined either directly or indirectly by septum formation, although it is compatible with a converse relationship as are the recent observations of Helmstetter & Cooper (1968), which will be considered in more detail later.

We are consequently led back to the simplest interpretation of the sequence of changes (reproduced in Fig. 3) which occur during the period of transition in a 'shift-up' experiment. This is that it is the rate of increase in cell mass which in some way determines the frequency of initiation.

In the following section we will consider a simple mechanism which would permit a cell to titrate the increment in cell mass after one act of initiation and ensure that a new act of initiation occurred after each doubling in mass, but not before. We are encouraged to discuss this mechanism in detail because it appears to be consistent with most existing data, because it makes a number of testable predictions, but particularly because it leads to a quite novel interpretation of the control of the frequency of replication and the *number* per cell of extrachromosomal particles (such as F), and provides a simpler explanation of some of their properties than existing models. A summary of this model has been presented elsewhere (Pritchard, 1968).

THE MECHANISM OF CONTROL

We will assume, as Jacob, Brenner & Cuzin (1963) have done, that initiation involves a modification of the structure of the chromosome origin mediated by an enzyme or enzyme system (the Initiator). Once initiation has occurred replication of the chromosome follows.

The following postulates would constitute a mechanism which would ensure that one act of initiation is followed by another when, and only when, the cell volume has doubled:

(1) Initiator (*I*) is produced constitutively. As a first approximation

we will assume that it constitutes a constant fraction of total protein at all growth rates.

(2) An inhibitor (H) of initiation is produced by a gene which has the following properties: (a) it is located adjacent to the chromosome origin or is part of the origin itself; (b) it is transcribed only at the time of its replication; (c) it produces a fixed number of messenger RNA molecules from which a fixed number of molecules of the inhibitor protein are translated *at all growth rates*.

(3) There is a cooperative interaction between H and either the chromosome origin or I such that a twofold change in concentration of H effects a transition between complete and zero inhibition of initiation.

Consider a situation in which the concentration of inhibitor in a cytoplasmic volume containing one chromosome origin is greater than the critical concentration (h_c) at which initiation occurs. An increase in cytoplasmic volume by growth will lead to a progressive dilution of inhibitor until the concentration h_c is reached. Initiation will follow but will be associated with production of a new pulse of inhibitor, raising its concentration above h_c. Further growth will again be necessary before initiation can occur once more and the cycle be repeated. (Implicit in this model is the assumption of a dead-time between successive acts of initiation which is longer than the time taken for the inhibitor concentration to rise significantly above its critical level after initiation.)

It can readily be shown that at equilibrium the concentration of inhibitor will oscillate over a twofold range irrespective of h_c or the number of inhibitor molecules produced at each pulse, the minimum concentration reached being h_c and the maximum $2h_c$. The frequency of initiation will therefore be determined by the dilution rate of inhibitor which will be the reciprocal of the growth rate, successive acts of initiation occurring after successive doublings in the volume of cytoplasm. This is shown diagrammatically in Fig. 4. The cytoplasmic volume (V_c) per chromosome origin at the time of initiation will be determined solely by h_c and the number of inhibitor molecules, a, produced by one chromosome at each initiation, the relationship being

$$V_c = \frac{a}{h_c}. \tag{1}$$

In the case of a bacterial cell which may contain more than one chromosome origin at the time of initiation, the cytoplasmic volume (V_i) at the time of initiation will be

$$V_i = \frac{a}{h_c} 2^n \quad (n = 0, 1, 2, 3, \ldots), \tag{2}$$

where 2^n is the number of chromosome origins present prior to initiation (the significance of the number n will become apparent later).

It is appropriate at this point to consider one important property of a control mechanism of this type.

Although it would ensure that the *mean* time interval between successive acts of initiation would be equal to the doubling time, the interval between successive acts of initiation in individual cells would be expected to show a distribution about this mean. The consequence of such a distribution can be seen by considering what follows if in a given instance initiation occurs before h_c is reached (dashed line in Fig. 4). The pulse of inhibitor resulting from this act of initiation will occur in a cytoplasmic volume less than V_i and the resulting concentration will therefore be greater than $2h_c$. Consequently the time interval before the concentration of inhibitor falls again to h_c will be correspondingly increased. Thus there will be a negative correlation between successive inter-initiation times, and the relationship given by equation (2) will be maintained by the population as a whole despite deviations from it by individual cells. In the same way it can be shown that deviations from the mean number of inhibitor molecules produced at each pulse will not disturb this relationship, and the system is therefore self-regulating. This is an essential property which any mechanism of control must have if the time of initiation in relation to any other growth parameter is to remain constant.

Equation (2) leads to the striking prediction that the cell volume per number of H genes at the time of initiation will be a constant at all growth rates. Since we are postulating that the H gene is adjacent to, or part of, the chromosome origin this ratio can be written as cell volume/number of chromosome origins. In order to determine whether this prediction holds in reality, it is necessary to know both the number of chromosome origins per cell and the cell volume at the time of initiation at different growth rates.

The first of these numbers as a function of the growth rate has been determined for one strain of *Escherichia coli* and a reasonable estimate of the second is possible.

Using an ingenious technique for fractionating the cells of a culture according to age, Helmstetter & Cooper (1968) were able to demonstrate that discrete changes in the rate of DNA synthesis occur at specific times in the cell cycle of *Escherichia coli* B/r. From the magnitude of these changes in rate (and from the times in the cell cycle at which they occur) in cultures with different growth rates, they came to the conclusion not only that over a wide range of growth rates the time (C)

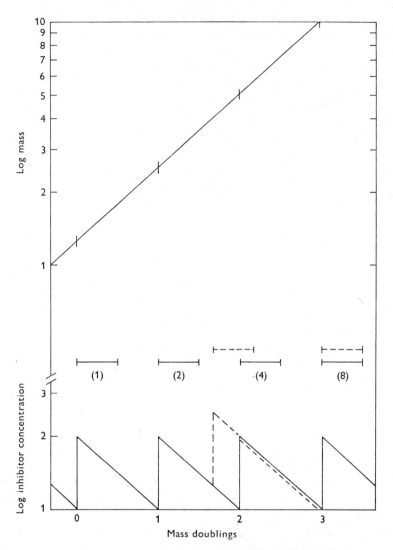

Fig. 4. Changes in inhibitor concentration due to initiation of chromosome replication and growth in volume of a protoplasmic mass. Vertical bars on the upper curve mark off mass doublings. Horizontal lines mark off periods of replication of chromosomes. Numbers in parentheses are the total numbers of chromosomes present in the protoplasmic mass shown. The lower curve represents the concentration of inhibitor. Initiation occurs at an arbitrary concentration 1·0 and at an arbitrary mass per chromosome of 1·25 units. The dashed curve and replication times show the consequences of initiation occurring before the inhibitor concentration has fallen to 1·0.

taken to replicate the chromosome is constant but, more remarkably, that there is a constant time (D) between the end of a cycle of replication and cell division. Having estimated these times they were able to predict not only the cell age at which initiation would occur at a given growth rate but also the amount of DNA per cell that would be present at that growth rate. Careful measurements of the quantity of DNA per cell at different growth rates were in remarkably close agreement with the amounts predicted and provided convincing evidence that their hypothesis, and their estimate of $C+D$ (approximately 70 min. for batch-grown cultures in the exponential phase of growth), was correct.[1] The pattern of DNA replication which their scheme implies is shown diagrammatically in Fig. 5.

Dr W. Donachie has pointed out to us that with this information it is possible to estimate the cell volume at the time of initiation (V_i) relative to the volume immediately after cell division (V_o). To determine this value it is necessary to make an assumption as to the rate of increase in cell volume of individual cells in a culture growing exponentially. Direct microscopic observations (Schaechter, Williamson, Hood & Koch, 1962) on single cells of *Escherichia coli* are consistent with an exponential increase in volume of individual cells. Recent data obtained from measurement of the distribution of cell sizes using an electronic particle counter (Harvey, Marr & Painter, 1967) suggest that the rate of increase in volume changes in a more complex fashion but approximates more closely to an exponential rate than, for example, to a linear rate.

We will assume an exponential increase in volume of individual cells and therefore write

$$V_i = V_o . 2^{t/g},\tag{3}$$

where t is the cell age at the time of initiation and g is the doubling time. The effect of slight deviations from this relationship will in any case be negligible.

Cooper and Helmstetter have shown that the cell age at the time of initiation is given by

$$t = (n+1)\,g - (C+D),\tag{4}$$

[1] The technique used by Helmstetter & Cooper depends on the ability of *Escherichia coli* B/r to bind to membrane filters. Newborn cells are released, however, and are eluted continuously by passage of prewarmed medium through the reverse side of the filter. Cultures were pulse labelled with [³H]thymidine before binding and the amount of label present in cells in the different age fractions collected was used to measure the rate of DNA synthesis in their ancestors at the same age. The value found for $C+D$ was about 63 min. but this was corrected to take account of the fact that the growth rate was faster on the filters than in a corresponding batch culture. However, further examination of their data suggests that this difference increases as the growth rate decreases. $C+D$ may therefore increase slightly over the range of growth rates studied.

where n is an integer such that $(n+1)g \geqslant C+D$. Combining (3) and (4) we get

$$V_i = V_0 \cdot 2^{[(n+1)\,g-(C+D)]/g} \tag{5}$$

If we now make use of our assumption that the mass per origin at the time of initiation is constant, we can combine (5) and (2) to get

$$\frac{a \cdot 2^{n+1}}{2h_c} = V_0 \cdot 2^{[(n+1)\,g-(C+D)]/g},$$

hence

$$V_0 = \frac{a}{2h_c} \cdot 2^{(C+D)/g}, \tag{6}$$

and the average cell volume is given by

$$\overline{V} = k\frac{a}{h_c} \cdot 2^{(C+D)/g} \tag{7}$$

where k is a constant.

Hence a plot of log $\overline{V}$ against $1/g$ will have a slope given by $(C+D)$ log 2 and an intercept which will be a function of a/h_c.

Reliable data on the relationship between mean cell volume and growth rate have not been obtained for *Escherichia coli*, but in *Salmonella typhimurium* the data reproduced in Fig. 2 are available (Schaechter *et al.* 1958). These relate to cell mass not volume, but as a first approximation we may assume that cell density does not change significantly with growth rate. These measurements are clearly consistent with a general relationship of the type predicted by equation (7). The value $C+D$ for *S. typhimurium* has not been measured directly, but two pieces of evidence suggest that it is about 70 min. as in *E. coli* B/r. First, as Cooper & Helmstetter have pointed out, the curve for DNA per cell as a function of the growth rate appears to have the same slope in the two strains, which would indicate similar values for $C+D$. Secondly, the time taken for the rate of cell division to accelerate to the new rate in a 'shift-up' experiment should be equal to $C+D$, and from the data reproduced in Fig. 3 this also seems to be about 70 min. If the curve superimposed on their data by Schaechter *et al.* (1958) is used to estimate $C+D$ from equation (7) a value of 63 min. is obtained. There is thus a good agreement between three separate estimates of $C+D$, one of which makes use of equation (7).

In comparing the predicted relationship between cell mass and growth rate with the data from *Salmonella typhimurium*, we have assumed a constant value for $C+D$. Cooper & Helmstetter (1968) have presented evidence that in *Escherichia coli* B/r $C+D$ increases if the doubling time of the culture is greater than about 70 min. and have suggested that at slower growth rates than this $C+D$ may be pro-

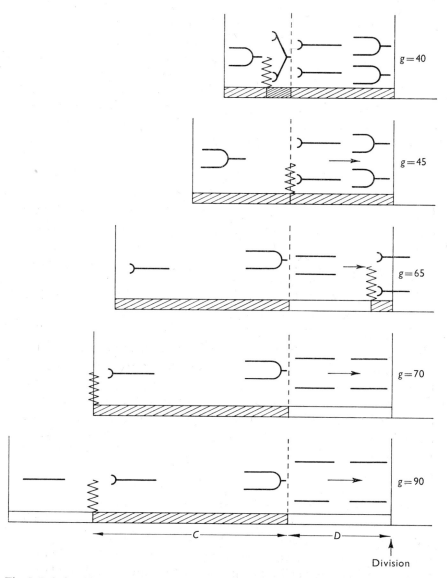

Fig. 5. Relationship between replication and cell division as envisaged by Cooper & Helm-stetter. The replication time (*C*) as indicated by diagonal hatching. *D* is the time between completion of replication and cell division. Denser diagonal hatching indicates the fraction of a cell cycle during which there is dichotomous replication. Zig-zag, time of initiation; dashed line, time of completion.

portional to the doubling time (see also footnote, p. 273). A full discussion of this question is given by these authors, and all that is necessary here is to point out that it does not affect the validity of the principle that the rate of DNA synthesis is determined by the frequency of initiation, and also that in equation (7) a constant value of $C + D$ is not assumed. If this term increases in proportion to the doubling time at slow growth rates the curve of mass against the generation rate will reach a minimum value and have zero slope at growth rates less than that at which $(C + D)/g$ becomes constant.

Data of Kjeldgaard (see Maaløe & Kjeldgaard, 1966) indicate that there is a continuous reduction in average cell mass down to generation rates as low as 0·2 per hour in *Salmonella typhimurium*. Data of Ecker & Schaechter (1963) show a similar reduction, although in this case growth rate was controlled by glucose limitation in a chemostat and the measurements are not necessarily comparable to corresponding measurements in batch cultures (for a discussion of this point see Schaechter *et al.* 1958). Nevertheless, it does seem clear that there is a continuous decrease in cell mass as the generation rate decreases towards zero and this in turn implies that there is not a proportionate increase in $C + D$.

On the other hand, Cooper & Helmstetter (1968) have shown that if $C + D/g$ becomes constant at slow growth rates the amount of DNA per cell will also become constant, and measurements of DNA per cell by Schaechter *et al.* (1958) suggest that this may happen at generation rates less than about 1·0 per hour. Thus there may be a discrepancy between the changes in mass per cell and DNA per cell at slow growth rates. The measurements of DNA per cell were only taken down to generation rates of 0·6 per hour, however, and the amount of scatter between the individual determinations is too great for any firm conclusions to be made.

A careful determination of the size and composition of cells at very slow growth rates in batch cultures will be necessary to provide a critical test of the relationship predicted by equation (7). If the average mass per cell decreases continuously down to zero growth rate but the amount of DNA per cell remains constant at rates less than about 1·0 per hour in the same strain, our model, in its simplest form, would be invalid; it might be necessary for example to assume that the value a is not constant.

OTHER MECHANISMS OF INITIATION

A number of authors have suggested mechanisms for the control of initiation which differ from the one discussed here. It is difficult to evaluate them since in most cases they have not been defined in sufficient detail, or were formulated without the benefit of data now available. They seem to fall into two classes, however, which can be outlined in a general way.

One class (cf. Jacob *et al.* 1963; Lark, 1966) makes use of the assumption that the chromosome origin is physically associated with the cell membrane or a cell organelle (such as a mesosome) which has a complex structure. Initiation of replication requires the synthesis of a new structure, or replication of an existing structure, or growth of the membrane. Once initiation has occurred it cannot recur until a new structure has been made or an appropriate development of the membrane has taken place. The other class (Maaløe & Kjeldgaard, 1966; see also footnote, p. 266) assumes that a fixed amount of an initiator is produced per generation. This substance has a high affinity for DNA (the chromosome origin?) and initiation occurs when this is 'saturated' with initiator.

The basic assumption behind both types of mechanism is that control is positive, involving the synthesis of a structure or substance necessary for initiation at a rate determined by the growth rate, and that initiation 'consumes' this structure or substance. Both classes seem to predict that the frequency of initiation would be determined by the growth rate but neither of them appears to make the essential prediction that there be a negative correlation between successive inter-initiation times—they are not self-regulating. Furthermore, neither class seems to account for some features of the behaviour of autonomous plasmids, such as F, as will be discussed in the following section.

There is one other possible mechanism which might determine the frequency of initiation which must be mentioned since it is similar in most respects to the inhibitor mechanism we have discussed. In another form it has been proposed by Ycas *et al.* (1965) as a general mechanism for control of cell division. The necessary assumptions are that under a given set of conditions a substance—in our case H—is synthesized at a constant rate proportional to the gene dose, and that H is unstable. The *amount* of H at equilibrium will be determined by the relative rates of synthesis and breakdown, but the *concentration* will depend on the cell volume. This concentration will therefore fall as the cell volume increases and at a critical value initiation will occur, resulting in a doubling of its rate of synthesis, and equilibrium concentration, if the gene producing H is adjacent to the chromosome origin.

This model and our model make the same predictions. They are both specific interpretations of a general mechanism of control based on changes in concentration of an inhibitor of initiation as a result of growth. There is no simple means of distinguishing between them (or other interpretations within this class) at present. We will therefore confine attention to the specific interpretation we have proposed since it seems to have the virtue of greater simplicity. With the type of mechanism based on the ideas of Ycas *et al.* (1965) additional assumptions would be necessary to take account of changes in the rate of protein synthesis which would presumably be associated with changes in growth rate. Additional assumptions would also be necessary to account for the inhibition of initiation by amino acid starvation (Maaløe & Hanawalt, 1961).

It seems appropriate to discuss here two observations relevant to control of initiation in *Escherichia coli* which appear to be inconsistent with the mechanism we have discussed. Lark & Lark (1965) have suggested that in cultures of *E. coli* (strain 15т⁻; 555-7) growing exponentially with succinate as sole carbon source (mean doubling time = 70 min.) individual cells have two chromosomes which replicate alternately. This is clearly inconsistent with a system of control based on the concentration of a cytoplasmic inhibitor. The initial observations which led to this interpretation of the pattern of replication in succinate-grown cells were based on the hypothesis—due initially to Maaløe & Hanawalt (1961) and substantially confirmed by Lark, Repko & Hoffman (1963), Cerdá-Olmedo, Hanawalt & Guerola (1968), and Wolf, Newman & Glaser (1968)—that when a culture is deprived of a required amino acid no further initiation of replication occurs but all cells replicating their DNA complete this process.[1] The theoretical increment in DNA expected if (1) all the cells in a culture are replicating their DNA, (2) the rate of replication is constant, and (3) each chromosome has only one replication fork, is about 40%. Lark & Lark (1965) found that amino acid deprivation of succinate-grown cells resulted in only a 20% increment in DNA, although the average DNA content of such cultures was about two genome equivalents and the majority of cells seemed to be replicating their DNA. They therefore concluded that only one of the two chromosomes was replicating at any one time.

---

[1] It has usually been assumed that amino acid starved cells do not initiate new rounds of replication because *de novo* synthesis of a protein specifically involved in initiation is necessary. An alternative possibility on our model is that further initiation does not occur because increase in cell volume ceases.

There are several potential sources of error in this type of experiment. The theoretical increment in DNA depends critically on the fraction of cells in the population which are not replicating their DNA. This was estimated by Lark & Lark (1965) from the fraction of unlabelled cells remaining after a culture had been pulse-labelled with tritiated thymine. As Cooper & Helmstetter (1968) have pointed out, the technical difficulties inherent in this technique can lead to a considerable under-estimate of this fraction and a corresponding overestimate of the predicted increment in DNA. Deviations from the predicted increment in DNA during amino acid starvation are also possible for a different reason. Thus Billen & Hewitt (1966) have shown that the increment in DNA observed during amino acid starvation of the polyauxotrophic strain used by Lark & Lark (1965) depends on which of the required amino acids are removed from the growth medium. In addition, Doudney (1966) has shown that inhibition of protein synthesis results in a progressive decline in the capacity to *replicate* DNA. He found that if a glucose-grown culture was deprived of an essential amino acid *and* thymine for 60 min. and thymine then restored, negligible synthesis of DNA occurred. It thus seems that the completion of replication observed by Maaløe & Hanawalt (1961) and Lark *et al.* (1963) during inhibition of protein synthesis was due to the fortunate circumstance that under the cultural conditions used the decay in the capacity to replicate DNA was slow enough to permit a substantial fraction of the cell population to complete replication of their chromosomes. It is therefore possible that the capacity to replicate DNA during amino acid deprivation of the succinate-grown cultures used by these authors decays more rapidly than it does in glucose-grown cultures resulting in a smaller increment in DNA than would otherwise be expected. Lark & Bird (1965a, b) provided further evidence to support the hypothesis of alternate replication of two chromosomes in succinate-grown cells, by examining the distribution of label among progeny of individual cells previously exposed to [³H]thymine. As Cooper & Helmstetter (1968) have pointed out, however, this evidence is open to more than one interpretation. Thus although the evidence against alternate replication in succinate-grown cells is not compelling it does seem that the replication behaviour of succinate-grown cells may not be of the alternate type suggested by Lark & Lark.

The second observation which appears to be incompatible with an inhibitor-dilution mechanism concerns the replication pattern of glucose-grown cultures (doubling time = 45 min) which have previously undergone a period of thymine starvation (Pritchard & Lark, 1964).

After a period of deprivation (which stops DNA synthesis) equivalent to about one doubling time, the subsequent rate of DNA synthesis was increased. Evidence was presented that this increase in rate was due to initiation of new rounds of replication during or immediately after thymine starvation, and this interpretation is supported by recent experiments of Cerdà-Olmedo et al. (1968) and Wolf et al. (1968). On the basis of an inhibitor-dilution model, the reinitiation of replication at the chromosome origin after a period of thymine starvation would be due to the continued increase in cell volume which occurs under these conditions, and we should expect that reinitiation would occur at both of the origins possessed by a replicating chromosome. Hence we should expect that the maximum increase in rate of DNA replication should be threefold in a glucose-grown culture if all chromosomes are replicating their DNA at the time thymine is removed and have only one replication fork viz:

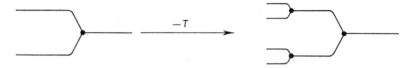

The observed maximum increase in rate (which, as expected, was reached after a period of starvation equivalent to about one doubling time—Fig. 6a, b) was closer to twofold than threefold. It was therefore suggested that either reinitiation occurs in only 50% of the cells in the culture, or it occurs in all cells but only one of the two chromosome origins of each replicating chromosome is affected. None of the experiments performed by Pritchard & Lark (1964) permitted a distinction between these alternatives, or a third possibility, that some cells fail to replicate their DNA at all after a period of thymine starvation. The

Legend to Figs. 6(a) and (b) opposite.

Fig. 6. Stimulation of DNA synthesis by prior treatment with nalidixic acid. (a) Exponential cultures of $15T^-$ (555–7) in glucose synthetic medium supplemented with the necessary nutritional factors (methionine, arginine, tryptophan and thymine) were exposed to nalidixic acid (50 μg./ml.) for the times (min.) indicated on the curves. They were then washed and resuspended in pre-warmed medium containing ^{14}C-thymine and samples taken at intervals to measure incorporation of label into acid precipitable material. Doubling time of the culture before treatment was 41 min. The horizontal axis (t, time of sampling; g doubling time) gives the theoretical increment in DNA in arbitrary units at the time of sampling assuming DNA synthesis to be exponential. (b) Data from Fig. 6 (a) and other similar experiments replotted to show the fractional increase in the rate of DNA synthesis as a function of the length of exposure to nalidixic acid. The horizontal axis gives the theoretical increase in mass of the culture in arbitrary units during the period of inhibition (t_i) if mass increases exponentially during inhibition of DNA synthesis. The point marked $-T$ shows the stimulation following thymine starvation in a parallel experiment.

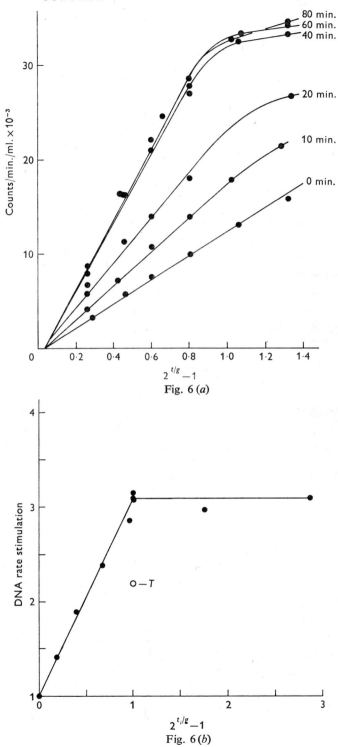

Fig. 6 (a)

Fig. 6 (b)

occurrence of reinitiation at only one of the two chromosome origins possessed by each replicating chromosome would be incompatible with an inhibitor-dilution model of initiation, but Lark & Bird (1965a) have presented data in support of this interpretation of the twofold increase in rate of DNA synthesis observed. We feel that this interpretation may be incorrect for two reasons, however. First, Lark & Bird's interpretation of their data depends on the validity of their hypothesis of alternate replication in succinate-grown cells. Secondly, although we have repeatedly confirmed that thymine starvation gives a maximum increase in the rate of DNA synthesis, closer to twofold than threefold, treatment of the same strain with nalidixic acid (a specific inhibitor of DNA synthesis) under the same conditions results in at least a threefold increase in rate (Fig. 6) which also seems to be due to reinitiation at the chromosome origin (Fig. 7).

There are several features of the curves shown in Figs. 6 and 7 which are notable in the present context. The decreasing rate of DNA synthesis observed in the 40–80 min. curves of Fig. 6a is transient. The 40 min. curve subsequently parallels the control curve. The rate stimulation of DNA synthesis, as a function of the theoretical increment in mass during treatment with nalidixic acid, is linear over one doubling time (Fig. 6b). Since in these experiments the increase in optical density of the treated and control cultures was not appreciably different over this period (data not shown), it follows that the rate stimulation is proportional to the mass increment during treatment as predicted by our model. The fact that the rate stimulation tends to a maximum after a period of treatment corresponding to one doubling period (although increase in optical density continues at a decreasing rate), is interesting, but the reason for this effect is not clear. It could mean that for some reason replication of a chromosome at three sites simultaneously (involving seven replication forks) is not possible. It could also mean that there is a limit to the rate of synthesis of deoxynucleotide triphosphate precursors which determines the maximum rate of DNA synthesis that can be achieved. The latter alternative might provide an explanation for the smaller rate stimulation that is obtained as a result of thymine starvation since this is known to disturb the control systems modulating the synthesis of the deoxynucleotide triphosphates (see Beacham, Barth & Pritchard, 1968). The difference in rate stimulation obtained by the two treatments does not appear to be due to a significant contribution from repair synthesis to the over-all rate of incorporation of labelled thymine after nalidixic acid treatment, since in experiments of the type shown in Fig. 7 little DNA of intermediate density is found.

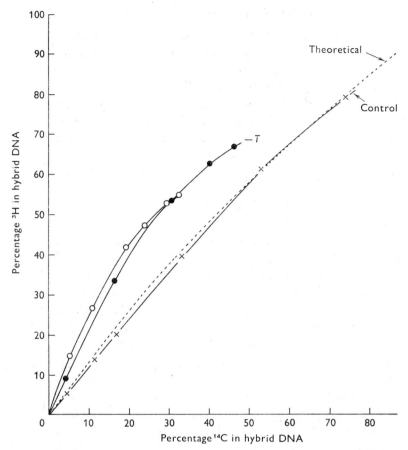

Fig. 7. Alteration in the sequence of replication induced by prior treatment with nalidixic acid or by thymine starvation. An exponential culture (15T⁻; 555-7) uniformly labelled with [¹⁴C] thymine in glucose synthetic medium was transferred to medium devoid of its required amino acids (but still with [¹⁴C]thymine) for 90 min. to allow replicating chromosomes to terminate. The culture was then washed to remove thymine, resuspended in medium with the required amino acids and [³H]thymine, and incubated for 29 min., when a 4·2% increment in DNA had occurred. (The increment in DNA was monitored frequently in a separate aliquot of the culture containing [¹⁴C]thymine at ten times the specific activity previously used.) The culture was again washed, transferred to medium identical to the initial growth medium and incubated for 2 hr. It was again washed and divided into portions which were treated as follows: ×, growth medium with 5-bromouracil in place of thymine; ○, growth medium plus nalidixic acid (50 μg./ml.) for 40 min., then into growth medium with 5-bromouracil; ●, growth medium without thymine for 40 min., 5-bromouracil then added. Samples were taken at intervals after addition of 5-bromouracil and analysed by density gradient separation according to Pritchard & Lark (1964) with minor modifications. The theoretical curve assumes that the distribution of replication forks is random with respect to the [³H]thymine labelled segments of DNA.

REPLICATION OF PLASMIDS

The number of episomes and other plasmids known to be capable of a stable co-existence with *Escherichia coli* continues to grow: it includes the classical sex factor F, drug-resistance transfer factors (R factors) of several distinct types, colicinogenic particles (Col factors) likewise of several distinct types, a haemolysin-producing factor (Hly factor), an enterotoxin factor (Ent factor), bacteriophages such as P1 (which does not become integrated into the bacterial chromosome)[1] and λ (which does). For further information about these plasmids the reviews of Bertani (1958), Jacob & Wollman (1961), Falkow, Johnson & Baron (1967), Watanabe (1967), and Meynell, Meynell & Datta (1968) can be consulted.

In this section we shall consider whether a mechanism of control of initiation similar to the one we have postulated for the bacterial chromosome might also control the replication of these particles.

The way in which the multiplication of λ prophage is coordinated with cell growth is understood in general terms. A protein specified by the λ genome can bind to a specific region of the λ DNA (Ptashne, 1967) thereby repressing the synthesis of other λ-specified proteins which mediate its replication. The repressor protein has also been shown to have a second function which is probably to inhibit prophage development directly when bound to it (Thomas & Bertani, 1964; Green, Gotchel, Hendershott & Kennel, 1967; Weisberg & Gallant, 1967). *Autonomous* replication of the prophage is consequently inhibited in cells containing repressor and its maintenance by the cell is only possible because it can become inserted into the bacterial chromosome (Campbell, 1962) where it is replicated *passively* by the replication functions of its host. A consequence of this system of repression is that lysogenic cells are immune to superinfection by λ, the repressor being produced by the prophage also inhibiting the replication of super-infecting particles which are consequently diluted out by subsequent growth of the culture (Wolf & Meselson, 1963). Repression can be abolished by agents which inhibit DNA synthesis. Such agents are called inducers. Loss of repression also occurs spontaneously in a small fraction of any population of lysogenic cells. In both cases the result will be excision of the prophage, autonomous replication of its DNA and, ultimately, lysis of the cell and release of mature λ particles.

[1] No chromosome location for P1 can be detected by genetic methods (Boice & Luria, 1961) but some recent experiments (Inselburg, 1968) involving DNA–DNA hybridization have been interpreted as indicating that in P1 lysogens the P1 DNA is an integral part of the bacterial chromosome.

sm of induction and spontaneous loss of repression is not

ect to one of its functions—direct blocking of phage
—the λ repressor behaves in a manner analogous to our
hibitor. We have therefore considered the possibility that
repressor might be dependent on replication of the gene
the amount of repressor synthesized being such that it does
s critical concentration at any time in the cell cycle. It
while to consider this possibility, since it could provide an
for the puzzling phenomenon of induction by inhibitors of
sis. Unfortunately for such a simple explanation of in-
is compelling evidence that synthesis of repressor occurs
ce of replication of the λ genome. Superinfection of an
with mutants of λ which synthesize repressors distin-
m wild-type repressor results in synthesis of the mutant
though replication of the superinfecting particle does not
1966; Tomizawa & Ogawa, 1967).
wn about the control of replication of plasmids which can
hase with the host cell without becoming integrated into the
, but there are striking similarities between the structure
ur of such plasmids and λ. In discussing these similarities
ine our attention to the sex factor F, since it has been the
ively studied. In most respects its properties are typical of
ids.
udies (Broda, Beckwith & Scaife, 1964) suggest that the F
the form of a ring, and a related particle (a Col VB factor)
own to have circular DNA (Hickson, Roth & Helinski,
become inserted into the bacterial chromosome (the Hfr
this state it appears to be replicated passively as part of the
osome at least in some strains (Espardellier-Joset, Harri-
& Marcovich, 1967; Berg & Caro, 1967; Abe & Tomizawa,
et al. 1968). It also seems to be capable of autonomous
since in newly infected cells it replicates more frequently
cterial chromosome (De Haan & Stouthamer, 1963). The
cant difference between λ and F in the context of this dis-
hat the latter, but not the former, can multiply along with
ell without integration into the chromosome. The most
imilarity is that cells harbouring F and other plasmids are
superinfection by homologous particles. The nature of the
on immunity exhibited by Hfr cultures was studied by
oss (1962). They showed that although Hfr cells are partially

resistant to infection by F (in that they are less efficient
F in F*lac*+ × Hfr matings than are F− cells in F*lac*+ ×
infection by F does occur. The progeny of infected cells
the F particle, however, F*lac* infection being detectable o
of the efficient rescue of the *lac*+ gene by recombination wit
of the recipient. The behaviour of F*lac*+ × Hfr mating
studied in more detail by Dubnau & Maas (1968) using
cells as recipients. Starvation converts Hfr cells into F−
(i.e. it lowers resistance to superinfection). These autho
the observations of Scaife & Gross (1962) and showed, in a
the infecting F particle fails to replicate and is diluted out
sequent growth of the recipient cultures. The superinfectic
shown by Hfr cells is thus remarkably similar to that
lysogens.

The response of F+ cultures to superinfection by F*lac*
different. F− phenocopies of the F+ cells also act as efficie
of F*lac* but in this case the progeny of the infected cell
exceptions,[1] carry *either* the F particle *or* the F*lac* particl
the majority of recipient colonies containing F*lac*+ cells ob
plating out single cells from an F*lac*+ × F+ mating mixtu
gated. Pure *lac*+ clones are infrequent. This behaviour, whi
of non-integrated plasmids, shows not only that two hom
genetically different F particles cannot co-exist indefinitely
clone, but also that the superinfecting particle can replace
homologue. Although accurate estimates have not been
published data suggest that a superinfected cell gives ris
F*lac* progeny with a relative frequency not grossly dif
equality.

It seems likely that the mutually exclusive behaviour
grated F particles has the same basis as the superinfectio
exhibited by Hfr recipients, the difference being that i
situation the incoming particle cannot replace the resident
which is inserted into the chromosome, except by recombi
behaviour of F+ recipients has been described as 'exclusi
1963). Since it probably has the same basis as the immunit
Hfr recipients, which is itself similar to that shown by λ,
tinue to use the term immunity to describe it, although neit
terms seems entirely appropriate.

---

[1] Exceptional clones in which two genetically distinct F primes (or an F
integrated F particle) persist, have been reported (Cuzin, 1962; De Haan
1963; Cuzin & Jacob, 1967c).

In view of the similarity in behaviour of λ lysogens and cells carrying non-integrated plasmids, it is not surprising that analogies should frequently be sought between the mechanisms of control which prevent their autonomous replication. There is a difficulty, however. If the functions controlling replication of a non-integrated particle are repressed, how is it to replicate? One possibility is that functions determining initiation specified by the host's genome can substitute for corresponding repressed functions specified by the plasmid. A mechanism of control of this sort could lead to synchronous replication of the plasmid and the host chromosome.

Although cellular control of initiation of non-integrated particles has not been ruled out[1] it does not by itself account for superinfection immunity; there would be no *a priori* reason why an incoming particle should not be replicated along with the resident homologue. In addition, as we shall show later, it does not account for the failure of an Hfr cell to initiate replication both at the chromosome origin and at the site of insertion of the F particle.

Jacob *et al.* (1963) suggested a mechanism of control quite different from the λ type, which accounts for many of the properties of extra-chromosomal particles. They proposed that such particles (and also the bacterial chromosome), are attached to specific sites on the cell membrane; that the number of such sites is limited, and that maintenance and replication of the particle requires that it be attached to such a site. Initiation of replication of non-integrated particles was assumed to be under *host* control but the functions (including 'initiator') which mediate initiation and replication were assumed to be specified by the particle itself. Thus this model also assumed cellular control of initiation but accounts for superinfection immunity (and the limitation on the number of particles per cell) by postulating a limited number of specific attachment sites. Jacob *et al.* (1963) also proposed that regular segregation of such particles (and the host chromosome) is ensured by growth of the membrane, and formation of the septum, between the sites occupied by sister particles (or chromosomes).

The following properties of F and similar plasmids are consistent with this hypothesis:

(1) The number per cell is small; in the case of F, indirect estimates are compatible with there being only one per chromosome origin—suggesting a small number of specific attachment sites.

[1] The existence of temperature sensitive mutants of the F factor which fail to be replicated at high temperature (Cuzin & Jacob, 1965, 1967*a*) is compatible with cellular control of replication. The mutations found might interfere with functions necessary for replication itself but not for initiation of replication.

(2) The fraction of cells which lose the particle is typically very small —indicating a reliable segregation mechanism.

(3) Superinfecting particles are either lost—because all available attachment sites are occupied—or replace the resident particle.

(4) Particles which can co-exist in the same cell do so because they occupy different attachment sites with different specificities.

There are two distinct aspects of this hypothesis. One is the notion of attachment to the cell membrane and a pattern of membrane growth leading to regular segregation of sister particles. The other is the notion of a limited number of attachment sites with appropriate specificity which control the number, and consequently the replication, of the particles.

There is considerable cytological (see Ryter, 1968) and chemical (Ganesan & Lederberg, 1965; Smith & Hanawalt, 1967) evidence for an association between bacterial DNA and the cell membrane. Cuzin & Jacob (1967b) have also provided striking evidence in support of such an association. They showed that when a culture of cells containing a mutant F particle whose replication was inhibited at high temperatures was uniformly labelled with [³H]thymidine at the permissive temperature and then allowed to grow through eight generations in unlabelled medium at the non-permissive temperature, the radioactive atoms incorporated into the polynucleotide strands of the chromosome and the F particle had largely remained associated together in the same daughter cells. This behaviour suggests that there is a physical association of the two ancestrally labelled structures which could result from attachment to a common area of the membrane. An equally remarkable observation of Dubnau & Stocker (1967) is also consistent with this suggestion. They found that when phage P22, grown on a strain carrying an R factor and three different Col factors, was used to transfer the R factor to a second strain by transduction, about 20 %, on average, of the R⁺ transductants also became Col⁺. This important observation suggests that these extrachromosomal particles must be located in proximity to each other inside the cell (it might even be taken to suggest a covalent linkage of their genomes despite contrary genetic evidence—cf. also footnote, p. 284).

Although the notion of an association between DNA and the cell membrane has considerable experimental support, the further idea that the membrane-attachment hypothesis provides a satisfactory explanation for the limited number of particles per cell, for the control of replication of extrachromosomal particles, and for the superinfection immunity, is less convincing. We have already discussed one difficulty inherent in models of this sort which assume that control of replication is positive. In the case of extrachromosomal particles there is an additional problem which is most easily demonstrated by a comparison of the different states of the F factor.

The configuration of an F+ cell in terms of the membrane-attachment hypothesis is shown in Fig. 8. Both chromosome and F particle are

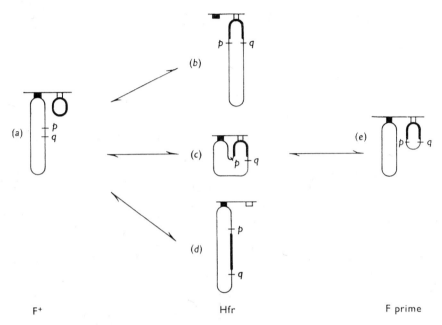

Fig. 8. Possible relationships between F and the cell membrane. F+ (*a*): chromosomes of F particle are shown as loops (chromosome, thin line; F particle, thick line) attached to different sites on the membrane. Letters *p* and *q* represent loci on the chromosome. Hfr (*b, c, d*): the F particle is shown integrated between the loci *p* and *q*. F-prime (*e*): detachment of the F particle has occurred by a cross-over between chromosome loci distal to *p* and *q*.

shown attached by their origins to their respective membrane sites, all of which are saturated. When the F particle becomes integrated into the bacterial chromosome three alternative configurations (Fig. 8*b–d*) seem possible. Configuration (*b*) would predict immunity to super-infecting F particles as is found (Scaife & Gross, 1962). It would also predict that vegetative replication of the chromosome would be under the control of the F particle and that the origin of vegetative replication should be located at the site of F insertion. This is not found (Abe & Tomizawa, 1967; Berg & Caro, 1967; Espardellier-Joset *et al.* 1967; Wolf *et al.* 1968), although a transposition of the vegetative origin to the site of F insertion may possibly occur in some strains or circumstances (Nagata, 1963; Wolf *et al.* 1968). Configuration (*d*), conversely, predicts no change in the location of the vegetative origin but loss of immunity. Configuration (*c*) is compatible with superinfection immunity, but it also predicts that the chromosome should have two origins of vegetative replication which could have opposite polarity (since genetic evidence indicates that the F particle can be inserted with either of the two possible orientations). Thus none of these three configurations as they stand provides a satisfactory interpretation both of superinfection immunity and the absence of initiation of replication at the site of insertion of F.

This paradox cannot be resolved by supposing that insertion of F into the chromosome so modifies its structure as to make it incapable of mediating its own replication during vegetative growth. Detachment of the F particle can occur leading to the formation of a new particle (F prime) which now includes part of the bacterial chromosome (Fig. 8*e*). F-prime particles are probably produced by a cross-over between two regions of the bacterial chromosome (spanning the particle) which, fortuitously, have limited homology (Broda *et al.* 1964). At least in some cases F-prime particles retain chromosomal genes (*p* and *q*) which spanned the F factor in the Hfr parent (Broda *et al.* 1964; Scaife, 1966; Berg & Curtiss, 1967). Thus detachment involving no change in the structure of the F factor itself permits the F-prime particle to replicate in step with growth of the host cell.

We can summarize the paradox raised by the reversible transition between the Hfr and F-prime states as follows. Since an F factor in the F-prime state mediates its own replication, why does it not do so in the Hfr state? If the answer to this question is to be found in detachment of the particle from the membrane in the Hfr state, why are Hfr cells immune to superinfection?

The difference in behaviour of F prime and Hfr cells presents an even more puzzling problem. An F-prime cell contains two replicons which replicate with the same frequency. A reciprocal cross-over between them will link them together, and the two origins present in the double replicon seem to be equivalent. If linking together two replicons prevents one of the two origins from initiating replication, what are the selection rules which determine which of the two is inactivated? The membrane attachment site model does not appear to provide an answer to any of these questions. We will therefore attempt to reinterpret them in terms of the inhibitor-dilution hypothesis.

We will assume that initiation of replication of F is controlled in the same fashion as initiation of replication of the chromosome. Let H_F be an inhibitor specified by a gene on the F particle. Let initiation occur when the inhibitor concentration falls to a critical value $[h_c]_F$. If inhibitor binds directly to the F particle we need not assume that the F particle specifies its own initiator. If inhibitor binds to initiator then each replicon must specify its own initiator. We need not distinguish between these two possibilities for the present since the consequences will be the same. We will also assume that in an Hfr cell the gene specifying H_F is transcribed at the time it is replicated even if this replication is passive—i.e. mediated by the chromosome origin of the host.

The number of F particles relative to the number of chromosome origins in an F$^+$ cell will be determined by the ratio of the values $[a/h_c]_F$ to $[a/h_c]_K$ (where K refers to inhibitor specified by the bacterial chromosome) and the absolute number of F particles per cell by the cell volume. If the ratio is unity the number of F particles will be equal to the number of chromosome origins and initiation of replication of the particle and the chromosome will be synchronous. If $[a/h_c]_F$ is larger than $[a/h_c]_K$ there will be fewer F particles than chromosome origins and replication of the F particle will occur after initiation of replication of the chromosome. In the converse case there will be more F particles than chromosome origins and they will replicate before initiation of replication of the chromosome. If the two ratios are not equal the numbers of F particles relative to the number of chromosome origins, and their relative times of initiation in the cell cycle, will be a function of the growth rate.

According to this interpretation of control of F replication, super-infection immunity of an F$^+$ cell occurs because the number of particles per cell at equilibrium is fixed by the cell volume and $[a/h_c]_F$. Infection of a cell by another particle will disturb this equilibrium temporarily (since the superinfecting particle will replicate when the resident particle does and thus raise the concentration of inhibitor to a value greater than $2h_c$), but the number of particles per cell will ultimately settle down to the equilibrium number again. If the products of replication of one particle always segregate to different daughter cells, genetically distinguishable particles will not remain together. This interpretation predicts that the incoming particle and the resident particle, or particles, will have an equal probability of appearing among the progeny. It also predicts that the incoming particle will replicate in synchrony with resident particles and in this respect is quite different from specific membrane attachment-site models. Immunity of an Hfr recipient has the same basis, but the incoming particle will be preferentially lost since persistence of the integrated particle is assured.

One way in which initiation of replication of the F particle might be prevented as a result of insertion into the chromosome can be demonstrated by a hypothetical example. Assume the following relationship:

$$[a/h_c]_F = 1 \cdot 5 [a/h_c]_K, \tag{8}$$

or

$$[V_c]_F = 1 \cdot 5 [V_c]_K, \tag{9}$$

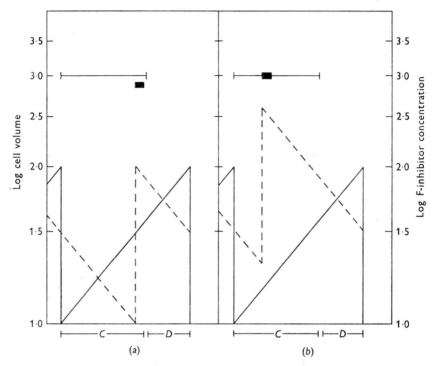

Fig. 9. Effect of insertion of an F particle into the chromosome on F inhibitor concentration. Solid curve represents changes in cell volume during the division cycle. For simplicity a doubling time equal to $(C+D)$ is shown. The replication periods of the chromosome (thin horizontal line) and F particle (thick line) are indicated above the curves. F inhibitor concentration is shown by a dashed curve. The relationship $[V_c]_F = 1 \cdot 5 [V_c]_K$ is assumed. (a) F+ cell; (b) Hfr.

then in an F+ cell the F particle will replicate later in the cell cycle than the time of initiation of replication of the chromosome (Fig. 9). If the F particle becomes integrated into the chromosome as shown in Fig. 9b its inhibitor will be synthesized earlier in the cell cycle, and at a smaller cell volume, than in an F+ cell. At equilibrium its concentration will therefore rise to a value greater than $2h_c$ and never fall to h_c. Consequently, initiation of replication of the F particle will normally be blocked.

It should be stressed that we have arbitrarily assumed that $[V_c]_F$ and $[V_c]_K$ are not equal solely to illustrate one type of selection rule which would ensure that only one of two replicons linked in series could initiate replication during vegetative growth. Since the time of F replication relative to the time of initiation of chromosome replication is not known, we cannot estimate what the true values are. It could be argued that the relationship $[V_c]_F \leqslant [V_c]_K$ would be favoured since if $[V_c]_F > [V_c]_K$ loss of the F particle might occur at slow growth rates (i.e. small cell volumes). In these circumstances the type of selection rule described above could not operate.

Different selection rules can be devised based on the inhibitor-dilution model but the example mentioned above will suffice to illustrate the general idea.

CONCLUSION

We have suggested that initiation of replication of the bacterial chromosome occurs at a specific cell volume, which is determined by the number of copies of the chromosome origin per cell, and have outlined a specific mechanism which would define the cell volume at which initiation occurs. We have also suggested that the replication of extra-chromosomal particles might be controlled in a similar way. The model contains one assumption for which there is no precedent—that of a gene whose transcription is linked to its replication. A variant of this model, however, which assumes continuous synthesis of an unstable gene product, has similar properties.

The main differences between models of this type and others which have been suggested are: (1) they assume that control of replication is negative—in the sense that the variable component in the system is inhibitory in its action; (2) the cell does not contain specific sites to which individual plasmids and the chromosome are attached; and (3) the number of copies of any plasmid present in one cell is determined by properties of the plasmid itself and the cell volume. It should be stressed that we do not wish to imply that replicons are not attached to the cell membrane, or that they are not attached to particular structures, or that attachment is not a necessary condition for replication, only that attachment sites, if they exist, are not replicon-specific and do not dictate the time of initiation.

This discussion of control of replication has intentionally been speculative. The justification we make for such speculation is that the mechanism of control that has been suggested leads to several predictions which can be tested. It has also helped us to see which are the important questions to be answered before the control of initiation of replication in bacteria can be understood even in a general way. Among these are whether or not components specified by the host are directly involved in initiation of replication of extrachromosomal particles, the basis of superinfection immunity, and, most important, the selection rules which determine the site of initiation of replication of a chromosome into which an F particle has been inserted.

ACKNOWLEDGEMENTS

We wish to thank our colleagues in Leicester, especially Dr T. H. Wood, for many helpful discussions during the preparation of this article. We are also grateful to Professors O. Maaløe and N. O. Kjeldgaard for permission to reproduce Fig. 2. J. Collins and P. T. Barth are grateful to the Medical Research Council for Studentships.

REFERENCES

ABE, M. & TOMIZAWA, J-I. (1967). Replication of the *Escherichia coli* K12 chromosome. *Proc. natn. Acad. Sci. U.S.A.* **58**, 1911.

BEACHAM, I. R., BARTH, P. T. & PRITCHARD, R. H. (1968). Constitutivity of thymidine phosphorylase in deoxyriboaldolase negative strains: dependence on thymine requirement and concentration. *Biochim. biophys. Acta*, **166**, 589.

BERG, C. M. & CARO, L. G. (1967). Chromosome replication in *Escherichia coli* I. Lack of influence of the integrated F factor. *J. molec. Biol.* **29**, 419.

BERG, C. M. & CURTISS, R. III. (1967). Transposition derivatives of an Hfr strain of *Escherichia coli* K-12. *Genetics*, **26**, 503.

BERTANI, G. (1958). Lysogeny. *Advanc. Virus Res.* **5**, 151.

BILLEN, D. & HEWITT, R. (1966). Influence of starvation for methionine and other amino acids on subsequent bacterial deoxyribonucleic acid replication. *J. Bact.* **92**, 609.

BOICE, L. B. & LURIA, S. E. (1963). Behaviour of prophage P1 in bacterial matings. I. Transfer of the defective prophage P1 d1. *Virology*, **20**, 147.

BRODA, P., BECKWITH, J. R. & SCAIFE, J. (1964). The characterization of a new type of F-prime factor in *E. coli* K12. *Genet. Res., Camb.* **5**, 489.

CAMPBELL, A. (1962). Episomes. *Advanc. Genet.* **11**, 101.

CERDÀ-OLMEDO, E., HANAWALT, P. C. & GUEROLA, N. (1968). Mutagenesis of the replication point by nitrosoguanidine: map and pattern of replication of the *Escherichia coli* chromosome. *J. molec. Biol.* **33**, 705.

COOPER, S. & HELMSTETTER, C. E. (1968). Chromosome replication and the division cycle of *Escherichia coli* B/r. *J. molec. Biol.* **31**, 519.

CUZIN, F. (1962). Multiplication autonome de l'épisome sexuel d'*Escherichia coli* K12 dans une souche Hfr. *C. r. hebd. Séanc. Acad. Sci., Paris*, **254**, 4211.

CUZIN, F. & JACOB, F. (1965). Analyse génétique fonctionnelle de l'épisome sexuel d'*Escherichia coli* K12. *C. r. hebd. Séanc. Acad. Sci., Paris*, **260**, 2087.

CUZIN, F. & JACOB, F. (1967a). Mutations de l'épisome F d'*Escherichia coli* K12. II. Mutants à réplication thermosensible. *Annls Inst. Pasteur, Paris*, **112**, 397.

CUZIN, F. & JACOB, F. (1967b). Existence chez *Escherichia coli* K 12 d'une unité génétique de transmission formée de différents réplicons. *Annls Inst. Pasteur, Paris*, **112**, 529.

CUZIN, F. & JACOB, F. (1967c). Association stable de deux épisomes F différents dans un clone d'*Escherichia coli. Annls Inst. Pasteur, Paris*, **113**, 145.

DE HAAN, P. G. & STOUTHAMER, A. H. (1963). F-prime transfer and multiplication of sexduced cells. *Genet. Res., Camb.* **4**, 30.

DOUDNEY, C. O. (1966). Requirement for ribonucleic acid synthesis for deoxyribonucleic acid replication in bacteria. *Nature, Lond.* **211**, 39.

DUBNAU, E. & MAAS, W. K. (1968). Inhibition of replication of an F'*lac* episome in Hfr cells of *Escherichia coli. J. Bact.* **95**, 53.

DUBNAU, E. & STOCKER, B. A. D. (1967). Behaviour of three colicine factors and an R (drug-resistance) factor in Hfr crosses in *Salmonella typhimurium. Genet. Res., Camb.* **9**, 283.

ECKER, R. E. & SCHAECHTER, M. (1963). Bacterial growth under conditions of limited nutrition. *Ann. N.Y. Acad. Sci.* **102**, 549.

ESPARDELLIER-JOSET, F., HARRIMAN, P. D., GOTS, J. & MARCOVICH, H. (1967). Localisation de l'origine de réplication végétative chez une souche Hfr d'*Escherichia coli* K12. *C. r. hebd. Séanc. Acad. Sci., Paris*, **264**, 1541.

FALKOW, S., JOHNSON, E. M. & BARON, L. S. (1967). Bacterial conjugation and extrachromosomal elements. *Ann. Rev. Genetics*, **1**, 87.

GANESAN, A. T. & LEDERBERG, J. (1965). A cell membrane bound fraction of bacterial DNA. *Biochem. biophys. Res. Commun.* **18**, 824.

GREEN, M. H., GOTCHEL, B., HENDERSHOTT, J. & KENNEL, S. (1967). Regulation of bacteriophage lambda DNA replication. *Proc. natn. Acad. Sci. U.S.A.* **58**, 2343.

HARVEY, R. J., MARR, A. G. & PAINTER, P. R. (1967). Kinetics of growth of individual cells of *Escherichia coli* and *Azotobacter agilis. J. Bact.* **93**, 605.

HELMSTETTER, C. E. & COOPER, S. (1968). DNA synthesis during the division cycle of rapidly growing *Escherichia coli* B/r. *J. molec. Biol.* **31**, 507.

HERBERT, D. (1959). Some principles of continuous culture. *Recent Progress in Microbiology*, p. 381. Ed. G. Tunevall. Stockholm: Almqvist and Wiksell.

HICKSON, F. T., ROTH, T. F. & HELINSKI, D. R. (1967). Circular DNA forms of a bacterial sex factor. *Proc. natn. Acad. Sci. U.S.A.* **58**, 1731.

INSELBURG, J. (1968). Physical evidence for the integration of prophage Pl into the Escherichia coli chromosome. *J. molec. Biol.* **31**, 553.

JACOB, F., BRENNER, S. & CUZIN, F. (1963). On the regulation of DNA replication in bacteria. *Cold Spring Harb. Symp. quant. Biol.* **28**, 329.

JACOB, F. & WOLLMAN, E. L. (1961). *Sexuality and the Genetics of Bacteria*. New York: Academic Press.

KJELDGAARD, N. O. (1961). The kinetics of ribonucleic acid and protein formation in *Salmonella typhimurium* during the transition between different states of balanced growth. *Biochim. biophys. Acta*, **49**, 64.

KJELDGAARD, N. O., MAALØE, O. & SCHAECHTER, M. (1958). The transition between different physiological states during balanced growth of *Salmonella typhimurium. J. gen. Microbiol.* **19**, 607.

KOCH, A. L. & SCHAECHTER, M. (1962). A model for statistics of the cell division process. *J. gen. Microbiol.* **29**, 435.

KUBITSCHEK, H. E. (1962). Normal distribution of cell generation rate. *Expl Cell Res.* **26**, 439.

LARK, K. G. (1966). Regulation of chromosome replication and segregation in bacteria. *Bact. Rev.* **30**, 3.

LARK, K. G. & BIRD, R. (1965 a). Premature chromosome replication induced by thymine starvation: restriction of replication to one of the two partically completed replicas. *J. molec. Biol.* **13**, 607.

LARK, K. G. & BIRD, R. E. (1965 b). Segregation of the conserved units of DNA in *Escherichia coli. Proc. natn. Acad. Sci. U.S.A.* **54**, 1444.

LARK, K. G. & LARK, C. (1965). Regulation of chromosome replication in *Escherichia coli*: Alternate replication of two chromosomes at slow growth rates. *J. molec. Biol.* **13**, 105.

LARK, K. G., REPKO, T. & HOFFMAN, E. J. (1963). The effect of amino acid deprivation on subsequent deoxyribonucleic acid replication. *Biochim. biophys. Acta*, **76**, 9.

LIEB, M. (1966). Studies of heat-inducible λ mutants. II. Production of C_1 product by superinfecting λ^+ in heat-inducible lysogens. *Virology*, **29**, 367.

MAALØE, O. (1961). The control of normal DNA replication in bacteria. *Cold Spring Harb. Symp. quant. Biol.* **26**, 45.

MAALØE, O. & HANAWALT, P. C. (1961). Thymine deficiency and the normal DNA replication cycle. I. *J. molec. Biol.* **3**, 144.

MAALØE, O. & KJELDGAARD, N. O. (1966). *Control of Macromolecular Synthesis.* New York: Benjamin.

MAAS, R. (1963). Exclusion of an *Flac* episome by an Hfr gene. *Proc. natn. Acad. Sci. U.S.A.* **50**, 1051.

MEYNELL, E., MEYNELL, G. G. & DATTA, N. (1968). Phylogenetic relationships of drug-resistance factors and other transmissible bacterial plasmids. *Bact. Rev.* **32**, 55.

NAGATA, T. (1963). The molecular synchrony and sequential replication of DNA in *Escherichia coli. Proc. natn. Acad. Sci. U.S.A.* **49**, 551.

OISHI, M., YOSHIKAWA, H. & SUEOKA, N. (1964). Synchronous and dichotomous replications of the *Bacillus subtilis* chromosome during spore germination. *Nature, Lond.* **304**, 1069.

PRITCHARD, R. H. (1965). Structure and replication of the bacterial chromosome. *Br. med. Bull.* **21**, 203.

PRITCHARD, R. H. (1966). Replication of the bacterial chromosome. *Proc. R. Soc. B,* **164**, 258.

PRITCHARD, R. H. (1968). Control of DNA synthesis in bacteria. *Heredity* (Abstract) **23**, 472.

PRITCHARD, R. H. & LARK, K. G. (1964). Induction of replication by thymine starvation at the chromosome origin in *Escherichia coli. J. molec. Biol.* **9**, 288.

PTASHNE, M. (1967). Specific binding of the λ phage repressor to λ DNA. *Nature, Lond.* **214**, 232.

ROSENBERGER, R. F. & KESSEL, M. (1967). Synchrony of nuclear replication in individual hyphae of *Aspergillus nidulans. J. Bact.* **94**, 1464.

RYTER, A. (1968). Association of the nucleus and the membrane of bacteria: A morphological study. *Bact. Rev.* **32**, 39.

SCAIFE, J. (1966). F-prime factor formation in *E. coli* K 12. *Genet. Res. Camb.* **8**, 189.

SCAIFE, J. & GROSS, J. D. (1962). Inhibition of multiplication of an F-lac factor in Hfr cells of *Escherichia coli* K–12. *Biochem. biophys. Res. Commun.* **7**, 403.

SCHAECHTER, M., MAALØE, O. & KJELDGAARD, N. O. (1958). Dependency on medium and temperature of cell size and chemical composition during balanced growth of *Salmonella typhimurium. J. gen. Microbiol.* **19**, 592.

SCHAECHTER, M., WILLIAMSON, J. P., HOOD, J. R. Jun. & KOCH, A. L. (1962). Growth, cell and nuclear divisions in some bacteria. *J. gen. Microbiol.* **29**, 421.

SMITH, D. W. & HANAWALT, P. C. (1967). Properties of the growing point region in the bacterial chromosome. *Biochim. biophys. Acta*, **149**, 519.

THOMAS, R. & BERTANI, L. E. (1964). On the control of the replication of temperate bacteriophages superinfecting immune hosts. *Virology*, **24**, 241.

TOMIZAWA, J-I. & OGAWA, T. (1967). Effect of ultraviolet irradiation on bacteriophage lambda immunity. *J. molec. Biol.* **23**, 247.

WATANABE, T. (1967). Evolutionary relationships of R factors with other episomes and plasmids. *Fedn Proc.* **26**, 23.

WEISBERG, R. A. & GALLANT, J. A. (1967). Dual function of the λ prophage repressor. *J. molec. Biol.* **25**, 537.

WOLF, B. & MESELSON, M. (1963). Repression of the replication of superinfecting bacteriophage DNA in immune cells. *J. molec. Biol.* **7**, 636.

WOLF, B., NEWMAN, A. & GLASER, D. A. (1968). On the origin and direction of replication of the *Escherichia coli* K 12 chromosome. *J. molec. Biol.* **32**, 611.

YCAS, M., SUGITA, M. & BENSAM, A. (1965). A model of cell-size regulation. *J. theoret. Biol.* **9**, 444.

THE DEVELOPMENT OF ORGANELLES
CONCERNED WITH ENERGY PRODUCTION

DAVID LLOYD

Medical Research Council Group for Microbial Structure and Function, Department of Microbiology, University College, Cathays Park, Cardiff

INTRODUCTION

The energy-yielding and conserving reactions of oxidation and photo-synthesis have many common features. In both processes the energy of electrons passing down a chain of redox carriers from a negative to a positive potential is harnessed as the potential free energy of hydrolysis of the terminal pyrophosphate bond of ATP. Both processes require that some of the reactants be embedded in highly hydrophobic micro-environments, and highly organized in chains or clusters. It is the 'coupling membranes' (Mitchell, 1966) of the energy-yielding organelles that provide the specialized locales for these processes. The most fully investigated system is the mammalian mitochondrion, where the sequence of the respiratory carriers is reasonably well established.

The electron transport system of the chloroplast is less well defined, and similar pathways of electron transport in bacteria are also far less well established. The nature of the coupling process is still the subject of a great debate because the identity of the several non-phosphorylated and phosphorylated high-energy intermediates postulated in the chemical theory has defied intensive investigation, and now their very existence has been questioned.

The energy-yielding organelles of eucaryotic micro-organisms are similar in structure to those of higher organisms. Thus yeasts, fungi, protozoa and algae all possess mitochondria, while members of the last named phylum also possess chloroplasts. Procaryotic micro-organisms do not possess specialized discrete energy-yielding organelles as such, and the functions of electron transport and phosphorylation in both oxidative and photosynthetic processes are performed by the cytoplas-mic membranes. While the bacterial membranes are responsible for a variety of functions other than these, differentiation to give specialized areas of membrane responsible for the energy-producing and conserving reactions has occurred in many species. Thus the 'electron transport particles' of *Azotobacter* and the 'chromatophores' of photosynthetic

bacteria are almost certainly artifacts of isolation produced by com-
minution of membranes; *in vivo* these are probably formed by invagina-
tions of the cytoplasmic membranes of these organisms. The term
mesosome has been used to describe any membraneous organelle produced by differentiation of the bacterial cell membrane. It now seems
likely that there are several different types of mesosome which perform
different functions, and that the structures seen in anaerobes such as the
Lactobacilli may be quite unrelated functionally to those of the Bacilli
and other highly aerobic organisms.

The emphasis on energy-producing processes of mitochondria and
chloroplasts until recent years led to a neglect of the other properties of
these organelles. The permeability properties of the inner mitochondrial
membrane, for example, are of key importance in regulation and control
of energy metabolism. The findings that both chloroplasts and mitochondria are capable of considerable biosynthetic feats such as protein
synthesis and lipid synthesis opened up new fields of investigation.
Studies of protein synthesis by isolated organelles have revealed that this
process is a genuine property of organelles which is distinct from that
occurring on hyaloplasmic ribosomes. The discovery, in chloroplasts
and mitochondria, of DNA as a likely candidate for the role of the genetic
material implicated in mechanisms of extra-chromosomal inheritance,
suggested that these organelles may be, at least, semi-autonomous.

Krebs (1967) has recently stressed the importance of asking the
right questions; perhaps this is a convenient point to highlight the
present state of our ignorance of the mechanisms of development of
energy-yielding organelles by formulating the following questions, of
relevance to future progress in this field:

(1) What is the lay-out of the components of the coupling membranes
and how are they integrated into a supra-molecular membrane structure?

(2) Where are the individual membrane components synthesized, and
how are the rates of synthesis of different components controlled?

(3) How are the various membrane components assembled to give a
functional energy-yielding organelle membrane?

(4) How do existing organelles grow, and can existing membranes be
extensively modified without net mass increase? If so, does turnover of
existing membrane components contribute to this process?

(5) What control mechanisms are involved in the assembly and
modification of organelle membranes? What is the extent of organelle
autonomy with regard to both membrane-bound and soluble components? At what level does correlation occur between chromosomal
and extrachromosomal information?

(6) What determines the growth rate of organelles? Do they divide and, if so, how is the division of organelles linked to the process of cell division?

While recent work throws some light on the processes of organelle development, many of these questions cannot yet be answered definitely.

THE MOLECULAR STRUCTURE OF THE MEMBRANES OF ENERGY-PRODUCING ORGANELLES

Mitochondria

While techniques of electron microscopy have yielded a great deal of information on the gross morphology of organelles, they are of only limited use for the elucidation of the ultrastructure of membranes at a molecular level, and give little information about the molecular packing arrangements of phospholipid, protein and polysaccharide components. The introduction of other physical approaches to the field of membrane structure has already indicated the potential value of techniques such as X-ray and electron diffraction (Parsons, 1966), reflectance measurements (Thompson & Huang, 1966), optical rotatory dispersion (Wallach 1966) and infra-red analysis (Maddy & Malcolm, 1965). The application of these techniques, together with chemical analyses of highly purified membrane fractions, has shown that membranes do not have a common structural pattern (Parsons, 1966; Korn, 1966). Although the Davson–Danielli model for myelin is consistent with all the available data on this much-studied membrane structure, it is not of general application and recent findings are contrary to the 'Unit membrane theory' (Robertson, 1959).

The membranes of mitochondria are probably the best characterized of all membrane structures other than myelin. Now that the inner and outer membranes of mammalian mitochondria can be separated (Sottocasa, Kuylenstierna, Ernster & Anders, 1967; Parsons *et al.* 1967) detailed analyses have been made of the enzymes and lipids of these two membranes. The outer membrane, which is 55Å thick, has a high lipid content, no cytochromes coupled to oxidative phosphorylation, cytochrome b_5 similar to that of the endoplasmic reticulum, NADH-cytochrome c reductase activity which is rotenone-insensitive, and (unlike endoplasmic reticulum) no NADPH-cytochrome c reductase. The inner mitochondrial membrane contains all the true respiratory chain components and the enzymes of oxidative phosphorylation. It has a low ratio of lipid to protein; much of the phospholipid is cardiolipin and it has very little cholesterol (Parsons, Williams & Chance, 1966).

About 20% of the mitochondrial membrane protein consists of cyto-chromes and flavoproteins; in the inner membrane the electron transport components are intimately associated with structural protein (Criddle, Bock, Green & Tisdale, 1962). It has been suggested that the phospho-lipid of the inner membrane (which is not present in sufficient quantity to give a bimolecular leaflet) is arranged as a flat micelle structure, 9 Å in thickness, with the individual molecules hexagonally packed. This leaves 45 Å available for membrane proteins but it is not yet clear whether these are present in a globular or extended form (Parsons, 1966).

The inner mitochondrial membrane is thrown into folds or cristae, and the inner surface of this membrane is covered with regular knob-like structures, the inner membrane subunits. These structures have been seen in both negatively stained material and in sections, and have been identified as the oligomycin-sensitive ATPase of mitochondria, the enzyme involved in the terminal step of oxidative phosphorylation (Racker et al. 1965). Similar inner membrane subunits have been found in other membranes having an energy-coupling function such as chloroplasts, 'chromatophores' and the plasma membranes of some bacteria.

The elegant spectrophotometric work of Chance has led to the con-clusion that the cytochromes are present in 1:1 ratios, and he has postulated that the entire function of electron transport could be accomplished by a unit respiratory assembly or 'oxysome' containing one molecule of each cytochrome associated with 2 or 3 molecules of flavoprotein and 10 to 20 molecules of reduced pyridine nucleotide. Kinetic data have suggested that the haem proteins may be closely enough associated so that only rotational movement is necessary for their interaction (Chance, 1967). The estimated molecular weight of this entity would be of the order of 2×10^6 and it would have a diameter of about 170 Å. The relationship of this entity to the organization of the mitochondrial membrane is shown in Fig. 1. The chemiosmotic hypo-thesis (Mitchell, 1966) requires that in all coupling membranes the electron transport chains zig-zag so that any two redox couples par-ticipating in a phosphorylation step are orientated transversely across the membrane, making possible a undirectional flow of protons. The anisotropic ATPase must also span the membrane so as to be accessible to OH^- ions from one side and to protons from the other. The elucida-tion of the molecular organization of coupling membranes in relation to the mechanism of energy conservation is still a formidable problem.

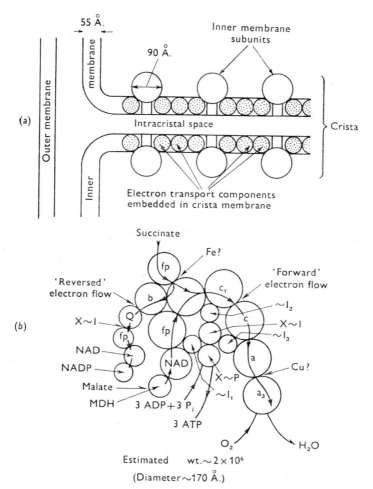

Fig. 1. Oxysome models proposed by Chance (1965). (a), Arrangement of the respiratory chain and coupling mechanism in relation to the organization of the mitochondrial membranes. (b), Pathways of electron transport and energy transfer in the 'oxysome'. I_1, I_2 and I_3 represent inhibitory ligands involved at the three coupling sites, and $X{\sim}I$ and $X{\sim}P$ are the high-energy intermediates proposed in the chemical theory of oxidative phosphorylation.

The cell membranes of non-photosynthetic bacteria

The structure and functions of the bacterial cytoplasmic membrane have been reviewed by Hughes (1962) and by Salton (1967). A number of cell-free systems have been prepared from both aerobic and facultative bacteria which carry out oxidative phosphorylation. While the bacterial systems have low P:O ratios and seldom show respiratory control, the respiratory carriers, like those of mitochondria, are localized in a membrane. The phosphorylating extracts from *Mycobacterium phlei*, for example, can be resolved into a particulate fraction (which contains the electron transport components) and a soluble fraction which contains the coupling factor proteins and the bulk of the enzymes which carry out the exchange reactions associated with oxidative phosphorylation (Asano & Brodie, 1965). The organization of these components *in vivo* remains to be investigated.

Photosynthetic membrane structures

The molecular structure of photosynthetic lamellar systems has been investigated by a wide variety of techniques including the newer freeze-etching process using the electron microscope, X-ray diffraction, and optical methods such as birefringence, dichroism and polarization of fluorescence. Many models have been proposed as a basis for understanding chloroplast and chromatophore function in relation to structure, and several different functional units of photosynthesis have been described.

The photosynthetic lamellae appear as flattened globular areas (thylakoids) with pebbled surfaces. Each of the pebbles has an area which could accommodate 200 chlorophylls stacked in close array. These correspond to the 'quantasomes' seen in shadowed preparations or to the 175Å × 90Å and 100Å × 90Å particles found embedded in the membranes of freeze-etched specimens by Park & Branton (1967). The occurrence of the 175Å particle seems to be restricted to membranes containing chlorophyll or bacteriochlorophyll, suggesting that it may be uniquely involved in quantum conversion but not uniquely associated with oxygen evolution. It seems likely that several quantasomes are associated with one oxygen-evolving site. Furthermore, the surfaces seen in freeze-etched preparations are non-polar in nature; so that here again the Davson–Danielli model of membrane structure is untenable.

Any functional unit of photosynthesis must contain the complete molecular recipe for the over-all processes of quantum conversion, photosynthetic electron transport and phosphorylation. The current theory of excitation transfer by resonance requires a favourable orientation of close-packed chlorophyll molecules, and much of the work in this field has been directed to the elucidation of the stacking of the planar porphyrins with their hydrophobic side chains. Little is known of the physical arrangement of the electron transport chains or of the phosphorylation mechanism.

Menke (1967) has sought a consistent interpretation of results obtained by various methods with thylakoids of bacteria. He suggests that these have an outer boundary layer consisting mainly of protein and that the major portion of the lipids is localized within the structure. The thylakoids of chloroplasts appear to be similar and have a protein layer consisting of 40Å particles which may correspond to a structural protein of molecular weight 25,000.

THE ASSEMBLY OF MEMBRANE STRUCTURES *IN VITRO*

The *in vitro* reconstitution of fractionated membrane components to give functional complexes may provide information relevant to problems of the mechanism of organelle assembly during growth. Reagents such as cholate, deoxycholate, *tert*–amyl alcohol, butanol, petroleum ether and cyclohexane which cleave protein–lipid and/or lipid–lipid bonds are effective in fragmenting the mitochondrial electron transport chain into 'complexes' (Green & Fleischer, 1962). When these isolated complexes are mixed under appropriate conditions they reassemble to form a product capable of integrated electron transport (Fowler & Hatefi, 1961).

'Structural protein' has a molecular weight of 22,500 and accounts for 40–50% of the mitochondrial protein (Criddle *et al.* 1962). It combines in a one-to-one ratio with each of the cytochromes; cytochrome c is bound less firmly than b- or a-type cytochromes. Structural protein also self-polymerizes and binds phospholipid; it appears to have hydrophobic surfaces and may form the backbone of the inner mitochondrial membrane. Indeed, up to 90% of the lipid of this membrane can be extracted without loss of gross mitochondrial structure (Richardson, Hultin & Green, 1963).

Another preparation of mitochondrial structural protein (F_4) (Zalkin & Racker, 1965) aggregates into sheets when mixed with phospholipid and the resulting structures resemble vesicles obtained on disruption of mitochondria (Stoeckenius, 1966). Racker suggests that this protein may form the framework for the association of the respiratory enzymes and coupling factors involved in the integrated system of the inner membrane. Kagawa & Racker (1966) have succeeded in reconstituting phosphorylating submitochondrial particles from their component fragments. Morphologically the reconstituted particles were indistinguishable from the starting material but complete restoration of function necessitated the inclusion of several coupling factors. The properties of individual membrane components may change when they are separated from the parent membrane, although reconstitution sometimes leads to restoration of these properties. Thus oligomycin sensitivity of ATPase is

20

dependent on the association of this enzyme with other membrane constituents.

Resolution of bacterial cell membranes by treatment with detergents or exposure to ultrasonically generated hydrodynamic shearing forces gives rise to 4S particles, which may not represent subunits in any functional sense (Brown, 1965; Salton & Netschey, 1965). Triple-layered aggregates are sometimes formed on reassociation (Razin, Morowitz & Terry, 1965).

Thus the formation of membranes *in vitro* can proceed by spontaneous combination of components; little or no directive influence other than the complementarity of surface is needed for the assembly process. Whether or not the actual assembly of organelle membranes *in vivo* is also an externally undirected process is not yet clear.

BIOSYNTHETIC CAPACITIES OF ORGANELLES

DNA and its synthesis

Mitochondrial DNA comprises less than 1 % of the total cell DNA, has a buoyant density which distinguishes it from nuclear DNA and is characterized by its rapid renaturation (suggesting extensive homogeneity with regard to base sequence). The circular (sometimes catenated or dimeric) molecules from mammalian mitochondria (Clayton & Vinograd, 1967), have a circumference of 5–6μ, and a molecular weight of 10–11 × 10^6 daltons (Borst, 1967), corresponding to a sequence of 15,000 base pairs. The genetic information carried by this DNA (assuming all the molecules are identical) would be sufficient to code for 5000 amino acids or about 25 proteins. There is no convincing evidence for the circularity of mitochondrial DNA from micro-organisms, but it is larger than that from mammalian mitochondria. For instance, DNA strands 47μ long have been obtained from yeast mitochondria (Borst, van Bruggen & Ruttenberg, 1968) and *Tetrahymena* mitochondria each contain seven or more 17·6μ strands of DNA (Suyama, 1966, and personal communication). Mitochondrial DNA has also been isolated from yeast by Sinclair & Stevens (1966) and by Avers (1967) and from *Euglena* (Edelman, Epstein & Schiff, 1966) and *Neurospora* (Luck & Reich, 1964).

Chloroplasts contain between 10^8 and 4 × 10^9 daltons of DNA per organelle but the molecular weight and structure of intact chloroplast DNA is not known. The kinetoplast (the large mitochondrion-like organelle of certain flagellates) also contains DNA which has a buoyant density distinct from that of nuclear DNA (Du Buy, Mattern & Riley,

1966). For further details of organelle DNAs see reviews of Borst, Kroon & Ruttenberg (1967) and Granick & Gibor (1967).

The incorporation of [³H]-thymidine into the mitochondria of whole cells of a slime mould (Guttes, Hanawalt & Guttes, 1967) and *Tetrahymena* (Parsons, 1965), and into mitochondria isolated from *Physarum polycephalum* (Brewer, de Vries & Rusch, 1967) and from yeast (Wintersberger, 1968) has been taken as indicative of the synthesis of mitochondrial DNA. The product was characterized as such in the case of yeast and incorporation was shown to be sensitive to actinomycin D and mitomycin C. Parsons & Simpson (1967) and Neubert, Bass & Oberdisse (1968) have investigated DNA synthesis in mammalian mitochondria. The latter have characterized a mitochondria-specific DNA polymerase and have measured the turnover rate of mitochondrial DNA (7–10 days).

DNA synthesis in isolated chloroplasts has been demonstrated by Spencer & Whitfeld (1967).

RNA and its synthesis

Ribosomal RNAs (23S and 16S) and transfer RNA (4S) from yeast mitochondria were first characterized by Wintersberger (1966). The similarities between ribosomal RNAs of yeast mitochondria and those from *Escherichia coli* have been emphasized by Rogers, Preston, Titchener & Linnane (1967) and these workers propose that the mitochondrial ribosomes, like those from bacteria, are 70S particles, while the eucaryotic hyaloplasmic ribosomes are of the 80S type. There is some evidence obtained from work with mammalian mitochondria to support the view that mitochondrial ribosomes are smaller than those of the hyaloplasm. However Rifkin, Wood & Luck (1967) have obtained a ribosome monomer (81S) from mitochondria of *Neurospora crassa* which gives subunits of 61S and 47S. They were able to distinguish between the RNAs of these particles and those of hyaloplasmic ribosomes by their differing sensitivities to Mg^{2+}, base compositions and sedimentation characteristics. Dure, Epler & Barnett (1967) have also studied the sedimentation properties of these RNAs and Küntzel & Noll (1967) have described the isolation of mitochondrial polysomes from this fungus.

Mitochondrial-specific amino-acyl-transfer RNAs and their synthetases have been characterized in *Neurospora* (Barnett & Brown, 1967; Barnett, Brown & Epler, 1967) and in *Tetrahymena* (Suyama & Eyer, 1967). However, as suggested by Suyama (1967) on the basis of hybridization studies, these transfer RNAs are probably synthesized outside the mitochondria and subsequently transported into the organelles.

Certainly the synthesis of these RNA species could not be demonstrated in isolated mitochondria (Suyama & Eyer, 1968). On the other hand, mitochondrial messenger RNA may be synthesized within the organelle. DNA-dependent RNA polymerase activity has been shown in mitochondria from yeast (Wintersberger, 1966) and from *Neurospora crassa* (Luck & Reich, 1964) and some mitochondrial RNA does have properties characteristic of a 'messenger' (Neubert, Helge & Merker, 1968).

The ribosomes isolated from chloroplasts (Svetailo, Philippovich & Sissakian, 1967; Sager & Hamilton, 1967), like those from mitochondria, are distinct from hyaloplasmic ribosomes and are also of the 70S type. Chloroplast polysomes have been studied by Chen & Wildman (1967), and by Stutz & Noll (1967). Hybridization studies suggest that chloroplast ribosomal RNA is coded for by chloroplast DNA (Steel-Scott & Smillie, 1967) and autoradiographs have shown that newly synthesized RNA first appears over regions of chloroplast DNA (Gibbs, 1967).

Protein synthesis

For some years there was doubt about the ability of mitochondria to synthesize proteins, but it is now quite clear that the incorporation of amino acids into the protein of isolated mitochondria is not dependent on the presence of contaminating particles or bacteria. Protein synthesis has been studied in mitochondria from *Tetrahymena* (Mager, 1960), *Neurospora* (Hall & Greenawalt, 1964) and yeast (Grivell, 1967; Lamb, Clark-Walker & Linnane, 1968), as well as from mammals. Some structural integrity of the mitochondria is necessary and there are requirements for Mg^{2+}, inorganic phosphate and an oxidizable substrate as energy source (Roodyn, 1966). Incorporation into soluble mitochondrial enzymes such as malate dehydrogenase and cytochrome *c* is negligible (Roodyn, Suttie & Work, 1962), but a major product of mitochondrial protein synthesis is associated with a membrane-derived fraction rich in phospholipid, RNA and respiratory enzymes (Roodyn, 1962). Further treatment of the labelled product with detergents and butanol resulted in the isolation of two proteins which were intimately associated with lipid, and Roodyn concluded that the major site of incorporation of amino acids, *in vitro*, is into insoluble lipoprotein, probably derived from the mitochondrial membrane. Labelled structural protein has been isolated by Haldar, Freeman & Work (1966), while Beattie, Basford & Koritz (1967a) and Neuport, Brdiczka & Bücher (1967) have shown that the incorporation of radioactive amino acids is predominantly into proteins of the inner mitochondrial membrane and not into the soluble proteins such as cytochrome c. Recent work in yeast indicates that cytochrome c is coded for by nuclear DNA (Sherman *et al.* 1967) and is synthesized on extra-mitochondrial ribosomes (Gonsález-Cadavid & Campbell, 1967; Freeman, Haldar & Work, 1967). Kadenbach (1968) has come to a similar conclusion from *in vitro* experiments with a reconstructed microsomal-mitochondrial system. The *in vitro* reconstitution of electron transport and oxidative phosphorylation in mitochondria from a cytochrome-c deficient yeast mutant by the addition of purified cytochrome c has been achieved by Matoon & Sherman (1966). This suggests that *in vivo* synthesis of cytochrome c in an extra-mitochondrial compartment of the cell may be followed by its spontaneous incorporation into its specific functional site.

Isolated chloroplasts also incorporate amino acids into proteins. Their amino acid activating enzymes were demonstrated by Bové & Raacke (1959) and by Marcus (1959), and the kinetics of labelling of amino acyl-transfer RNA in chloroplasts suggests that it is an early product of amino acid incorporation (Spencer & Wildman, 1964). Protein synthesis has also been demonstrated in chloroplasts isolated from *Euglena* (Eisenstadt & Brawerman, 1964) and from *Acetabularia* (Goffeau & Brachet, 1965). The proteins synthesized, however, may include more than the limited number of structural proteins synthesized by mitochondria, since Smillie *et al.* (1967) have indirect evidence for the *in vivo* synthesis of *both* soluble *and* lamellar-bound enzymes inside the chloroplast.

Lipid synthesis

The formation of malonyl-CoA and the synthesis of long-chain fatty acids in beef heart mitochondria was demonstrated by Hülsmann (1962) and the respiration-dependent incorporation of inorganic phosphate into total phospholipid of isolated mitochondria was studied by Garbus, De Luca, Loomans & Strong (1963). However some of the enzymes involved in the synthesis of complex mitochondrial lipids are known to be extramitochondrial (Wilgram & Kennedy, 1963); for example, lecithin synthesis occurs on microsomes (Schneider, 1963). Phospholipid metabolism has been reviewed by Dawson (1966) and factors affecting lipid synthesis in mitochondria have been discussed by Hülsmann, Wit-Peeters & Benckhuysen (1966).

Chloroplasts, like mitochondria, apparently synthesize lipids by an energy-requiring process. Stumpf, Bové & Goffeau (1963) have been able to show the light-dependent incorporation of acetate into lipids of isolated chloroplasts and the synthesis of long-chain fatty acids.

GROWTH OF ORGANELLES

Growth of the energy-yielding organelle in non-photosynthetic bacteria

Growth of *Haemophilus parainfluenzae* requires the formation of a membrane-bound electron transport system consisting of six cytochrome components (White & Smith, 1962). Changes of oxygen tension, growth with a limiting amount of haemin, or growth in the presence of secobarbital leads to modifications of the cytochrome oxidase composition and the proportions of the cytochromes under conditions where there is no cell division (White, 1962). Seven different membrane-bound flavoprotein dehydrogenases can be formed in varying proportions during different growth conditions and the great variability in the composition of the cytochrome and flavoprotein components of the electron transport systems during conditions of limited net growth suggests that each pigment is added individually to the membrane (White, 1964). Wright & White (1966) have also shown that various components of the electron transport system are formed at different rates during the growth of spheroplasts of *Haemophilus*. Thus the synthesis of each cytochrome, flavoprotein, and also that of 2-demethyl-vitamin K_2 and phospholipid appears to be controlled individually but changes in phosphatidyl-glycerol metabolism appear to be closely correlated with cytochrome synthesis (D. C. White & A. Tucker, personal communication). Addition of oxygen to anaerobically grown *Staphylococcus aureus* results in the formation of a complex membrane-bound electron transport system. Frerman & White (1967) have investigated the wide range of respiratory pigment and lipid changes that occur in this transition and suggest that the membrane of this organism exists as a mosaic of domains with differing compositions. This concept is diametrically opposed to that of Green & Silman (1967) and of Green *et al.* (1967), regarding the formation of the mitochondrial membrane from identical multienzyme packets and further work will be needed to resolve the problem.

Cohen-Bazire, Kunisawa & Poindexter (1966) were able to correlate an increase in the number of small, peripheral mesosomes in oxygen-limited cultures of *Caulobacter crescentus* with an abnormally high content of haem pigments, suggesting that mesosomes are the respiratory organelles of this organism. The formation and function of mesosomes is the subject of much current research (see Salton, 1967).

Development of the bacterial photosynthetic organelle

The growth and photosynthetic pigment content of phototrophic bacteria is a function of light intensity and Sistrom (1962) has high-lighted the difficulties of measurement of pigment content of growing cells. In order to distinguish between the growth of new membrane and the addition of pigment to existing membrane it is necessary to use a chemostat in which growth rate and light intensity can be independently varied. In *Rhodospirillum rubrum*, *Rhodopseudomonas spheroides* and in *Chloropseudomonas ethylicum* both the complexity of the photo-synthetic apparatus and the content of bacteriochlorophyll per unit mass of membrane are modified under different light intensities (Cohen-Bazire & Sistrom, 1966; Holt, Conti & Fuller, 1966). However, Sistrom (1962) has shown that the bacteriochlorophyll content of the membrane system, once formed, is independent of light intensity. This agrees with the results reported by Bull & Lascelles (1963), which show that bacterio-chlorophyll synthesis is closely linked with the protein synthesis necessary for the formation of new membrane. Photopigment synthesis is also controlled by oxygen concentration (see Wimpenny, this Sym-posium).

The organelles of growing eucaryotic micro-organisms; growth and division

The elegant experiments of Luck (1962, 1965 *a*, *b*) with *Neurospora crassa* opened a new chapter in the biology of organelle development, and gave the first biochemical evidence for the mechanism of organelle growth. Prior to this work cytological investigations had indicated that mitochondria might originate: (*a*) from many, diverse and unrelated membrane structures, (*b*) *de novo*, or (*c*) by the processes of growth and division of pre-existing mitochondria (for review see Novikoff, 1961). It should be stressed that these three alternative mechanisms are not mutually exclusive; while the evidence for growth and division of organelles in growing cells is now quite impressive, '*de novo*' synthesis of mitochondria from 'premitochondrial' structures clearly does occur in rather special physiological situations which will be considered separately in a later section.

Luck (1962) studied the incorporation of labelled choline into phos-phatidyl choline of the mitochondrial membrane of the choline-requiring strain of *Neurospora crassa* (chol-1, 34486). The choline containing lipid was sufficiently stable to serve as a satisfactory marker through three mass doubling cycles. Radiographic analysis showed that a random distribution of radioactivity occurred in mitochondrial populations, isolated as mitochondrial fractions from fully labelled cells, and also from cells undergoing three subsequent doubling cycles in unlabelled choline. Control experiments indicated that the random distribution did not result from fractionation procedures; it was also unlikely that it was effected *in vivo* by repeated mitochondrial fusions and fissions. The data favoured the hypothesis that the mitochondrial mass increases by a continuous process of addition of new choline-containing lipid units to the existing mitochondrial structure and that the number of individuals in the mitochondrial population increases by division of organelles. This division process would distribute the label at random so that the pre-existing membrane would be transmitted nearly uniformly to all progeny. Confirmation of this hypothesis came from the observation that, by altering the choline concentration in the medium, the buoyant density of the *Neurospora* mitochondria subsequently isolated could be altered (Luck, 1965a). All the mitochondria showed a change of density as a single population, suggesting that mitochondria grow by accretion at random of new lecithin into existing mitochondrial structures, and that the mitochondrial population increases by division. Subsequently Luck (1965b) was able to show that mitochondrial composition is determined by the relative rates of incorporation of protein and phospholipid, and that these rates can vary independently in response to precursor concentration in the culture medium.

The relationships between mitochondrial development and cell division have been investigated in a number of organisms. Kahn & Blum (1967) measured the rate of mitochondrial protein synthesis during synchronized division of *Astasia longa* by pulse-labelling experiments with $^{35}SO_4$ and found that the rate of mitochondrial protein synthesis was highest just before cell division; they failed to detect any significant change in the ratio of synthesis of mitochondrial protein to that of total cell protein at any time of the division cycle. Unfortunately these authors presented no data on mitochondrial numbers. However, Elliott & Bak (1964) have made an interesting cytological study of the changes in position, number and structure of mitochondria during the growth of *Tetrahymena pyriformis* through the exponential phase and into the stationary phase. They obtained evidence for mitochondrial division and showed that it kept pace with cytokinesis so that the population of mitochondria remained essentially at the minimal level during logarithmic growth. As the ciliates entered the stationary phase, the mitochondria increased in number, became oval and migrated throughout the cytoplasm. There then followed a phase in which many of the mitochondria were broken down in acid-phosphatase-containing vacuoles.

Synchronous division of mitochondria behind the growing hyphal tips of *Neurospora crassa* has been reported by Hawley & Wagner (1967), who observed 'cupping' of mitochondria followed by division to give closely associated organelles. A number of unicellular algae have a single chloroplast; several species are also known which have only one mitochondrion per cell and it would be interesting to know whether their replication is linked. Manton & Parke (1960) have presented elegant electron micrographs which show the stages of cell division of a small marine flagellate; division of both the chloroplast and the mitochondrion were clearly defined.

It is more difficult to define precisely the relationship between replication of mitochondria and of DNA, but some progress has been made. Reich & Luck (1966) have produced strong evidence for the replication of the mitochondrial DNA of *Neurospora*. They showed that mitochondrial DNA possesses physical continuity during several cycles of replication during vegetative growth. The synthesis of nuclear and mitochondrial DNA were partly independent, as the precursors involved in the formation of the two species were drawn from two distinct pools. It was also shown that the pattern of inheritance of mitochondrial DNA species themselves was uniparental by way of the maternal parent. Although these properties are to be expected of any extra-chromosomal hereditary material, Reich & Luck stressed that their data do not prove that mitochondrial DNA has a genetic function. Parsons (1965) studied *Tetrahymena* radioautographically after pulse labelling with [³H]-thymidine; his results show that mitochondrial DNA became labelled even when no labelling of nuclear material occurred. The incorporation into mitochondrial DNA was a continuous process which therefore was not synchronous with the synthesis of nuclear DNA. Corneo *et al.* (1966) have demonstrated the semiconservative replication of mitochondrial DNA in yeast and shown this replication was asynchronous with respect to that of the nuclear DNA.

Inhibition of organelle development

Chloramphenicol inhibits the growth of blue-green algae and most bacteria in which it inhibits protein synthesis. Higher concentrations are required to inhibit growth and protein synthesis in fungi and some yeasts and protozoa, while cell-free protein synthesizing systems from higher animals are apparently resistant to chloramphenicol (Vazquez,

1966). Cycloheximide, on the other hand, inhibits protein synthesis in higher animals but not in bacteria. Evidence is now accumulating which suggests that organelles such as mitochondria and chloroplasts resemble bacteria in their sensitivity to chloramphenicol regardless of their source.

Chloramphenicol (25–150 μg./ml.) inhibits the growth of *Tetrahymena* (Mager, 1960): and selectively inhibits mitochondrial protein synthesis. Linnane, Biggs, Huang & Clark-Walker (1968) showed that growth of an obligately aerobic and of a facultative yeast on a nonfermentable substrate was inhibited in the presence of high concentrations of chloramphenicol. The drug inhibited the formation of cristae and led to reduction in the mitochondrial content of cytochrome oxidase and cytochrome *b*, and of NADH and succinate dehydrogenases. *In vitro* protein synthesis by isolated mitochondria was sensitive to chloramphenicol and the investigators concluded that *in vivo* the protein synthetic capacity of mitochondria may be restricted to the membrane-bound enzymes and proteins of the inner membrane. The high concentration of chloramphenicol (4 mg./ml.) required for inhibition was surprising, but Wilkie, Saunders & Linnane (1967) have since shown that some strains of yeast are sensitive to as little as 0·05 μg./ml. The concentration of antibiotic required to inhibit cytochrome synthesis by yeast is genetically determined and a number of genes determining sensitivity and resistance to several antibiotics have been recognized (Linnane, 1968). A genetic change in an erythromycin-resistant mutant has been traced to a biochemical change in the protein synthesis system of the mitochondrion. Clark-Walker & Linnane (1966) showed that the cytoplasmic ribosomal system of *Saccharomyces cerevisiae* is sensitive to cycloheximide and is largely unaffected by inhibitors of bacterial protein synthesis. On the other hand, the mitochondrial system is sensitive to chloramphenicol, tetracycline, oxytetracycline, erythromycin, carbomycin, spiramycin, oleandomycin and lincomycin but not cycloheximide. It was suggested that the differential inhibition of the two systems was correlated with the different ribosomal sizes of the mitochondrial and cytoplasmic systems. Further work with a cytoplasmic yeast mutant, and with glucose-repressed and derepressed normal yeast, led Clark-Walker & Linnane (1967) to suggest that cytochrome *c* and readily solubilized mitochondrial enzymes as well as proteins of the outer membrane are synthesized by non-mitochondrial systems.

The effects of chloramphenicol on the formation of mitochondrial enzymes of yeast are paralleled in the flagellate *Polytomella caeca*

(Evans & Lloyd, 1967b). The synthesis of cytochromes b and $(a+a_3)$ was inhibited during growth with chloramphenicol, whereas cytochrome c and several enzymes of the TCA cycle and β-oxidation of propionate were not affected. Rotenone-insensitive NADH-cytochrome c reductase activity, which may be an outer membrane function (Lloyd & Chance, 1968), was also not affected by growth with chloramphenicol. Similar effects have been noted with the mitochondrial enzymes of *Tetrahymena* (G. Turner & D. Lloyd, unpublished results) but in this case the inhibition of mitochondrial synthesis also results in a decrease in mitochondrial size as well as disorganization of the cristae membranes (Plate 1); the decreased size is uniform through the entire mitochondrial population. On removing the chloramphenicol, restoration of normal mitochondrial structure and enzymic activity occurs. Again the change affects the whole mitochondrial population uniformly; these data favour the concept of Luck (1962) that mitochondria grow by accretion of new membrane components and increase their numbers by division.

Studies with *Euglena gracilis* have shown that chloramphenicol inhibits the light-dependent growth of the alga; this effect appears to be due to a preferential inhibition of protein synthesis in the chloroplasts (Smillie, Evans & Lyman, 1963; Pogo & Pogo, 1965). The protein synthesis system of isolated chloroplasts is more sensitive to chloramphenicol than the corresponding cytoplasmic systems (App & Jagendorf, 1963; Eisenstadt & Brawerman, 1964) and Aaronson, Ellenbogen, Yellen & Hutner (1967) have shown that chloramphenicol, and also DL-ethionine, may be used to differentiate between cytoplasmic and chloroplast protein synthesis in *Euglena*. Furthermore, synthesis of some of the specific chloroplast components including chlorophyll (Linnane & Stewart, 1967) and the Calvin cycle and electron transport enzymes (Smillie *et al.* 1967) is inhibited by antibiotics which affect bacterial protein synthesis including chloramphenicol. Similarly, Schrader, Beevers & Hageman (1967) have shown that the induction of nitrite reductase, a chloroplast enzyme in higher plants, is inhibited by chloramphenicol, whereas that of nitrate reductase, an enzyme located outside chloroplasts, is not sensitive to this antibiotic.

Other inhibitors of plastic differentiation include actinomycin D (Pogo & Pogo, 1964) and *O*-acyl derivatives of oleandomycin and erythromycin (Celmer & Ebringer, 1967). On the other hand, puromycin amino nucleoside markedly affects the growth of *Euglena*, but the chloroplasts or proplastids of this organism are relatively insensitive to this inhibitor. Loss of chloroplasts and chlorophyll can be induced by a variety of treatments including ultraviolet irradiation, incubation with streptomycin and culture at elevated temperatures, but the literature on these effects is too extensive to review here. Lyman (1967) has recently reported that nalidixic acid specifically inhibits chloroplast replication in *Euglena*.

Differentiation of organelles

The classical studies of Linnane, Vitols & Nowland (1962) and Polakis, Bartley & Meek (1964) have established that the formation of mitochondria can occur in the absence of net cell growth when anaerobically grown yeast is aerated. Shatz (1965) has isolated a membrane fraction from anaerobically grown yeast which contained several of the enzymes which are associated with mitochondrial membranes in aerobically grown yeast; it was suggested that this fraction contained 'premitochondrial particles'. The cytology of anaerobically grown yeast depends on medium composition (Wallace, Huang & Linnane, 1968), and the process of mitochondrial formation in yeast (Jayaraman, 1966) is subject to catabolite repression as well as control by oxygen concentration (Lukins, Tham, Wallace & Linnane, 1966) (see also review by Wimpenny, this Symposium). Anaerobically grown yeast does contain mitochondrial DNA; thus the genetic continuity of the mitochondria is preserved even when their characteristic structures are lost (Fukuhara, 1967a).

Several biochemical studies of development of fungal spores have been made (Sussmann, 1965). The conidia of *Neurospora crassa* possess mitochondria (Weiss, 1965) and parallel increases in their capacity for oxidative phosphorylation during hyphal development have been shown by Hall & Greenawalt (1964).

Griffiths, Lloyd, Roach & Hughes (1967) were unable to demonstrate mitochondria in cysts of the soil amoeba *Hartmanella castellanii* or to detect the cytochromes or tightly coupled oxidations characteristic of mitochondria isolated from the trophic amoeba. Presumably excystment should provide another differentiating system suitable for the study of mitochondrial biogenesis.

Chloroplast development on the illumination of etiolated dark-grown photosynthetic systems has been reviewed by Kirk & Tilney-Bassett (1967). The plastic precursors (proplastids) in dark-grown *Euglena* are similar to structures found in higher plants (Epstein, Boy de la Tour & Schiff, 1960). On illumination these structures elongate and their inner membrane invaginates to give fully fledged lamellae which increase in number linearly with time. The appearance of chlorophyll is associated with an increased turnover of RNA and protein and accompanied by a shift of protein from cytoplasm to chloroplast (Brawerman & Chargaff, 1959). The lamellar structure develops with the formation of chlorophyll (Wolken, 1959) and newly synthesized chlorophyll which lacks its phytol side chain is absorbed in precursor bodies which are 'structureless'. Layers form as the side chains are attached (Butler, 1961) and Smillie (1962) was able to show changes of enzyme activities as the chloroplasts developed.

Turnover of organelle components in resting cells

Since the original experiments of Fletcher & Sanadi (1961) on the turnover of rat-liver mitochondria, there have been several studies on various mammalian systems including those by Taylor, Bailey & Bartley (1967) and by Beattie, Basford & Koritz (1967b). While the original workers concluded that the mitochondria were turned over as an entity with a half life of 10 days, it is now apparent that phospholipid, membrane protein and soluble protein components have different half lives. Neubert, Bass & Oberdisse (1968) have shown that mitochondrial DNA of tissues with a low rate of mitosis has a half life of 7–8 days. There are no reports of the measurement of mitochondrial turnover in resting cells of eucaryotic micro-organisms other than that of Fukuhara (1967a) on the respiratory system in anaerobically grown yeast.

Modification of existing organelles

Experiments with the colourless alga, *Prototheca zopfii*, suggested that adaptation of non-proliferating acetate-grown cells to propionate involved the induction of enzymes which were partly mitochondrial in location (Lloyd & Venables, 1967). An extension of this work with the flagellate, *Polytomella caeca* (Evans & Lloyd, 1967a; 1968), has confirmed that some of the enzymes of a β-oxidation pathway of propionate utilization are inducible in the mitochondria of resting cells. The newly formed enzymes catalyse the oxidation of the intermediates of the pathway, but mitochondria isolated in the early stages of adaptation are not capable of tightly coupled phosphorylation with these substrates. After 18 hours' adaptation a fully functional energy-producing mechanism has been formed; P/O ratios and respiratory control ratios[1] are then as high as in mitochondria from propionate-grown cells. Chloramphenicol does not inhibit this adaptation but cycloheximide and actinomycin D are both inhibitory. It thus seems likely that the

[1] Respiratory control ratio $= \dfrac{\text{respiration rate in the presence of added ADP}}{\text{respiration rate after exhaustion of ADP}}$

inducible mitochondrial enzymes are synthesized on extramitochondrial ribosomes. If the assumption is correct that there is little turnover of the mitochondria of resting cells during adaptation, then the transport of nascent enzymes into existing mitochondria and their integration with the inner mitochondrial membrane electron transport system must be postulated.

Adaptation of *Pseudomonas fluorescens* KB1 to the oxidation of nicotinic acid involves modification of the cell membrane of this organism (Hunt, Rodgers & Hughes, 1959). The inducible nicotinic acid hydroxylase appears to be located in discrete patches on the membrane. Whether the adaptation involves the insertion of the new enzyme into an existing site or the extensive turnover and rearrangement of many membrane components is not yet clear.

Control of organelle development: informational role of organelle DNA

Mechanisms of extranuclear inheritance have been studied extensively in fungi (Jinks, 1965), yeasts (Wilkie, 1964) and in algae (Sager, 1965). While the biochemical genetics of cytoplasmic mutants of yeast, *Neurospora* and *Chlamydomonas* have been intensively studied there was little convincing evidence for the informational role of the DNA of mitochondria and chloroplasts until very recently. Thus it was thought that cytoplasmic 'petite' mutants of yeast contained no mitochondrial DNA until Mounolou, Jakob & Slonimski (1966) showed that in fact this DNA is present in a physically modified form. The mitochondrial DNA from a cytoplasmic mutant had a different buoyant density from that of the wild-type; furthermore, two extreme genetic states of the cytoplasmic hereditary determinants, the recessive and the dominant one, differ in the densities of their mitochondrial DNAs. It is as yet uncertain whether the change of density corresponds to a change in base-composition, differences in the extent of methylation, or the presence of unknown components. These workers also posed the question of whether different cytoplasmic mutations result from an induced change in the density of a single molecular species of mitochondrial DNA or from a selection (either mitochondrial or extra mitochondrial) among a population of non-identical mitochondrial DNA molecules originally present in the wild-type cell.

If all the mitochondrial DNAs within the cell are identical the amount of information carried must be relatively small. If different mitochondria contain different DNAs, then more information may be carried and the genetics would become very complicated; population genetics between mitochondria of the same cell becomes a possibility. There is some evidence for non-equivalence of mitochondrial DNA molecules with regard to their susceptibility to mutagenesis by ethidium bromide (Slonimski, Perrodin & Croft, 1968) and complementation by mitochondria has been suggested as the basis for heterosis in maize hybrids (McDaniel & Sarkissian, 1966). Although renaturation studies favour extensive homogeneity of mitochondrial DNA, single base-pair differences would be sufficient to make the DNA of different mitochondria genetically distinct.

Fukuhara (1967b) has used hybridization techniques to investigate the informational role of mitochondrial DNA. Both aerobically and anerobically grown cells contain an RNA species that hybridizes with mitochondrial DNA and the total amount of this RNA is significantly greater in aerobic cells. This RNA is not cytoplasmic ribosomal or transfer RNA and is a metabolically stable species associated with membranes. He suggested that this may be ribosomal RNA of the mitochondria and has evidence that it does not hybridize with nuclear DNA. A preferential transcription of mitochondrial DNA occurs during respiratory adaptation (Mounolou, Perrodin & Slonimski, 1968). The onset of adaptation to oxygen of anaerobically grown $\rho + grande$ yeast corresponds to an extremely rapid synthesis of a specific and small part of mitochondrial DNA (about 10^5 base pairs/ cell). This process precedes the biosynthesis of respiratory enzymes and the development of mitochondrial structures, and the proportion of RNA hybridizing with mitochondrial DNA also increases during this respiratory adaptation.

Suyama (1967) has shown that the mitochondria of *Tetrahymena* contain a membrane-bound (ribosomal) RNA which hybridizes with mitochondrial DNA and has sedimentation coefficients of 18 and 28S. Further evidence for cytoplasmic coding for the mitochondrial ribosomes has been obtained by Linnane, Saunders, Gingold & Lukins (1968). Erythromycin resistance in *Saccharomyces cerevisiae* mutants is extrachromosomally inherited and Linnane, Lamb, Christodoulou & Lukins (1968) suggest that the synthesis of yeast mitochondrial ribosomes is under the control of a cytoplasmic genetic determinant which may prove to be mitochondrial DNA.

The only direct evidence for a gene–protein relationship involving mitochondrial DNA has been provided by Woodward & Munkres (1966). They were able to show that two different extrachromosomal mutations of the DNA of respiratory deficient mutants of *Neurospora* lead to alterations of the primary structure and function of the structural protein of the mitochondria. The amino acid compositions of mitochondrial structural protein from three other mutants was indistinguishable from the wild-type protein. Two of these were respiratory deficient as a result of nuclear mutations while the third was not respiratory deficient and had an alteration in a non-chromosomal determinant. The amino acid replacement in the mutant structural proteins led to an altered affinity for binding of nucleotide coenzymes and malate dehydrogenase. Woodward & Munkres postulated that the expression of the mutations as pleiotropic respiratory deficiencies results from the role of the structural protein in the organization and assembly of the respiratory chain. They observed that cytochrome *c* accumulated by 16-fold in a non-particulate cell fraction from the respiratory mutants, riboflavin occurred in twofold excess, and there was 15-fold excess of long chain unsaturated fatty acids. Deficiencies in cytochrome *a*, *b* and cytochrome oxidase were also observed. As a result of these experiments Woodward & Munkres (1967) have proposed that the structural protein might better be called 'organizer protein'. Since the gene which controls the primary structure of this protein is located in mitochondrial DNA and there are at least as many copies of this gene as there are mitochondria per cell, many copies of structural protein m-RNA are produced and much more structural protein is formed, than is formed in the case of proteins encoded by nuclear DNA (assuming equivalent rates of transcription and translation). Thus, whereas most of the cellular protein is required in catalytic amounts, the regulation of synthesis of a protein required in such quantities as structural protein may be more easily accomplished outside the nuclear environment. After complexing specifically with phospholipid and enzymes it may then be incorporated into a growing membrane.

Further evidence that mitochondrial DNA codes for specific mitochondrial membrane proteins has been obtained by acrylamide gel electrophoresis of proteins from mutant yeasts by Slonimski, Poglase & Peel (cited by Work, 1967).

The transmission of cytoplasmically determined characters from one cell to another by purified mitochondrial preparations has been demonstrated with two systems. Tuppy & Wildner (1965) used spheroplasts of a stable acriflavin-induced 'petite' mutant of *Saccharomyces cerevisiae*, auxotrophic for adenine and thymine, and incubated these with mitochondria isolated from a leucine auxotroph with normal respiration. 'Recombinants' (2·6 %) were obtained with normal respiration and without leucine requirement, but still auxotrophic for adenine and thymine. Diacumakos, Garnjobst & Tatum (1965) used a micro injection technique to introduce abnormal mitochondria of a mutant into a normal recipient; it had previously been shown (Garnjobst, Wilson & Tatum, 1965) that the cytoplasmic character became pheno-typically dominant in a heterocaryon. A high proportion of the injected *Neurospora* cells acquired the growth characteristics and abnormal cyto-chrome spectrum of the mutant donor strain. Controls showed that this situation could not be produced by transmission of purified nuclei, nuclear DNA or mitochondria isolated from a normal strain.

Avers, Rancourt & Lin (1965) have observed that diverse mito-chondrial populations persist in the common nucleocytoplasmic milieu of uninucleate yeast cells for many thousands of generations. Thus the unique phenotypes of genetically distinct systems are maintained in a common environment. Federman & Avers (1967) have studied the perpetuation of a heterogeneous mitochondrial population through many generations following the production of hybrid zygotes by conjugation of wild-type and 'petite' strains. The gradual alteration of the proportions of cristate and non-cristate mitochondria appeared to be a consequence of selective advantages, and led these investigators to support the hypothesis that the mitochondrion is a semi-autonomous organelle endowed with a limited genetic capacity for phenotype control.

CONCLUDING REMARKS

The surface area/volume ratio of bacteria is high, and sufficient space on the cytoplasmic membrane for the organization of energy-yielding functions is obtained by its invagination in highly aerobic chemotrophs and also in phototrophs. The association of an episome with this primitive organelle may have been an event that led to the evolution of the discrete semi-autonomous energy-producing organelles in eucaryotic protists, and more recently in the metazoa. The alternative proposal that DNA-containing organelles are descendants of autonomous symbionts has many followers, especially in the light of the remarkable similarity between the ribosomes of bacteria, chloroplasts and mito-chondria (with respect to both size and sensitivity to antibiotics). It is also clear that the bacterial cytoplasmic membrane and the inner mito-chondria membrane are similar with regard to the localization of polyglycerophosphatides (cardiolipin) and electron transport and oxidative phosphorylation enzymes. Both types of membrane have a low total lipid content (20–30 %) and little or no cholesterol (Parsons, 1966).

Progressive loss of functions controlled by symbiont DNA may have led to the present situation in which only a modest contribution to the information necessary for organelle growth is provided by organelle DNA. It is now clear, at least in the case of the mitochondrion, that this information is restricted to that required for the synthesis of certain hydrophobic proteins and ribosomal RNA; the latter may be of paramount importance for the reading of messages from the nucleus and thus for the control of organelle synthesis. Because there is more than one source of genetic information in eucaryotes, the problems of organelle growth are also the problems of nucleocytoplasmic relationships. While *in vitro* reconstitution experiments suggest a certain spontaneity of self-assembly of specialized energy-yielding membranes on a backbone of structural protein (when the 'soluble' components are provided in the correct stoichiometry), nothing is known of this process *in vivo* (Green & Hechter, 1965). Valuable clues to this process should come from the study of mutants unable to make specific membrane components; indeed, a start has already been made by the important observations of Woodward & Munkres (1966). Mutants which have defective oxidative phosphorylation systems (Kovač, Lachowcz & Slonimski, 1967; Kovač & Hrusouska, 1968; Kovač & Weissova, 1968) should prove especially valuable.

Finally there are the problems of the control of organelle division during growth, modification of existing organelles, and '*de novo*' origin of organelles during differentiation. It seems likely that many of the difficulties of reconciling the growth processes at a molecular level with the macroscopic phenomena of organelle development observed by cell physiologists can be attributed to the dynamic state of membrane systems. The continual changes in membrane systems seen on phase-contrast observation of living cells must reflect the ceaseless activity of their molecular components. Roodyn, Reis & Work (1960) have previously cited the observations of Ritchie & Hazeltine (1953) in this context: '...under optimum conditions they resemble an aggregation of vigorous earthworms. In addition to twisting, turning and progressive movement...a mitochondrion in *Allomyces* can shorten and thicken or lengthen and become slender or a lump may pass along its length like a rat being swallowed by a snake. Mitochondria can be seen to swell, coil, branch, fragment, coalesce, put out pseudopods, even to get tied in knots. They can change their form until they resemble bubbles, strings of beads, dumb-bells, lemons or snow shoes'.

ACKNOWLEDGEMENTS

The author wishes to express his gratitude to Professor D. E. Hughes for valuable discussions concerning ideas proposed in this article and to workers who have kindly provided unpublished data.

REFERENCES

AARONSON, S., ELLENBOGEN, B. B., YELLEN, L. K. & HUTNER, S. H. (1967). *In vivo* differentiation of *Euglena* cytoplasmic and chloroplast protein synthesis with chloramphenicol and DL-ethionine. *Biochem. biophys. Res. Commun.* **27**, 535.

APP, A. A. & JAGENDORF, A. T. (1963). Incorporation of labelled amino acids into isolated spinach chloroplasts. *Pl. Physiol.* **39**, 772.

ASANO, A. & BRODIE, A. F. (1965). Phosphorylation coupled to different segments of the respiratory chains of *Mycobacterium phlei. J. biol. Chem.* **240**, 4002.

AVERS, C. J. (1967). Heterogeneous length distribution of circular DNA filaments from yeast mitochondria. *Proc. natn. Acad. Sci. U.S.A.* **58**, 620.

AVERS, C. J., RANCOURT, M. W. & LIN, F. H. (1965). Intracellular mitochondrial diversity in various strains of *Saccharomyces cerevisiae. Proc. natn. Acad. Sci. U.S.A.* **54**, 527.

BARNETT, W. E. & BROWN, D. H. (1967). Mitochondrial transfer RNA's. *Proc. natn. Acad. Sci. U.S.A.* **57**, 452.

BARNETT, W. E., BROWN, D. H. & EPLER, J. L. (1967). Mitochondria specific amino acyl-RNA synthetase. *Proc. natn. Acad. Sci. U.S.A.* **57**, 1775.

BEATTIE, D. S., BASFORD, R. E. & KORITZ, S. B. (1967a). The inner membrane as the site of *in vivo* incorporation of L-[14C] leucine into mitochondrial protein. *Biochemistry, N.Y.* **6**, 3099.

BEATTIE, D. S., BASFORD, R. E. & KORITZ, S. B. (1967b). The turnover of the protein components of mitochondria from rat liver, kidney and brain. *J. biol. Chem.* **242**, 4584.

BORST, P. (1967). Mitochondrial DNA. *Biochem. J.* **105**, 37P.

BORST, P., KROON, A. M. & RUTTENBERG, G. J. C. M. (1967). Mitochondrial DNA and other forms of cytoplasmic DNA. In *Symposium on Structure and Function of Genetic Elements*, p. 81. Ed. D. Shugar. London: Academic Press.

BORST, P., VAN BRUGGEN, E. F. J. & RUTTENBERG, G. J. C. M. (1968). In *Round Table Discussion on Biochemical Aspects of the Biogenesis of Mitochondria*. Ed. E. C. Slater, J. M. Tager, S. Papa and E. Quagliariello. Bari: Adriatica Editrice. (In the Press.)

BOVÉ, J. & RAACKE, I. D. (1959). Amino acid-activating enzymes in isolated chloroplasts from spinach leaves. *Archs Biochem. Biophys.* **85**, 521.

BRAWERMAN, G. & CHARGAFF, E. (1959). Factors involved in the development of chloroplasts in *Euglena gracilis. Biochim. biophys. Acta,* **31**, 178.

BREWER, E. N., DE VRIES, A. & RUSCH, H. P. (1967). DNA synthesis in isolated mitochondria of *Physarum polycephalum. Biochim. biophys. Acta,* **145**, 686.

BROWN, J. W. (1965). Evidence for a magnesium-dependent dissociation of bacterial cytoplasmic membrane particles. *Biochim. biophys. Acta,* **94**, 97.

BULL, M. J. & LASCELLES, J. (1963). The association of protein synthesis with the formation of pigments in some photosynthetic bacteria. *Biochem. J.* **87**, 15.

BUTLER, W. L. (1961). Chloroplast development: energy transfer and structure. *Archs Biochem. Biophys.* **92**, 287.

CELMER, W. D. & EBRINGER, L. (1967). Effects of certain O-acyl derivatives of olandeomycin and erythromycin on chloroplasts of *Euglena gracilis. J. Protozool.* **14**, 263.

CHANCE, B. (1965). Control of energy metabolism in mitochondria. In *Control of Energy Metabolism*, p. 415. Ed. B. Chance, R. W. Estabrook and J. R. Williamson. New York: Academic Press.

CHANCE, B. (1967). The reactivity of haemoproteins and cytochromes. *Biochem. J.* **103**, 1.

CHEN, J. L. & WILDMAN, S. G. (1967). Functional chloroplast ribosomes from tobacco leaves. *Science*, **155**, 127.

CLARK-WALKER, G. D. & LINNANE, A. W. (1966). *In vivo* differentiation of yeast cytoplasmic and mitochondrial protein synthesis with antibiotics. *Biochém. biophys. Res. Commun.* **25**, 8.

CLARK-WALKER, G. D. & LINNANE, A. H. (1967). Biogenesis of mitochondria in *Saccharomyces*. A comparison between the cytoplasmic respiratory deficient mutant yeast and chloramphenicol inhibited wild type cells. *J. Cell Biol.* **34**, 1.

CLAYTON, D. A. & VINOGRAD, J. (1967). Circular dimer and catenate forms of mitochondrial DNA in human leucocytic leucocytes. *Nature, Lond.* **216**, 652.

COHEN-BAZIRE, G., KUNISAWA, R. & POINDEXTER, J. S. (1966). The internal membranes of *Caulobacter crescentus*. *J. gen. Microbiol.* **42**, 301.

COHEN-BAZIRE, G. & SISTROM, R. W. (1966). The procaryotic photosynthetic apparatus. In *The Chlorophylls*, p. 313. Ed. L. P. Vernon and G. R. Seely. New York: Academic Press.

CORNEO, G., MOORE, C., SANADI, D. G., GROSSMAN, L. I. & MARMUR, J. (1966). Mitochondrial DNA in yeast and some mammalian species. *Science, N.Y.* **151**, 687.

CRIDDLE, R. S., BOCK, R. M., GREEN, D. E. & TISDALE, H. (1962). Physical characteristics of proteins of the electron transfer system and interpretation of the structure of the mitochondria. *Biochemistry, N.Y.* **1**, 827.

DAWSON, R. M. C. (1966). The metabolism of animal phospholipids and their turnover in cell membranes. In *Essays in Biochemistry*, vol. II, p. 62. Ed. P. N. Campbell and G. D. Greville. London: Academic Press.

DIACUMAKOS, E. G., GARNJOBST, L. & TATUM, E. L. (1965). A cytoplasmic character in *Neurospora crassa*. Role of nuclei and mitochondria. *J. Cell Biol.* **26**, 427.

DU BUY, H. G., MATTERN, C. F. T. & RILEY, F. L. (1966). Comparison of the DNA's obtained from brain nuclei and mitochondria of mice and from the nuclei and kinetoplasts of *Leishmania enrettii*. *Biochim. biophys. Acta*, **123**, 298.

DURE, L. S., EPLER, J. L. & BARNETT, W. E. (1967). Sedimentation properties of mitochondrial and cytoplasmic ribosomal RNA's from *Neurospora*. *Proc. natn. Acad. Sci. U.S.A.* **58**, 1883.

EDELMAN, M., EPSTEIN, H. T. & SCHIFF, J. A. (1966). Isolation and characterization of DNA from the mitochondrial fraction of *Euglena*. *J. molec. Biol.* **17**, 463.

EISENSTADT, J. M. & BRAWERMAN, G. (1964). The protein-synthesizing systems from the cytoplasm and the chloroplasts of *Euglena gracilis*. *J. molec. Biol.* **10**, 392.

ELLIOTT, A. M. & BAK, I. J. (1964). The fate of mitochondria during aging in *Tetrahymena pyriformis*. *J. Cell Biol.* **20**, 113.

EPSTEIN, H. T., BOY DE LA TOUR, E. & SCHIFF, J. A. (1960). Fluorescence studies of chloroplast development in *Euglena*. *Nature, Lond.* **185**, 825.

EVANS, D. A. & LLOYD, D. (1967*a*). Inducible enzymes in mitochondria of *Polytomella caeca*. *Biochem. J.* **103**, 21–22P.

EVANS, D. A. & LLOYD, D. (1967*b*). The effect of chloramphenicol on the mitochondria of *Polytomella caeca*. *Biochem. J.* **103**, 22P.

EVANS, D. A. & LLOYD, D. (1968). Adaptive formation of phosphorylation coupled oxidations in *Polytomella caeca* mitochondria. *Fedn Europ. Biochem. Soc. Proc.* (A 820.)

FEDERMAN, M. & AVERS, C. J. (1967). Fine structural analysis of intracellular mitochondrial diversity in *Saccharomyces cerevisiae*. *J. Bact.* **94**, 1236.

FLETCHER, M. J. & SANADI, D. R. (1961). Turnover of rat liver mitochondria. *Biochim. biophys. Acta*, **51**, 356.

FREEMAN, K. B., HALDAR, D. & WORK, T. S. (1967). The morphological site of synthesis of cytochrome *c* in mammalian cells (Krebs cells). *Biochem. J.* **105**, 947.

FRERMAN, F. E. & WHITE, D. C. (1967). Membrane changes during formation of a functional electron transport system in *Staphylococcus aureus*. *J. Bact.* **94**, 1868.

FOWLER, L. R. & HATEFI, Y. (1961). Reconstitution of DPNH oxidase, succinic oxidase and DPNH succinic oxidase. *Biochem. biophys. Res. Commun.* **5**, 203.

FUKUHARA, H. (1967*a*). Protein synthesis in non-growing yeast. Respiratory adaptation system. *Biochim. biophys. Acta*, **134**, 143.

FUKUHARA, H. (1967*b*). Informational role of mitochondrial DNA studied by hybridization with different classes of RNA in yeast. *Proc. natn. Acad. Sci. U.S.A.* **58**, 1065.

GARBUS, J., DE LUCA, H. F., LOOMANS, M. E. & STRONG, F. M. (1963). The rapid incorporation of phosphate into mitochondrial lipids. *J. biol. Chem.* **238**, 59.

GARNJOBST, L., WILSON, J. F. & TATUM, E. L. (1965). Studies on a cytoplasmic character in *Neurospora crassa*. *J. Cell Biol.* **26**, 413.

GIBBS, S. P. (1967). Synthesis of chloroplast RNA at the site of DNA. *Biochem. biophys. Res. Commun.* **28**, 653.

GOFFEAU, A. & BRACHET, J. (1965). DNA-dependent incorporation of amino acids into proteins of chloroplasts isolated from anucleate *Acetabularia* fragments. *Biochim. biophys. Acta*, **95**, 302.

GONSÁLEZ-CADAVID, N. F. & CAMPBELL, P. N. (1967). The biosynthesis of cytochrome *c*. Sequence of incorporation *in vivo* of [^{14}C] lysine into cytochrome *c* and total proteins of rat liver subcellular fractions. *Biochem. J.* **105**, 443.

GRANICK, S. & GIBOR, A. (1967). The DNA of chloroplasts, mitochondria, and centrioles. In *Progress in Nucleic Acid Research and Molecular Biology*, vol. 5, p. 143. Ed. J. N. Davidson and W. E. Cohn. New York: Academic Press.

GREEN, D. E., ALLMANN, D. W., BACHMANN, E., BAUM, H., KOPACZYK, K., KORMAN, E. F., LIPTON, S., MACLENNAN, D. H., McCONNELL, D. G., PERDUE, J. F., RIESKE, J. S. & TZAGOLOFF, A. (1967). Formation of membranes by repeating units. *Archs Biochem. Biophys.* **119**, 312.

GREEN, D. E. & FLEISCHER, S. (1962). On the molecular organisation of biological transducing systems. In *Horizons in Biochemistry*, p. 381. Ed. M. Kasha and B. Pullman. New York: Academic Press.

GREEN, D. E. & HECHTER, O. (1965). Assembly of membrane subunits. *Proc. natn. Acad. Sci. U.S.A.* **53**, 318.

GREEN, D. E. & SILMAN, I. (1967). Structure of the mitochondrial electron transfer chain. *A. Rev. Pl. Physiol.* **18**, 147.

GRIFFITHS, A. J., LLOYD, D., ROACH, G. I. & HUGHES, D. E. (1967). Metabolic changes in *Hartmanella castellanii* during encystment. *Biochem. J.* **103**, 21P.

GRIVELL, L. A. (1967). Amino acid incorporation by mitochondria isolated, essentially free of micro-organisms from *Saccharomyces carlsbergenesis*. *Biochem. J.* **105**, 44C.

GUTTES, E. W., HANAWALT, P. C. & GUTTES, S. (1967). Mitochondrial DNA synthesis and the mitotic cycle in *Physarum polycephalum*. *Biochim. biophys. Acta*, **142**, 181.

HALDAR, D., FREEMAN, K. B. & WORK, T. S. (1966). Biogenesis of mitochondria. *Nature, Lond.* **211**, 9.

HALL, D. O. & GREENAWALT, J. W. (1964). Oxidative phosphorylation by isolated mitochondria of *Neurospora crassa*. *Biochem. biophys. Res. Commun.* **17**, 565.

HAWLEY, E. S. & WAGNER, R. P. (1967). Synchronous mitochondrial division in *Neurospora crassa*. *J. Cell Biol.* **35**, 489.

HOLT ST. C., CONTI, S. F. & FULLER, R. C. (1966). Effect of light intensity on the formation of the photochemical apparatus in the green bacterium *Chloropseudomonas ethylicum*. *J. Bact.* **91**, 349.

HUGHES, D. E. (1962). The bacterial cytoplasmic membrane. *J. gen. Microbiol.* **29**, 39.

HÜLSMANN, W. C. (1962). Fatty acid synthesis in heart sarcosomes. *Biochim. biophys. Acta*, **58**, 417.

HÜLSMANN, W. C., WIT-PEETERS, E. M. & BENCKHUYSEN, C. (1966). Factors influencing fatty acid metabolism in mitochondria. In *Regulation of Metabolic Processes in Mitochondria*, p. 460. Ed. J. M. Tager, S. Papa, E. Quagliariello and E. C. Slater. Amsterdam: Elsevier.

HUNT, A. L., RODGERS, A. & HUGHES, D. E. (1959). Sub-cellular particles and the nicotinic acid hydroxylase system in extracts of *Pseudomonas fluorescens* KB1. *Biochim. biophys. Acta*, **34**, 354.

JAYARAMAN, J. (1966). Biochemical correlation of respiratory deficiency. Glucose respiration and mitochondrial synthesis in yeast. *Archs Biochem. Biophys.* **116**, 224.

JINKS, J. L. (1965). Mechanisms of inheritance. 4. Extranuclear inheritance. In *The Fungi*, vol. II, p. 619. Ed. G. C. Ainsworth and A. S. Sussmann. New York: Academic Press.

KADENBACH, W. (1968). Transfer of proteins from microsomes into mitochondria. Biosynthesis of cytochrome *c*. In *Round Table Discussion of Biochemical Aspects of the Biogenesis of Mitochondria*. Ed. E. C. Slater, J. M. Tager, S. Papa and E. Quagliariello. Bari: Adriatica Editrice. (In the Press.)

KAGAWA, Y. & RACKER, E. (1966). Partial resolution of the enzymes catalyzing oxidative phosphorylation X. Correlation of morphology and function in sub-mitochondrial particles. *J. biol. Chem.* **241**, 2475.

KAHN, V. & BLUM, J. J. (1967). The rate of mitochondrial protein synthesis during synchronized division of *Astasia*. *Biochemistry, N.Y.* **6**, 817.

KIRK, J. T. O. & TILNEY-BASSETT, R. A. E. (1967). *The Plastids, their Chemistry, Structure Growth and Inheritance.* London: Freeman and Co.

KORN, E. D. (1966). Structure of biological membranes. *Science, N.Y.* **153**, 1491.

KOVAČ, L. & HRUSOUSKA, E. (1968). Oxidative phosphorylation in yeast. II. An oxidative phosphorylation-deficient mutant. *Biochim. Biophys. Acta*, **153**, 43.

KOVAČ, L., LACHOWCZ, T. M. & SLONIMSKI, P. P. (1967). Biochemical genetics of oxidative phosphorylation. *Science, N.Y.* **158**, 1564.

KOVAČ, L. & WEISSOVA, K. (1968). Oxidative phosphorylation in yeast III. ATPase activity of the mitochondrial fraction from a cytoplasmic respiratory-deficient mutant. *Biochim. Biophys. Acta*, **153**, 55.

KREBS, H. A. (1967). The making of a scientist. *Nature, Lond.* **215**, 1441.

KÜNTZEL, H. & NOLL, H. (1967). Mitochondrial and cytoplasmic polysomes from *Neurospora crassa*. *Nature, Lond.* **215**, 1340.

LAMB, A. J., CLARK-WALKER, G. D. & LINNANE, A. W. (1968). The biogenesis of mitochondria. 4. The *in vitro* differentiation of mitochondrial and cytoplasmic protein synthesizing systems by antibiotics. *Biochim. biophys. Acta*, **161**, 415.

LINNANE, A. W. (1968). Some characteristics of the protein synthesizing system of yeast mitochondria. In *Round Table Discussion on Biochemical Aspects of the Biogenesis of Mitochondria.* Ed. E. C. Slater, J. M. Tager, S. Papa and E. Quagliariello. Bari: Adriatica Editrice. (In the Press.)

LINNANE, A. W., BIGGS, D. R., HUANG, M. & CLARK-WALKER, G. D. (1968). The effect of chloramphenicol on the differentiation of the mitochondrial organelle. In *Aspects of Yeast Metabolism*. Ed. R. K. Mills. Oxford: Blackwell. (In the Press.)

LINNANE, A. W., LAMB, A. S., CHRISTODOULOU, C. & LUKINS, H. B. (1968). The biogenesis of mitochondria. 6. The biochemical basis of the resistance of *Saccharomyces cerevisiae* toward antibiotics which specifically inhibit mitochondrial protein synthesis. *Proc. natn. Acad. Sci. U.S.A.* **59**, 1288.

LINNANE, A. W., SAUNDERS, G. W., GINGOLD, E. B. & LUKINS, H. B. (1968). The biogenesis of mitochondria, 5. Cytoplasmic inheritance of erythromycin resistance in *Saccharomyces cerevisiae*. *Proc. natn. Acad. Sci. U.S.A.* **59**, 903.

LINNANE, A. W. & STEWART, P. R. (1967). The inhibition of chlorophyll formation in *Euglena* by antibiotics which inhibit bacterial and mitochondrial protein synthesis. *Biochem. biophys. Res. Commun.* **27**, 511.

LINNANE, A. W., VITOLS, E. & NOWLAND, P. G. (1962). Studies on the origins of yeast mitochondria. *J. Cell Biol.* **13**, 345.

LLOYD, D. & CHANCE, B. (1968). Electron transport in mitochondria isolated from the flagellate, *Polytomella caeca*. *Biochem. J.* **107**, 829.

LLOYD, D. & VENABLES, S. E. (1967). The regulation of propionate oxidation in *Prototheca zopfii*. *Biochem. J.* **104**, 639.

LUCK, D. J. L. (1962). Genesis of mitochondria in *Neurospora crassa*. *Proc. natn. Acad. Sci. U.S.A.* **49**, 233.

LUCK, D. J. L. (1965a). Formation of mitochondria in *Neurospora crassa*: a study based on mitochondrial density changes. *J. Cell Biol.* **24**, 461.

LUCK, D. J. L. (1965b). Influence of precursor pool size on mitochondrial composition in *Neurospora crassa*. *J. Cell Biol.* **24**, 445.

LUCK, D. & REICH, E. (1964). DNA in mitochondria of *Neurospora crassa*. *Proc. natn. Acad. Sci., U.S.A.* **52**, 931.

LUKINS, H. B., THAM, S. H., WALLACE, P. G. & LINNANE, A. W. (1966). Correlation of membrane-bound succinate dehydrogenase with the occurrence of mitochondrial profiles in *Saccharomyces cerevisiae*. *Biochem. biophys. Res. Commun.* **23**, 363.

LYMAN, H. (1967). Specific inhibition of chloroplast replication in *Euglena gracilis* by nalidixic acid. *J. Cell Biol.* **35**, 726.

MADDY, A. H. & MALCOLM, B. R. (1965). Protein conformations in the plasma membrane. *Science, N.Y.* **150**, 1616.

MAGER, J. (1960). Chloramphenicol and chlortetracycline inhibition of amino acid incorporation into proteins in a cell-free system from *Tetrahymena pyriformis*. *Biochim. biophys. Acta*, **38**, 150.

MANTON, I. & PARKE, M. (1960). Further observations on small green flagellates with special reference to possible relatives of *Chromuline pusilla* Butcher. *J. mar. biol. Ass. U.K.* **39**, 275.

MARCUS, A. (1959). Amino acid dependent exchange between pyrophosphate and ATP in spinach preparations. *J. biol. Chem.* **234**, 1238.

MATOON, J. R. & SHERMAN, F. (1966). Reconstitution of phosphorylating electron transport in mitochondria from a cytochrome c-deficient yeast mutant. *J. biol. Chem.* **241**, 4330.

McDANIEL, R. G. & SARKISSIAN, I. V. (1966). Heterosis: complementation by mitochondria. *Science, N.Y.* **152**, 1640.

MENKE, W. (1967). The molecular structure of photosynthetic lamellar systems. In *Energy Conversion by the Photosynthetic Apparatus*, p. 328. Upton, New York: Brookhaven National Laboratory.

MITCHELL, P. (1966). Chemiosmotic coupling in oxidative and photosynthetic phosphorylation. *Biol. Rev.* **41**, 445.

MOUNOLOU, J. C., JAKOB, H. & SLONIMSKI, P. P. (1966). Mitochondrial DNA from yeast 'petite' mutants. Specific changes of buoyant density corresponding to different cytoplasmic mutations. *Biochem. biophys. Res. Commun.* **24**, 218.

MOUNOLOU, J. C., PERRODIN, G. & SLONIMSKI, P. P. (1968). Specific synthesis of a small part of mitochondrial DNA corresponding to the onset of the O_2-induced development of mitochondria. In *Round Table Discussion on Biochemical Aspects of the Biogenesis of Mitochondria*. Ed. E. C. Slater, J. M. Tager, S. Papa and E. Quagliariello. Bari: Adriatica Editrice. (In the Press.)

NEUBERT, D., BASS, R. & OBERDISSE, E. (1968). Biosynthesis and degradation of mammalian mitochondrial DNA. In *Round Table Discussion on Biochemical Aspects of the Biogenesis of Mitochondria*. Ed. E. C. Slater, J. M. Tager, S. Papa and E. Quagliariello. Bari: Adriatica Editrice. (In the Press.)

NEUBERT, D., HELGE, H. & MERKER, H. (1968). Biosynthesis of mammalian mitochondrial RNA. In *Round Table Discussion on Biochemical Aspects of Biogenesis of Mitochondria*. Ed. E. C. Slater, J. M. Tager, S. Papa and E. Quagliariello. Bari: Adriatica Editrice. (In the Press.)

NEUPERT, W., BRDICZKA, D. & BÜCHER, T. H. (1967). Incorporation of amino acids into the outer and inner membrane of isolated rat liver mitochondria. *Biochem. biophys. Res. Commun.* **27**, 488.

NOVIKOFF, A. G. (1961). Mitochondria (Chrondriosomes). In *The Cell*, vol. II, p. 299. Ed. J. Brachet and A. E. Mirsky. New York: Academic Press.

PARK, R. B. & BRANTON, D. (1967). Freeze-etching of chloroplasts from glutaraldehyde-fixed leaves. In *Energy Conversion by the Photosynthetic Apparatus*, p. 341. Upton, New York: Brookhaven National Laboratory.

PARSONS, J. A. (1965). Mitochondrial incorporation of tritiated thymidine in *Tetrahymena pyriformis*. *J. Cell Biol.* **25**, 641.

PARSONS, D. F. (1966). Ultrastructure and molecular aspects of cell membranes. *Proc. 7th Canadian Cancer Conf.*, p. 193. Oxford: Pergamon.

PARSONS, D. F., WILLIAMS, G. R. & CHANCE, B. (1966). Characteristics of isolated and purified preparations of outer and inner membranes of mitochondria. *Ann. N.Y. Acad. Sci.* **137**, 643.

PARSONS, D. F., WILLIAMS, G. R., THOMPSON, W., WILSON, D. & CHANCE, B. (1967). Improvements in the procedure for purification of outer and inner mitochondrial membrane. Comparison of outer-membrane with smooth endoplasmic reticulum. In *Mitochondrial Structure and Compartmentation*, p. 29. Ed. E. Quagliariello, S. Papa, E. C. Slater and J. M. Tager. Bari: Adriatica Editrice.

PARSONS, P. & SIMPSON, M. V. (1967). Biosynthesis of DNA by isolated mitochondria. *Science*, **155**, 91.

POGO, B. G. T. & POGO, A. O. (1964). DNA dependence of plastid differentiation. Inhibition by actinomycin D. *J. Cell Biol.* **22**, 296.

POGO, B. G. T. & POGO, A. O. (1965). Inhibition by chloramphenicol of chlorophyll, protein synthesis and growth in *Euglena gracilis*. *J. Protozool.* **12**, 96.

POLAKIS, E. S., BARTLEY, W. & MEEK, G. A. (1964). Changes in structure and enzyme activity of *Saccharomyces cerevisiae in* response to changes in the environment. *Biochem. J.* **90**, 369.

RACKER, E., TYLER, D. D., ESTABROOK, R. W., CORNOVER, T. E., PARSONS, D. F. & CHANCE, B. (1965). Correlations between electron transport activity, ATPase and morphology of submitochondrial particles. In *Oxidases and Related Redox Systems*, vol. 2, p. 1077. Ed. T. E. King, H. S. Mason and M. Morrison. New York: Wiley.

RAZIN, S., MOROWITZ, H. J. & TERRY, T. M. (1965). Membrane subunits of *Mycoplasma laidlawii* and their assembly to membrane like structures *Proc. natn. Acad. Sci. U.S.A.* **54**, 219.

REICH, E. & LUCK, D. J. L. (1966). Replication and inheritance of mitochondrial DNA. *Proc. natn. Acad. Sci. U.S.A.* **55**, 1600.

RICHARDSON, S. H., HULTIN, H. O. & GREEN, D. E. (1963). Structural proteins of membrane systems. *Proc. natn. Acad. Sci. U.S.A.* **50**, 821.

RIFKIN, M. R., WOOD, D. D. & LUCK, D. J. L. (1967). Ribosomal RNA and ribosomes from mitochondria of *Neurospora*. *Proc. natn. Acad. Sci. U.S.A.* **58**, 1025.

RITCHIE, D. & HAZELTINE, P. (1953). Mitochondria in *Allomyces* under experimental conditions. *Expl Cell Res.* **17**, 58 S.

ROBERTSON, J. D. (1959). The ultrastructure of cell membranes and their derivatives. *Biochem. Soc. Symp.* **16**, 3.

ROGERS, P. J., PRESTON, B. N., TITCHENER, E. B. & LINNANE, A. W. (1967). Differences between the sedimentation characteristics of RNA's prepared from yeast cytoplasmic ribosomes and mitochondria. *Biochem. biophys. Res. Commun.* **27**, 405.

ROODYN, D. B. (1962). The controlled disruption and subfractionation of mitochondria labelled *in vitro* with radioactive valine. *Biochem. J.* **85**, 177.

ROODYN, D. B. (1966). Factors affecting the incorporation of amino acids into protein by isolated mitochondria. In *Regulation of Metabolic Processes in Mitochondria*, p. 383. Ed. J. M. Tager, S. Papa, E. Quagliariello and E. C. Slater. Amsterdam: Elsevier.

ROODYN, D. B., REIS, P. J. & WORK, T. S. (1960). Protein synthesis in isolated mitochondria: its relationship to oxidative phosphorylation and its bearing upon theories of mitochondrial replication. In *Protein Biosynthesis*, p. 44. London: Academic Press.

ROODYN, D. B., SUTTIE, J. W. & WORK, T. S. (1962). Rate of incorporation *in vitro* of radioactive amino acids into soluble proteins in the mitochondrial fraction including catalase, malic dehydrogenase and cytochrome *c*. *Biochem. J.* **83**, 29.

SAGER, R. (1965). On non-chromosomal heredity in micro-organisms. In *Function and Structure in Micro-organisms*, p. 324. Ed. M. R. Pollock and M. H. Richmond. Cambridge University Press.

SAGER, R. & HAMILTON, M. G. (1967). Cytoplasmic and chloroplast ribosomes; ultracentrifugal characterization. *Science*, **157**, 709.

SALTON, M. R. J. (1967). Structure and function of bacterial cell membranes. *A. Rev. Microbiol.* **21**, 417.

SALTON, M. R. J. & NETSCHEY, A. (1965). Physical chemistry of isolated bacterial membranes. *Biochim. biophys. Acta*, **107**, 539.

SCHNEIDER, W. C. (1963). Intracellular distribution of enzymes, XIII. Enzymatic synthesis of deoxycytidine diphosphatecholine and lecithin in rat liver. *J. biol. Chem.* **238**, 3572.

SCHRADER, L. E., BEEVERS, L. & HAGEMAN, R. H. (1967). Differential effects of chloramphenicol on the induction of nitrate and nitrite reductase in green leaf tissue. *Biochem. biophys. Res. Commun.* **26**, 14.

SHATZ, G. (1965). Subcellular particles carrying mitochondrial enzymes in anaerobically-grown cells of *Saccharomyces cerevisiae*. *Biochim. biophys. Acta*, **96**, 342.

SHERMAN, F., STEWART, J. W., MARGOLIASH, E., PARKER, J. & CAMPBELL, W. (1967). The structural gene for yeast cytochrome *c*. *Proc. natn. Acad. Sci. U.S.A.* **55**, 1498.

SINCLAIR, J. H. & STEVENS, B. J. (1966). Circular DNA filaments from mouse mitochondria. *Proc. natn. Acad. Sci. U.S.A.* **56**, 508.

SISTROM, R. W. (1962). The kinetics of the synthesis of photopigments in *Rhodopseudomonas spheroides. J. gen. Microbiol.* **28**, 607.

SLONIMSKI, P. P., PERRODIN, G. & CROFT, J. H. (1968). Ethidium bromide induced mutation of yeast mitochondria: complete transformation of cells into respiratory deficient non-chromosomal 'Petites'. *Biochem. biophys. Res. Commun.* **30**, 232.

SMILLIE, R. M. (1962). Photosynthetic and respiratory activities of growing pea leaves. *Pl. Physiol.* **37**, 716.

SMILLIE, R. M., EVANS, W. R. & LYMAN, H. (1963). Meristems and differentiation. *Brookhaven Symp. Biol.* **16**, 89.

SMILLIE, R. M., GRAHAM, D., DWYER, M. R., GRIEVE, A. & TOBIN, N. F. (1967). Evidence for the synthesis *in vivo* of proteins of the Calvin cycle and of the photosynthetic electron-transfer pathway on chloroplast ribosomes. *Biochem. biophys. Res. Commun.* **28**, 604.

SOTTOCASA, G. L., KUYLENSTIERNA, B., ERNSTER, L. & ANDERS, B. (1967). In *Methods in Enzymology*, vol. x, p. 448. Ed. R. W. Estabrook and M. E. Pullman. New York: Academic Press.

SPENCER, D. & WHITFELD, P. R. (1967). DNA synthesis in isolated chloroplasts. *Biochem. biophys. Res. Commun.* **28**, 538.

SPENCER, D. & WILDMAN, S. G. (1964). The incorporation of amino acids into protein by cell-free extracts from Tobacco leaves. *Biochemistry, N.Y.* **3**, 954.

STEELE-SCOTT, W. & SMILLIE, R. M. (1967). Evidence for the direction of chloroplast ribosomal RNA synthesis by chloroplast DNA. *Biochem. biophys. Res. Commun.* **28**, 598.

STOECKENIUS, W. (1966). Morphological observations on mitochondria and related structures. *Ann. N.Y. Acad. Sci.* **137**, 641.

STUMPF, P. K., BOVÉ, J. M. & GOFFEAU, A. (1963). Relation of fatty acid synthesis and photophosphorylation in lettuce chloroplasts. *Biochim. biophys. Acta*, **70**, 260.

STUTZ, E. & NOLL, H. (1967). Characterization of cytoplasmic and chloroplast polysomes in plants: evidence for three classes of ribosomal RNA in nature. *Proc. natn. Acad. Sci. U.S.A.* **57**, 774.

SUSSMAN, A. S. (1965). In *The Fungi*, vol. II, p. 733. Ed. G. C. Ainsworth and A. S. Sussman. New York: Academic Press.

SUYAMA, Y. (1966). Mitochondrial DNA of *Tetrahymena*. Its partial physical characterization. *Biochemistry, N.Y.* **5**, 2214.

SUYAMA, Y. (1967). The origin of mitochondrial RNAs in *Tetrahymena pyriformis. Biochemistry, N.Y.* **6**, 2829.

SUYAMA, Y. & EYER, J. (1967). Leucyl-tRNA and leucyl-tRNA synthetase in mitochondria of *Tetrahymena pyriformis. Biochem. biophys. Res. Commun.* **28**, 746.

SUYAMA, Y. & EYER, J. (1968). RNA synthesis in isolated mitochondria from *Tetrahymena. J. biol. Chem.* **243**, 320.

SVETAILO, E. N., PHILIPPOVICH, I. I. & SISSAKIAN, N. M. (1967). Sedimentation properties of chloroplast and cytoplasmic ribosomes from pea seedlings. *J. molec. Biol.* **24**, 405.

TAYLOR, C. B., BAILEY, E. & BARTLEY, W. (1967). Studies on the biosynthesis of protein and lipid components of rat liver mitochondria. *Biochem. J.* **105**, 605.

THOMPSON, T. E. & HUANG, C. (1966). The water permeability of lipid bilayer membranes. *Ann. N.Y. Acad. Sci.* **137**, 740.

TUPPY, H. & WILDNER, G. (1965). Cytoplasmic transformation. Mitochondria of wild type Baker's yeast restoring respiratory capacity in a respiratory deficient 'Petite' mutant. *Biochem. biophys. Res. Commun.* **20**, 733.

VAZQUEZ, D. (1966). Mode of action of chloramphenicol and related antibiotics. *Symp. Soc. gen. Microbiol.* **16**, 169.

WALLACE, P. A., HUANG, M. & LINNANE, A. W. (1968). The influence of medium composition on the cytology of anaerobically grown *Saccharomyces cerevisiae. J. Cell Biol.* **37**, 207.

WALLACH, D. F. H. (1966). Protein conformations in cellular membranes, *Proc. natn. Acad. Sci. U.S.A.* **56**, 1552.

WEISS, B. (1965). An electron microscope and biochemical study of *Neurospora crassa* during development. *J. gen. Microbiol.* **39**, 85.

WHITE, D. C. (1962). Respiratory systems in the haemin-requiring *Haemophilus* species. *J. Bact.* **85**, 84.

WHITE, D. C. (1964). Differential synthesis of five primary electron transport dehydrogenases in *Haemophilus parainfluenzae. J. biol. Chem.* **239**, 2055.

WHITE, D. C. & SMITH, L. (1962). Haematin enzymes of *Haemophilus parainfluenzae. J. Bact.* **83**, 851.

WILGRAM, G. F. & KENNEDY, E. P. (1963). Intracellular distribution of some enzymes catalyzing reactions in the biosynthesis of complex lipids. *J. biol. Chem.* **238**, 2615.

WILKIE, D. (1964). *The Cytoplasm in Heredity.* London: Methuen.

WILKIE, D., SAUNDERS, G. & LINNANE, A. W. (1967). Inhibition of respiratory enzyme synthesis in yeast by chloramphenicol. Relationship between chloramphenicol tolerance and resistance to other antibacterial antibiotics. *Genet. Res.* **10**, 199.

WINTERSBERGER, E. (1966). Synthesis and function of mitochondrial RNA. In *Regulation of Metabolic Processes in Mitochondria*, p. 439. Ed. J. M. Tager, S. Papa, E. Quagliariello and E. C. Slater. Amsterdam: Elsevier.

WINTERSBERGER, E. (1968). Synthesis of DNA in isolated yeast mitochondria. In *Round Table Discussion on Biochemical Aspects of the Biogenesis of Mitochondria* Ed. E. C. Slater, J. M. Tager, S. Papa and E. Quagliariello. Bari: Adriatica Editrice. (In the Press.)

WOLKEN, J. J. (1959). The structure of the chloroplast. *A. Rev. Pl. Physiol.* **10**, 71.

WOODWARD, D. O. & MUNKERS, K. D. (1966). Alterations of a maternally inherited mitochondrial structural protein in respiratory deficient strains of *Neurospora. Proc. natn. Acad. Sci. U.S.A.* **55**, 872.

WOODWARD, D. O. & MUNKRES, K. D. (1967). Genetic control, function and assembly of a structural protein in *Neurospora.* In *Organizational Biosynthesis*, p. 489. Ed. H. J. Vogel, J. O. Lampen and V. Bryson. New York: Academic Press.

WORK, T. S. (1967). The function of mitochondrial nucleic acids in protein synthesis. *Biochem. J.* **105**, 38P.

WRIGHT, E. A. & WHITE, D. C. (1966). Formation of a functional electron transport system during growth of penicillin-induced spheroplasts of *Haemophilus parainfluenzae. J. Bact.* **91**, 1356.

ZALKIN, H. & RACKER, E. (1965). Partial resolution of the enzymes catalysing oxidative phosphorylation. V. Properties of coupling factor 4. *J. biol. Chem.* **240**, 4017.

EXPLANATION OF PLATE

Fig. 1. Electron micrograph of section of *Tetrahymena pyriformis* grown on proteose–peptone medium in the absence of chloramphenicol, showing normal mitochondria with outer membranes (*om*) and densely packed tubular cristae (*c*).

Fig. 2. Electron micrograph of section of *T. pyriformis* grown on proteose–peptone medium containing 200 μg./ml. chloramphenicol. Note the presence of small abnormal mitochondria (*m*) with intact outer membranes but deficient in cristae. (Preparations by G. Turner and G. I. Roach.)

PLATE 1

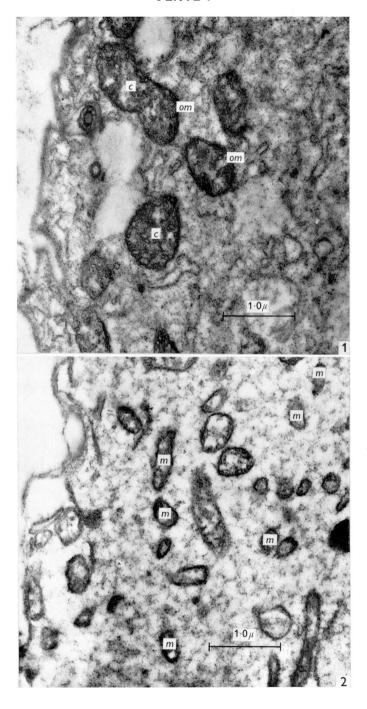

GROWTH AND DIFFERENTIATION OF PLANT CELLS IN CULTURE

D. H. NORTHCOTE

Department of Biochemistry, University of Cambridge

INTRODUCTION

By inserting a segment of a stem or root into a block of agar containing a nutrient medium the orderly growth of the tissues can be disorganized so that a callus is formed. The nutrient solution contains a simple salt mixture, a carbon source such as sucrose and small amounts of growth substances such as indolyl-3-acetic acid, or 2,4-dichlorophenoxy acetic acid, kinetin, amino acids, myoinositol, thiamine and pantothenic acids (White, 1943; Gautheret, 1959; Gamborg, Miller & Ojima, 1968).

Table 1. *Composition of a medium for the growth of callus tissue from* Acer pseudoplatanus

Substance	mg./l.	Substance	mg./l.
$MgSO_4 . 7H_2O$	250·0	$CuSO_4 . 5H_2O$	0·03
$CaCl_2 . 2H_2O$	75·0	$AlCl_3$	0·03
$NaNO_3$	600·0	$NiCl_3 . 6H_2O$	0·03
KCl	750·0	KI	0·01
$NaH_2PO_4 . H_2O$	125·0	Sucrose	20000
$FeCl_3 . 6H_2O$	1·0	Thiamine	1·0
$ZnSO_4 . 7H_2O$	1·0	Calcium D-pantothenate	1·0
H_3BO_3	1·0	2,4-Dichlorophenoxyacetic acid	6·0
$MnSO_4 . 4H_2O$	0·1	Coconut milk	20%, v/v

Usually, stimulation of callus formation is not possible on a completely defined medium and some of the growth factors are added by the inclusion of a liquid endosperm (coconut milk, 10–20%, v/v), yeast extract or casein hydrolysate (Table 1) (van Overbeck, Conklin & Blakeslee, 1942; Caplin & Steward, 1948; Steward & Caplin, 1951; Caplin, 1956; Torrey & Shigemura, 1957). In this type of medium most tissues will produce a callus and some of these will be capable of continued transfer away from the parent tissue if they are supplied with exogenous nutrients (Hildebrandt, Riker & Duggar, 1945; Hildebrandt & Riker, 1949; Nickell, 1956; Gautheret, 1959; Blakely & Steward, 1961). By agitating the solid callus in a solution of the nutrient that stimulates rapid growth of the callus, small clumps and some single cells are shed from the solid mass and in some cases the cultures can be maintained in the liquid medium as a fine cell suspension which contains single cells and small aggregates of about 100–1000 cells (Nickell, 1956; Torrey & Shigemura, 1957; Nickell & Tulecke, 1960; Blakely & Steward, 1961; Lamport, 1964).

GROWTH CHARACTERISTICS

When a transfer of a callus tissue is made to fresh medium, growth of the tissue occurs after a lag period of up to about 12–15 h. (Henshaw *et al.* 1966; Givan & Collin, 1967). Growth takes place during the next 24–72 hr. mainly by cell division and the mitotic index of the culture rises to approximately 1–4 % over this period. The first two or three divisions of the cells are usually in the same plane and at right angles to the long axis of the cell so that a column of cells of more or less equal volume are produced; subsequent divisions are more random and small aggregates of cells are formed (Vasil & Hildebrandt, 1965*a*, *b*, 1967; Henshaw *et al.* 1966). After about 7 days from the transfer, the cells no longer divide and growth occurs by an increase in area of the cell surface (Street & Henshaw, 1963), and usually the cells change from spherical to elongate. The growth characteristics of the culture depend not only upon the medium and the inherent properties of the cells but also upon the frequency of subculture and the size of the inoculum. The cells do not grow independently of one another and generally the culture has a heterogeneous population. In any one culture maintained by standard growth and transfer techniques the majority of the cells resemble one another but a fair proportion of the cells usually differ from the mean size, shape and ploidy of the population.

The general cytological features of the cells from any tissue callus are similar, even though the cultures were obtained from different species and produced from different organs. In young cells small vacuoles are present but the older cells are characterized by a large vacuole so that the cytoplasm and organelles are restricted to a thin layer around the wall and to transvacuolar strands (Blakely & Steward, 1961; Mitra & Steward, 1961) (Pl. 1, figs. 1, 2; Pl. 2, figs. 1, 2, 3). The nucleus is usually closely applied to the wall or suspended within the vacuole by strands (Pl. 1, figs. 1, 2). Amyloplasts, mitochondria, Golgi bodies and the endoplasmic reticulum are congregated around the nucleus, sparsely distributed in the peripheral cytoplasm and in the transvacuolar strands (Pl. 1, figs. 1, 2; Pl. 2, figs. 2, 3). Microtubules are found lying just under the plasmalemma and they encircle the cell (Pl. 2, fig. 2); in the peripheral cytoplasm they are often found between the plasmalemma and a mitochondrion or plastid or fat droplet (Pl. 3, figs. 1–4; Pl. 4, figs. 1–3). Generally a suspension culture of a tissue such as *Acer pseudoplatanus* contains cells with a dry weight of about 3×10^{-9} g., a diameter of between 50 and 100 μ and a Q_{O_2} of about 3·0 μl./mg. dry wt./hr.

WALL STRUCTURE AND FORMATION

The wall of a callus cell makes up 40–46 % of the dry weight of the cell (approx. $1·2 \times 10^{-9}$ g.). It is non-lignified; its chemical composition resembles that of a cambial cell wall both in the quantities of the main polysaccharide fractions and in the nature and amounts of the individual sugars of which they are composed (Lamport & Northcote, 1960 a; Stoddart, Barrett & Northcote, 1967) (Table 2).

Table 2. *Composition of polysaccharide components of sycamore callus and cambial tissues*

Substance	Callus, % cell-wall material	Cambium, % cell-wall material
Pectin	15	15
Hemicellulose	40	45
α cellulose	30	37
	% whole dry tissue extracted with 80% ethanol	
Glucose (from α cellulose)	7·6	11·5
Galactose	6·4	8·1
Mannose	0·8	0·8
Arabinose	8·2	6·3
Xylose	4·0	2·9
Rhamnose	1·0	1·1
Galacturonic acid	10·3	11·6

The pectic polysaccharides of the wall are particularly interesting and the nature and metabolism of these compounds can be readily studied by the use of plant tissue cultures. These compounds are laid down in the wall of the cells of the intact plant only during primary growth (Northcote, 1963). The pectic polysaccharides make up part of the matrix material of the wall and any change in their properties that depends on a change in their pattern of formation could greatly influence the physical nature of the wall and consequently have a direct effect on the growth of the cell. Pectic substances can be extracted from the cell with sodium hexametaphosphate (2 %, pH 4·0, 100° C., 4 hr.) and the starch present in the extract can be removed by α-amylase (Stoddart *et al.* 1967). The resultant solution contains four groups of pectic substances which can be separated by electrophoresis on glass fibre paper at pH 6·5 (Barrett & Northcote, 1965; Stoddart & Northcote, 1967*b*). One of these is a neutral fraction composed of arabinose and galactose and the other three are acidic polymers composed of galacturonic acid and varying amounts of neutral sugars (mainly arabinose and galactose but also rhamnose and xylose) (Barrett & Northcote, 1965; Stoddart *et al.* 1967). The acidic nature of the polymers can be metabolically varied by the formation of varying amounts of the methyl ester of the polyuronic acid chains and by the incorporation of blocks of neutral sugars into or on to the chains (Fig. 1).

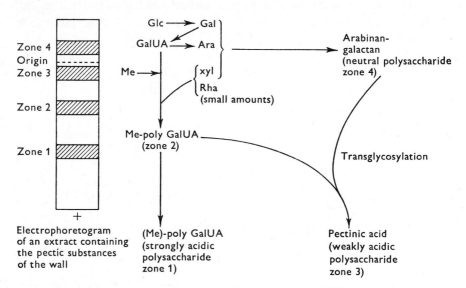

Fig. 1. Diagram to illustrate the possible metabolic relationships of the pectic polysaccharides that are present in an actively growing cell. The separation of the zones of the pectic material extracted with sodium hexametaphosphate (2%; pH, 4·0; 100°C.) and separated on a glass fibre strip by electrophoresis (pH, 6·5; 44 V./cm.; 30 min.) is depicted diagrammatically on the left of the scheme. Glc, glucose; Gal UA, galacturonic acid; ara, arabinose; xyl, xylose.

Electrophoretic analysis of the pectic substances in a mature tissue such as an apple fruit and in actively growing and dividing tissues such as apple cambium and a tissue culture prepared from apple fruit, shows that the amount and type of each pectic fraction varies with the growth conditions of the tissue. In apple-fruit pectin, the neutral fraction is present together with a weakly acidic fraction (pectinic acid) (Fig. 1). Only a trace of the more strongly acidic polymers is found. The pectinic acid contains chains of polygalacturonic acid which are combined with quite large blocks of neutral polysaccharide material containing arabinose and galactose. In the rapidly growing tissue, on the other hand, the more strongly acidic fractions are present. These fractions contain relatively less methylated polygalacturonic acid and less neutral sugars (Barrett & Northcote, 1965; Stoddart *et al.* 1967). When radioactive glucose and arabinose are fed to growing sycamore suspension cultures it can be shown that in all probability the polygalacturonic acid chains are synthesized separately from the neutral polymers containing arabinose and galactose. The neutral material can then be incorporated into the acidic polymers by a transglycosylation reaction which occurs either during the processing of the molecules in the Golgi bodies and vesicles before they are passed into the wall or after they have been deposited within the matrix of the wall (Stoddart & Northcote, 1967*a*).

Proteins that contain hydroxyproline are found in plant cells and it has been shown that the bulk of the hydroxyproline which occurs in the callus cells of sycamore and bean is located within the structure of the wall (Lamport & Northcote, 1960*a, b*). The function of the wall peptides is unknown, but it has been suggested that glycopeptides are present (Boundy *et al.* 1967) and that one of the linkages between the peptide and the sugar portion of the molecule could be an arabinosyl bond with the hydroxyl group of the hydroxyproline (Lamport, 1967). It is possible therefore that this type of compound could take part in the mechanism of a transglycosylation type of reaction which could alter the nature of the pectic polysaccharides in the wall. Free hydroxyproline added to a medium in which intact tissue is growing acts as a specific inhibitor of growth in area of the cell wall (Cleland, 1967*a*) and although the mechanism of this inhibition is unknown it is likely that either the hydroxyproline inhibits the formation of the peptides which contain hydroxyproline and which are formed by hydroxylation of incorporated proline in the peptide (Cleland, 1967*b*; Cleland & Olson, 1968), or that at the high concentrations of hydroxyproline used to achieve growth inhibition abnormal proteins may be synthesized (Holleman, 1967).

GROWTH FACTORS

The disorganized growth of the callus tissue is brought about by the supplied exogenous growth factors and depends upon the ratio of the concentrations of a hormone such as indolyl-3-acetic acid or 2,4-dinitrophenoxy acetic acid on the one hand and kinetin and the other growth factors in coconut milk on the other (Steward & Caplin, 1951; Nétien, 1957; Skoog & Miller, 1957; Torrey & Shigemura, 1957;

Steward, Mapes & Mears, 1958; Reinert, 1959; Steward & Shantz, 1959; Dougall & Shimbayashi, 1960; Blakely & Steward, 1961). With an imbalance of these supplied materials random cell division and growth in cell wall area occur; even cytokinesis and mitosis become out of step and daughter cells with abnormal ploidy arise (Gautheret, 1959; Torrey, 1959; Mitra, Mapes & Steward, 1960; Mitra & Steward, 1961; Cooper, Cooper, Hildebrandt & Riker, 1964).

Callus tissue can be induced in wound tissue of a plant by bacterial or viral infection and the resultant gall tissue is capable of maintenance on an external medium with limited amounts of various growth factors or with only a carbon source and a simple salt solution (Braun, 1957). The gall tissue can be maintained in this way even though the original bacterial or other infection has been removed from the tissue. If the tissue had not been infected at all then a range of growth factors must be supplied to the explant medium in order to grow a callus from it, and if it is to be maintained as a callus, the growth factors must be added to the medium at each subculture. It appears therefore as if the bacterial or viral infection has brought about a permanent increase in the activity or amount of the endogenous synthetic systems, in particular those for the production of indolyl-3-acetic acid and kinetin, which supply the growth factors to the callus mass of cells. The sites of indolyl-3-acetic acid and kinetin synthesis in the intact plant are unknown, but the work with the gall tissue indicates that the disorganized cells of a callus are capable of synthesizing such growth factors so that callus material can be used to investigate these processes.

Tobacco internode explants cultured on a basic medium which contains salts, glycine, nicotinic acid, thiamine, pyridoxamine and sucrose grow a callus at their morphological basal end (Nierdergang-Kamien & Skoog, 1956). The growth of callus depends upon the movement of indolyl-3-acetic acid within the stem segment to the basal end of the internode. Callus can be stimulated at the apical end of the stem segment only if a medium containing auxin and coconut milk is applied to that end. On the basic medium the excess of indolyl-3-acetic acid which stimulates the callus production is supplied by a movement of the hormone that is already present or that is being produced by the tissues of the stem segment, and by its accumulation at the basal end by a polar transport mechanism. The callus which is formed develops in an organized sequence. Within 8 days a definite wedge-shaped area of cambial activity, the dynamic wedge, is apparent and by 14 days an outgrowth of the tissue occurs and tracheids differentiate in the new tissue. Later an arching vascular cambium is formed which gives rise to lignified pitted tracheids on the inside and phloem with sieve elements on the outside (Sterling, 1950; Sheldrake & Northcote, 1968 a).

The auxin present in the internode at the time of excision accumulates at the basal end of the section within 3 hr. (Nierdergang-Kamien & Skoog, 1956). Yet such internode sections can be bisected from 1 to 23 days or even 4, 6 or 9 months after the initial explant and the apical halves will still retain the ability to form a callus at their basal end if they are reset in the basic medium. During a period of up to at least 74 days from the time the stem segment was placed in the agar medium, after excision, the lateral cambium of the cultured internode section is active and produces xylem and phloem within the stem segment (Sheldrake & Northcote, 1968 a).

The tobacco stem explants can be stripped, at the level of the cambium, to give two preparations. One of these consists of the inner layers of the original stem which at the newly exposed surface carry cells already differentiating into xylem or ray parenchyma, and the other consists of the outer tissues which at the newly exposed surface have the thin-walled cambial cells and on the outside of these the phloem but which carry no xylem cells.

The inner tissues if explanted on to the basic medium produced at the exposed outer surface a few cell divisions which gave rise to a few large wound cells and a callus formed at the basal end of the section. However, if the explant was bisected after 7 days no new callus formed on the basal end of the apical half if this was recultured in the medium. When the outer tissues were cultured they continued to give cell divisions all along the exposed surface of the explant. A callus developed at the basal end and a cambial zone appeared along the whole length of the explant which cut off new phloem elements and xylem tracheids. The new cambial zone was in continuity with an arch of cambium which formed in the callus. Ordered rows of differentiated phloem and xylem cells were formed and shoots sometimes formed on the callus. By the third day lignified, reticulately thickened xylem tracheids appeared at the basal end and from these new differentiation occurred acropetally until at the fourth day several files of lignified xylem cells were found stretching down to the apical end of the explant. The explant of the outer tissue could be bisected some 4 months after the original explant had been formed and the apical halves would still form fresh callus at their basal ends (Sheldrake & Northcote, 1968 a).

The polar transport of auxin is inhibited by 2,3,5-tri-iodo-benzoic acid (Nierdergang-Kamien & Skoog, 1956) and if this substance is put into the basic medium and explants of the intact stem segment or the stripped outer tissues are placed horizontally on to the medium, after they had been initially cultured for several days in a normal manner on the untreated basic medium, then a great many rows of xylem cells developed in the explants (Sheldrake & Northcote, 1968*a*).

These experiments with the tobacco internode segments suggest that differentiating xylem and/or phloem cells produce either auxin or its immediate metabolic precursors. Auxin is known to stimulate cambial activity and xylem differentiation (Torrey, 1953; Wetmore, 1955; Wareing, 1958). If auxin is produced as a consequence of xylem differentiation then in the absence of a polar transport mechanism for removing it a positive feed-back effect would stimulate the formation of serried ranks of xylem cells such as those obtained in the presence of the triodobenzoic acid. It is therefore possible that auxin is produced as a consequence of xylem rather than phloem differentiation (Sheldrake & Northcote, 1968*a*).

Autolysing plant and animal tissues produce auxin and maybe other growth factors (Sheldrake & Northcote, 1968*b*). The breakdown of the cytoplasmic contents of differentiating xylem and phloem cells represents a controlled and organized autolytic process which occurs within an intact plant. These dying cells could therefore be an important source of auxin and other growth factors, the supply of which to other tissues could be further regulated by transport mechanisms and enzymes which bring about the destruction of the hormones.

Within a mass of tissue culture cells at any one time not all the cells are alive and areas of necrosis can be found. These sites may therefore be important sources of growth materials for the living areas of the callus and may give rise to variable factors in the growth inocula and in the degree of spontaneous differentiation occurring within the callus at any particular time in any one type of growth medium.

CONTROL OF DIFFERENTIATION

A callus tissue such as that prepared from the stem of *Phaseolus vulgaris* and maintained on a medium containing 2,4-dichlorophenoxyacetic acid (6 p.p.m.) and 20 % coconut milk does not contain any cells which resemble phloem or xylem (Jeffs & Northcote, 1966). Differentiated cells can be induced in such a tissue if it is transferred to a medium containing less 2,4-dichlorophenoxyacetic acid or none at all and with varying quantities of sucrose (Torrey & Shigemura, 1957). Differentiation can also be induced in the tissue maintained on a basic medium that contains neither growth factors nor coconut milk if the cells are irrigated with a solution of growth factors and sucrose by the implantation into the callus mass of an agar wedge containing them (Wetmore & Rier, 1963; Jeffs & Northcote, 1966, 1967). Micropipettes containing solutions of growth factors have also been used to irrigate the tissue callus (Clutter, 1960). The purpose of the techniques is to bring about a variation in the ratio of the concentrations of exogenous and endogenous growth factors and nutrients in regions of the callus tissues (Skoog & Miller, 1957; Vasil & Hildebrandt, 1965b). It is possible that growth factors applied to the cultures in varying sequences may have different effects from those treatments where they are applied separately or concurrently (Steward, Kent & Mapes, 1967).

These experiments show that with certain ratios of concentrations of applied indolyl-3-acetic acid, sucrose and kinetin, xylem or xylem and phloem are differentiated within the callus tissue usually in the form of small nodules which contain, in addition, an active meristematic region in which the cell divisions seem to occur with a definite orientation to give a more organized tissue (Wetmore & Rier, 1963; Jeffs & Northcote, 1966, 1967; Rier & Beslow, 1967). Within these nodules the xylem occurs towards the centre and the phloem, as several groups of cells, at the periphery and separated from the xylem by the meristematic region (Pl. 5, Figs. 1–3). Thus the general pattern of development resembles in some respects that which occurs in the intact tissues from which the callus was formed.

Generally it is found that indolyl-3-acetic acid (0·1 %) and low concentrations of sucrose (1 % or less) bring about the induction of xylem tracheids, whereas higher concentrations of sucrose (2 % and above) together with the indolyl-3-acetic acid stimulate the formation of nodules and phloem. The induction of phloem is dependent upon the presence of the intact sucrose molecules and similar concentrations of glucose or fructose or both monosaccharides together in the induction medium does not stimulate phloem formation (Jeffs & Northcote, 1967). Some other disaccharides such as maltose and $\alpha\alpha$ trehalose also induce phloem formation but cellobiose does not, so that it seems essential to have an α-glucosyl linkage in the disaccharide in order to bring about the induction (Jeffs & Northcote, 1967). Thus sucrose has a physiological action in addition to its role for the transport of carbon within the plant. Since xylem can probably produce indolyl-3-acetic acid and phloem conducts the sucrose formed from photosynthesis, these vascular tissues can themselves provide some of the necessary growth factors to stimulate further differentiation. A system is therefore possible whereby the continuity of the vascular tissues in the growing plant can be maintained.

The induced differentiation within the callus tissue can be detected by optical microscopy either by the use of stains such as phloroglucinol which detects the lignin present in the secondarily thickened xylem walls (Jensen, 1962) or by the use of polarizing optics which clearly show the organized secondary thickening of the xylem (Jeffs & Northcote, 1966). The use of ultraviolet light fluorescent optics in the presence of a fluorescent dye such as aniline blue enables the callose of the sieve plates of the phloem tissue to be made visible (Pl. 5, Figs. 2, 3) (Currier & Strugger, 1956; Jeffs & Northcote, 1967; Northcote & Wooding, 1968).

It is also possible to use the induced differentiated cells in the callus tissue for ultrastructural studies of the differentiation processes which are going on and to compare them with those that occur within an intact stem or root. The fine structure of the cells during differentiation in both types of tissue is very similar. This is especially interesting for the formation of the sieve tubes since in the phloem of a cut stem it is always possible that a particular cytological characteristic seen in the sieve tube may have been brought about by the release of osmotic tension in the whole phloem conducting system when the stem was cut. Many structural features of the phloem tissues of stems and roots have been doubted because of the possibility of artifacts arising in this manner, especially since any continuous strands which might pass from one sieve tube to the next throughout the system would be damaged when the tissue was cut. However, the phloem which occurs as discrete nodules in induced callus tissue cannot be subjected to this type of artifact when it is removed from the callus mass and fixed for electron microscopic study. Nevertheless, the essential fine structural changes which occur during the differentiation of this induced phloem are almost identical to those which are seen with the intact plant (Laflèche, 1966; Northcote & Wooding, 1968) (Pl. 6, Figs. 1–4; Pl. 7, Figs. 1–4).

The measurement of xylem and phloem differentiation can be expressed quantitatively by comparing the composition of the cambial cell wall with that of the derived tissues (Jeffs & Northcote, 1966). Some of these differences can be stated in the form of the xylose/arabinose concentration ratio which is low in the primary wall and high in the secondary thickenings and reflects the changes in the composition of the polysaccharides of the matrix material of the wall (Table 3). These polysaccharides are synthesized within the cytoplasm and are transported to the wall within the Golgi vesicles (Northcote, 1968). The change in ratio is therefore dependent on a change in polysaccharide synthesis which accompanies the change of primary to secondary growth and consists of a repression of pectin synthesis during the secondary stage. At the time of wall thickening lignin synthesis is initiated and a measurement of the amount of this substance in a tissue will also give a quantitative indication of the extent of the differentiation (Jeffs & Northcote, 1966) (Table 4). The measurement of these indices of morphogenesis in intact stems and in callus tissue indicates the similarity between the induced differentiation and that which occurs in the growing plant.

Table 3. *Ratio of concentrations of xylose and arabinose in polysaccharide hydrolysates of ethanol (80 %)-extracted bean callus and sycamore stem tissues*

Tissue	Xylose/arabinose ratio
Control callus	0.409 ± 0.053
Differentiated callus (1-month induction)	0.527 ± 0.045
Differentiated callus (2-month induction)	0.682 ± 0.038
Sycamore cambium	0.46
Sycamore xylem	20.5

Table 4. *Concentration of phenolic aldehydes derived from oxidation of lignin by nitrobenzene of bean callus and sycamore stem tissue*

Tissue	p-Hydroxy benzaldehyde (μg./mg. of dry tissue)	Vanillin (μg./mg. of dry tissue)	Syringaldehyde (μg./mg. of dry tissue)
Control callus	85	27	0
Differentiated callus (1-month induction)	95	69	0
Differentiated callus (2-month induction)	187	118	0
Sycamore cambium	20	50	30
Sycamore xylem	70	125	195

The measurement of the changes which are observed indicates the possible sites at which the growth hormones responsible for the induction of the development exert their control within the cells. Thus during the initial stages of polysaccharide synthesis they may influence the enzymes (epimerases) which bring about the interconversion of the donor uridine diphosphate sugar molecules of the glucose series (glucose, glucuronic acid and xylose) with those of the galactose series (galactose, galacturonic acid and arabinose). These donor molecules give rise to the various pectic substances and hemicelluloses of the matrix material of the wall (Northcote, 1963). Also, the growth factors

344   D. H. NORTHCOTE

may influence the initiation of lignin synthesis at the enzymic changes which convert phenylalanine into cinnamic acid (phenylalanine-ammonia-lyase) (Rubery & Northcote, 1968).

CELL DIVISION IN A CALLUS CELL

Cell divisions of a callus are stimulated by the addition of kinetin to the medium (Digby & Wareing, 1966) and in some tissue suspensions the mitotic index can be raised for a short period from 1–2 % to about 10 % and over. The cells are especially useful for a microscopic study of cell division in a living cell since the suspension cells can be cultured in a micro-chamber which can be viewed directly under the optical microscope and the division cycle recorded by time lapse or continuous cinematography (Jones, Hildebrandt, Riker & Wu, 1960; Das, Hildebrandt & Riker, 1966) (Pl. 1, Figs. 1, 2).

The division is that of a vacuolated cell and the nucleus enters mitosis either while it is applied to the wall or more usually while it is suspended by cytoplasmic strands within the vacuole. Just before prophase the starch grains leave the vicinity of the nucleus and move to one of two positions within the cell. Cytoplasmic streaming, which is very active in the interphase cell, slows down and a cytoplasmic strand is usually apparent where the future cell plate will form. Generally the stages of mitosis are normal and at telophase the cell plate is established in the usual manner, although it may fuse with one side of the mother cell wall before it has been completely extended to the rest of the wall. Normally the two daughter nuclei remain close together on either side of the new wall for some time after the completion of cytokinesis, but move away to a more central position in the daughter cells before the next division.

CONCLUSIONS

Tissue cultures either in the form of solid callus or as liquid suspensions offer unique advantages as experimental material for the study of the biochemistry, cytology and development of plants.

Large amounts of tissue can be grown under sterile conditions and in a controlled environment at all times of the year. The conditions of culture and the supply of nutrients and growth factors can be varied in a manner which is quite impossible in an intact plant and which resembles that used for experiments with bacterial cultures. In addition, radioactive tracer compounds and metabolic variations of different cell clones obtained by the culture of natural or artificial mutations can be

EXPLANATION OF PLATES

PLATE 1

Fig. 1. Differential interference microscopy of a living bean (*P. vulgaris*) callus suspension cell. The nucleus is applied to the cell wall. Amyloplasts and mitochondria can be seen aggregated around the nucleus. Transvacuolar strands can be seen which also carry mitochondria and other organelles. × 1300.

Fig. 2. Differential interference microscopy of a living bean callus suspension cell. The nucleus is suspended in the vacuole by thin transvacuolar strands connected to the peripheral cytoplasm. × 1300.

PLATE 2

Fig. 1. Electron micrograph (glutaraldehyde, osmium fixation) of the cytoplasm near the nucleus of a bean callus suspension cell. × 30,000.

Fig. 2. Electron micrograph (glutaraldehyde, osmium fixation) of the peripheral cytoplasm of a bean callus suspension cell. Microtubules indicated by arrows are present just under the cytoplasm near the wall. × 45,000.

Fig. 3. Electron micrograph (glutaraldehyde, osmium fixation) of a portion of the tip of a strand of cytoplasm in the large vacuole of a bean callus suspension cell. × 45,000.

PLATE 3

Figs 1–4. Electron micrographs of bean callus suspension cells. The relationship of mitochondria in the peripheral cytoplasm to microtubules which lie just under the plasmalemma near the cell wall can be seen. The microtubules are indicated by arrows. (Glutaraldehyde osmium fixation.) Fig. 1, × 65,000; fig. 2, × 70,000; fig. 3, × 40,500; fig. 4, × 60,000.

PLATE 4

Figs. 1–3. Electron micrographs of bean callus suspension cells. The relationship of plastids (figs. 1, 2) and a fat droplet (fig. 3) to microtubules which lie just under the plasmalemma near the cell wall can be seen. The microtubules are indicated by arrows. (Glutaraldehyde osmium fixation.) Fig. 1, × 70,000; fig. 2, × 49,000; fig. 3, × 50,000.

PLATE 5

Fig. 1. Nodule containing actively dividing cells, xylem (at centre) and phloem tissue (at periphery) which had been induced in a block of bean callus by irrigation with IAA (0·1 mg./l.) and sucrose (2 %) from an agar wedge inserted in the block. Section stained with safranin; picric aniline blue. × 120.

Fig. 2. Xylem tracheid induced in a block of bean callus tissue (for conditions see caption to Pl. 5, fig. 1). Section stained with aniline blue and photographed with ultraviolet light. × 450.

Fig. 3. Nodule containing xylem (x) and phloem tissue (Ph) induced in a block of bean callus tissue (for conditions see caption to Pl. 5, fig. 1). The section was stained with aniline blue and photographed with ultraviolet light. × 160.

PLATES 6 AND 7

Pl. 6, figs. 1–4 and Pl. 7, figs. 1–4. Electron micrographs of developing phloem sieve tubes induced in a block of bean callus tissue (for conditions see caption to Pl. 5, fig. 1). Compare the inclusions shown with those illustrated by Laflèche (1966). (Glutaraldehyde, osmium fixation.) Pl. 6, fig. 1. A crystalline inclusion of the sieve tube. The cell plate and membranous elements can be seen. × 17,000. Pl. 6, fig. 2. Pore at the sieve plate lined by callose with a membranous inclusion in the lower cell. × 40,000. Pl. 6, figs. 3, 4. Pl. 7, figs. 1, 4. Stages in the formation and dispersal of the slime body. Pl. 6, fig. 3, × 15,000; Pl. 6, fig. 4, × 30,000; Pl. 7, fig. 1, × 15,000; Pl. 7, fig. 4, × 30,000. Pl. 7, fig. 2. Pore at the sieve plate (SP) with membranous inclusion. × 40,500. Pl. 7, fig. 3. Fully formed sieve plate (SP) between two sieve tubes. × 15,000.

(*Facing page* 344)

PLATE 1

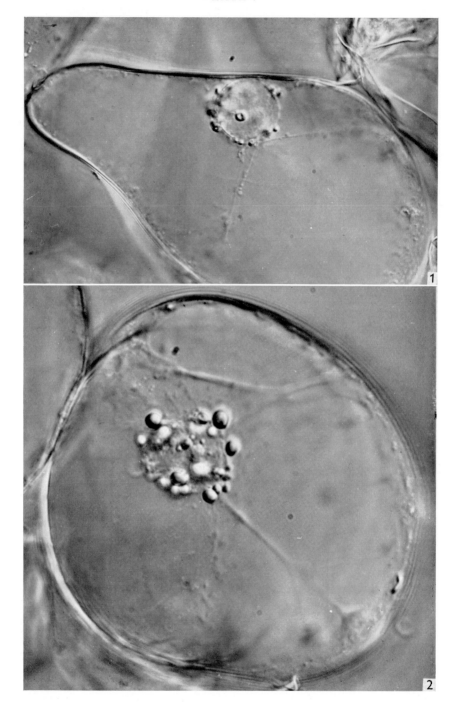

PLATE 2

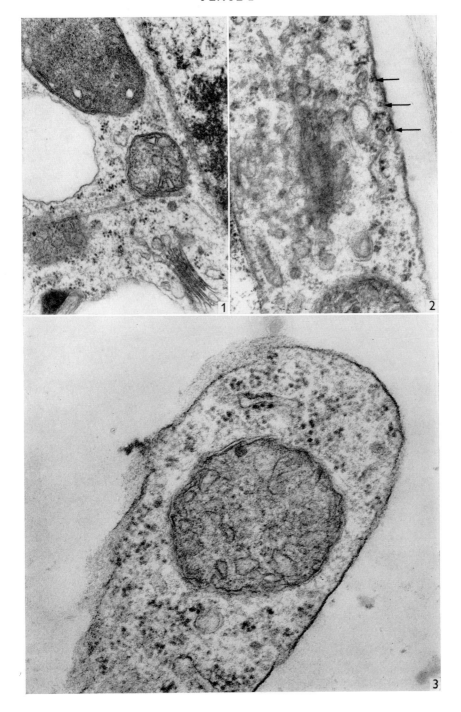

PLATE 3

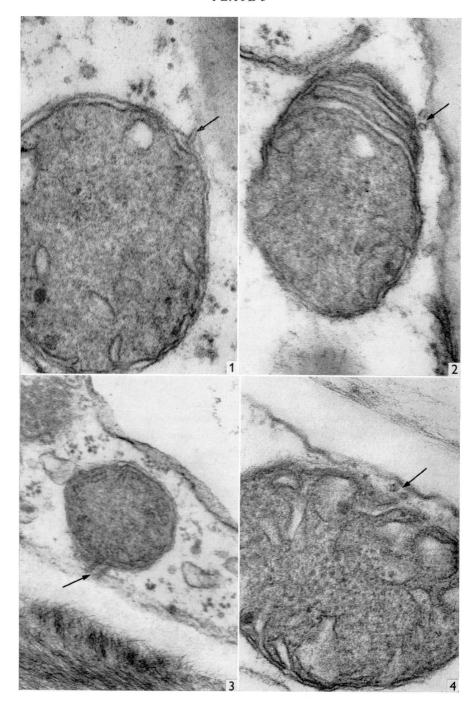

PLATE 4

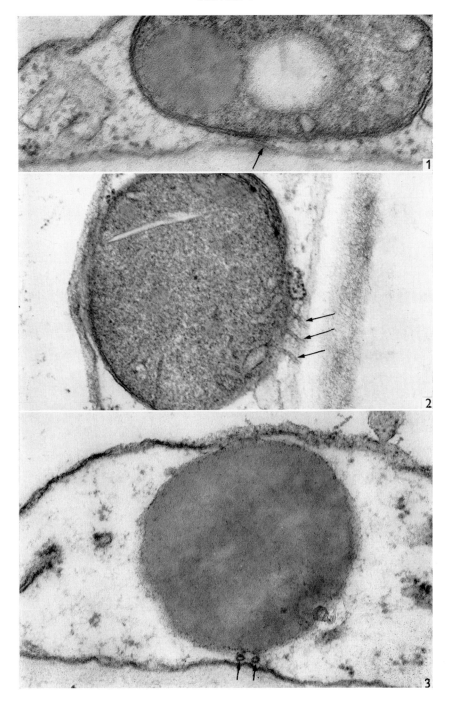

PLATE 5

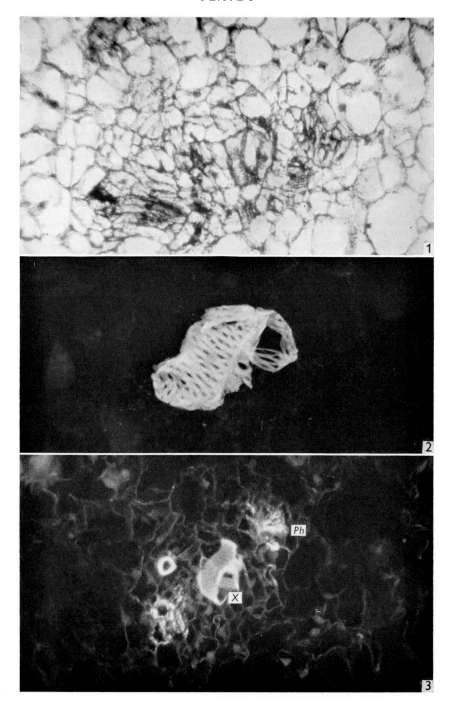

PLATE 6

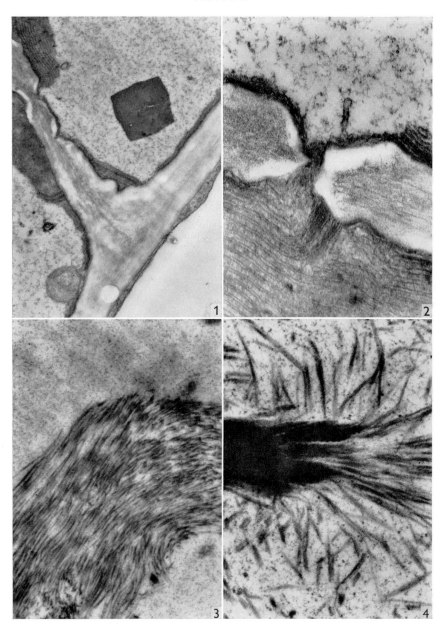

PLATE 7

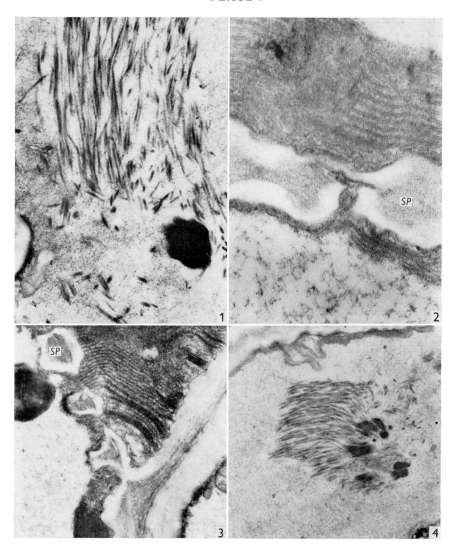

used to study biochemical reactions by techniques analogous to those already well established for micro-organisms.

Plant tissue cultures have however important differences from cultures of bacteria and other micro-organisms. The cells are derived from organisms in which the tissues normally differentiate and, in culture, the cells retain this property. Part of the mechanism for the differentiation and development of distinct tissues within an organism, where there is a separation of form and function of the various tissues, is that the cells can interact with one another to bring about a sequential and coordinated development of the plant. In the tissue culture the cells do not grow independently of one another, but they produce substances which influence the growth and development of other cells in the culture. This phenomenon is of extreme importance for the study of the mechanism of cell differentiation. In particular, tissue-culture experiments can be used to show the influence of growth factors such as indolyl-3-acetic acid, kinetin and sucrose on morphogenesis and the sites of cellular synthesis of these substances in a growing tissue can be related to the development of xylem and phloem within the intact plant.

During differentiation of a plant cell different materials are synthesized at various stages of the development; pectic polysaccharides for example are laid down in the cell wall during cell division and primary growth while lignin is deposited in the thickenings of the wall during secondary growth. These processes can be induced in plant tissue cultures and the amount of synthesis of the various substances can be measured during the differentiation of the cell. It is possible therefore to relate the induced differentiation to the synthesis or activation of the enzyme systems which are responsible for the formation of the substances laid down in the wall at the various stages of growth.

It is possible in some cases to induce differentiation within a tissue callus or cell suspension, not only to give different cell types but to produce an organized growth of cells which can subsequently be cultured so that a complete plant is obtained. Cell suspensions can be subjected to mutagenic reagents and by simple cloning techniques the mutants may be cultured. These methods therefore indicate important experimental procedures for the production of new plant varieties for agriculture and horticulture. Although tissue cultures have some potential use in other important industrial applications such as the production of plant secondary compounds (e.g. alkaloids) and as specific substrates for certain fungi (e.g. *Claviceps purpurea* for the manufacture of ergotamine), there are very difficult technical problems to be overcome before these can be exploited.

The cultured cells are very suitable for examination under the light microscope in a drop of the culture medium, where they divide, grow and develop for periods of some 3 weeks under sterile conditions. These cells therefore can be studied for the cytological features and changes which accompany cell differentiation especially since the growth of the cell may be controlled by the nutritional medium in which it is grown. The observations on the living cell are readily related to an electron microscope study of fixed material which can be prepared at various stages of the induced development in experiments which are run in parallel with the optical microscopic work.

Cultures of plant cells provide an important experimental system for the study of some of the most fundamental problems of biology and they also have a potentially important application to agriculture, horticulture and the pharmaceutical industry.

REFERENCES

BARRETT, A. J. & NORTHCOTE, D. H. (1965). Apple fruit pectic substances. *Biochem. J.* **94**, 617.

BLAKELY, L. M. & STEWARD, F. C. (1961). Growth induction in *Haplopappus gracilis*. 1. The behaviour of the cultured cells. *Am. J. Bot.* **48**, 351.

BOUNDY, J. A., WALL, J. S., TURNER, J. E., WOYCHIK, J. H. & DIMLER, R. J. (1967). A mucopolysaccharide containing hydroxyproline from corn pericarp. *J. biol. Chem.* **242**, 2410.

BRAUN, A. C. (1957). A physiological study of the nature of autonomous growth in neoplastic plant cells. *Symp. Soc. exp. Biol.* **11**, 132.

CAPLIN, S. M. (1956). Variability of carrots to auxin, casein hydrolysate and coconut milk. *Am. J. Bot.* **43**, 749.

CAPLIN, S. M. & STEWARD, F. C. (1948). Effect of coconut milk on the growth of excised plant tissue in liquid media under aseptic conditions. *Science, N. Y.* **108**, 655.

CLELAND, R. (1967a). Inhibition of cell elongation in Avena coleoptile by hydroxyproline. *Pl. Physiol.* **42**, 271.

CLELAND, R. (1967b). Inhibition of formation of protein bound hydroxyproline by free hydroxyproline in *Avena* coleoptile. *Pl. Physiol.* **42**, 1165.

CLELAND, R. & OLSON, A. C. (1968). Direct incorporation of hydroxyproline into *Avena* coleoptile proteins. *Biochemistry*, **7**, 1745.

CLUTTER, M. E. (1960). Hormonal induction of vascular tissue in tobacco pith. *Science, N.Y.* **132**, 548.

COOPER, L. S., COOPER, D. C., HILDEBRANDT, A. C. & RIKER, A. J. (1964). Chromosome numbers in single cell clones of tobacco tissue. *Am. J. Bot.* **51**, 284.

CURRIER, H. B. & STRUGGER, S. (1956). Aniline-blue and fluorescence microscopy of callose in bud-scales of *Allium cepa* L. *Protoplasma*, **45**, 552.

DAS, N. K., HILDEBRANDT, A. C. & RIKER, A. J. (1966). Cine-photo-micrography of low temperature effects on cytoplasmic streaming, nucleolar activity and mitosis in single tobacco cells in microculture. *Am. J. Bot.* **53**, 253.

DIGBY, J. & WAREING, P. F. (1966). The effect of growth hormones on cell division and expansion in liquid suspension cultures of *Acer pseudoplatanus. J. exp. Bot.* **17**, 718.

DOUGALL, D. K. & SHIMBAYASHI, K. (1960). Factors affecting growth of tobacco callus tissue and its incorporation of tyrosine. *Pl. Physiol.* **35**, 396.

GAMBORG, O. L., MILLER, R. A. & OJIMA, K. (1968). Nutrient requirements of suspension callus of soybean root cells. *Expl Cell Res.* **50**, 151.

GAUTHERET, R. J. (1959). *La culture des tissus végétaux.* Paris: Masson.

GIVAN, C. V. & COLLIN, H. A. (1967). Changes in respiration rate and nitrogen content associated with the growth of *Acer pseudoplatanus* L cells in suspension culture. *J. exp. Bot.* **18**, 321.

HENSHAW, G. C., JHA, K. K., MEHTA, A. R., SHAKESHAFT, D. J. & STREET, H. E. (1966). Studies on the growth in culture of plant cells. 1. Growth patterns in batch propagated suspension cultures. *J. exp. Bot.* **17**, 362.

HILDEBRANDT, A. C. & RIKER, A. J. (1949). The influence of various carbon compounds on the growth of marigold, paris-daisy, periwinkle, sunflower and tobacco tissue *in vitro. Am. J. Bot.* **36**, 74.

HILDEBRANDT, A. C., RIKER, A. J. & DUGGAR, B. M. (1945). Growth *in vitro* of excised tobacco and sunflower tissue with different temperatures, hydrogen-ion concentrations and amounts of sugar. *Am. J. Bot.* **32**, 357.

HOLLEMAN, J. (1967). Direct incorporation of hydroxyproline into protein of sycamore cells incubated at growth-inhibitory levels of hydroxyproline. *Proc. natn. Acad. Sci. U.S.A.* **57**, 50.

JEFFS, R. A. & NORTHCOTE, D. H. (1966). Experimental induction of vascular tissue in an undifferentiated plant callus. *Biochem. J.* **101**, 146.

JEFFS, R. A. & NORTHCOTE, D. H. (1967). The influence of indol-3yl acetic acid and sugar on the pattern of induced differentiation in plant tissue culture. *J. Cell Sci.* **2**, 77.

JENSEN, W. A. (1962). *Botanical Histochemistry*, p. 55. San Francisco: Freeman.

JONES, L. E., HILDEBRANDT, A. C., RIKER, A. J. & WU, J. H. (1960). Growth of somatic tobacco cells in microculture. *Am. J. Bot.* **47**, 468.

LAFLÈCHE, D. (1966). Ultrastructure et cytochimie des inclusions flagellées des cellules criblées de *Phaseolus vulgaris. J. Microscopie*, **5**, 493.

LAMPORT, D. T. A. (1964). Cell suspension cultures of higher plants: isolation and growth energetics. *Expl Cell Res.* **33**, 195.

LAMPORT, D. T. A. (1967). Evidence for a hydroxyproline-*O*-glycosidic cross link in the plant cell wall glycoprotein-extensin. *Nature, Lond.* **216**, 1322.

LAMPORT, D. T. A. & NORTHCOTE, D. H. (1960*a*). The use of tissue cultures for the study of plant-cell walls. *Biochem. J.* **76**, 52P.

LAMPORT, D. T. A. & NORTHCOTE, D. H. (1960*b*). Hydroxyproline in primary cell walls of higher plants. *Nature, Lond.* **188**, 665.

MITRA, J., MAPES, M. O. & STEWARD, F. C. (1960). Growth and organised development of cultured cells. 4. The behaviour of the nucleus. *Am. J. Bot.* **47**, 357.

MITRA, J. & STEWARD, F. C. (1961). Growth induction in *Haplopappus gracilis.* 2. The behaviour of the nucleus. *Am. J. Bot.* **48**, 358.

NÉTIEN, G. (1957). Action des gibberellins sur la culture des tissus végétaux cultivés *in vitro. C. r. hebd. Séanc. Acad. Sci., Paris* **244**, 2732.

NICKELL, L. G. (1956). The continuous submerged cultivation of plant tissue as single cells. *Proc. natn. Acad. Sci. U.S.A.* **42**, 848.

NICKELL, L. G. & TULECKE, W. (1960). Submerged growth of cells of higher plants. *J. biochem. microbiol. Technol. Engng*, **2**, 287.

NIERDERGANG-KAMIEN, E. & SKOOG, F. (1956). Studies on polarity and auxin transport in plants. 1. Modification of polarity and auxin transport by triiodobenzoic acid. *Physiologia Pl.* **9**, 60.

NORTHCOTE, D. H. (1963). Changes in the cell walls of plants during differentiation. *Symp. Soc. exp. Biol.* **17**, 157.

NORTHCOTE, D. H. (1968). Structure and function of plant-cell membranes. *Br. med. Bull.* **24**, 107.

NORTHCOTE, D. H. & WOODING, F. B. P. (1968). The structure and function of phloem tissue. *Sci. Progr.* **56**, 35.

OVERBECK, J. VAN, CONKLIN, M. E. & BLAKESLEE, A. F. (1942). Cultivation *in vitro* of small *Datura* embryos. *Am. J. Bot.* **29**, 472.

REINERT, J. (1959). Über die Kontrolle der Morphogenese und die Induction von adventivembryonen an gewebekulturen aus Karotten. *Planta*, **53**, 318.

RIER, J. P. & BESLOW, D. T. (1967). Sucrose concentration and the differentiation of xylem in callus. *Bot. Gaz.* **128**, 73.

RUBERY, P. H. & NORTHCOTE, D. H. (1968). Site of phenylalanine ammonia-lyase activity and synthesis of lignin during xylem differentiation. *Nature, Lond.* **219**, 1230.

SHELDRAKE, A. R. & NORTHCOTE, D. H. (1968*a*). The production of auxin by tobacco internode tissues. *New Phytol.* **67**, 1.

SHELDRAKE, A. R. & NORTHCOTE, D. H. (1968*b*). The production of auxin by autolysing tissues. *Planta*, **80**, 227.

SKOOG, F. & MILLER, C. O. (1957). Chemical regulation of growth and organ formation in plant tissues cultured *in vitro*. *Symp. Soc. exp. Biol.* **11**, 118.

STERLING, C. (1950). Histogenesis in tobacco stem segments cultured *in vitro*. *Am. J. Bot.* **37**, 464.

STEWARD, F. C. & CAPLIN, S. M. (1951). A tissue culture from potato tubers: the synergistic action of 2,4-D and coconut milk. *Science, N.Y.* **113**, 518.

STEWARD, F. C., KENT, A. E. & MAPES, M. O. (1967). Growth and organisation in cultured cells: sequential and synergistic effects of growth regulating substances. *Ann. N.Y. Acad. Sci.* **144**, 326.

STEWARD, F. C., MAPES, M. O. & MEARS, K. (1958). Growth and organised development of cultured cells. 2. Organisation in cultures grown from freely suspended cells. *Am. J. Bot.* **45**, 705.

STEWARD, F. C. & SHANTZ, E. M. (1959). The chemical regulation of growth: some substances and extracts which induce growth and morphogenesis. *A. Rev. Pl. Physiol.* **10**, 379.

STODDART, R. W., BARRETT, A. J. & NORTHCOTE, D. H. (1967). Pectic polysaccharides of growing plant tissues. *Biochem. J.* **102**, 194.

STODDART, R. W. & NORTHCOTE, D. H. (1967*a*). Metabolic relationships of the isolated fractions of the pectic substances of actively growing sycamore cells. *Biochem. J.* **105**, 45.

STODDART, R. W. & NORTHCOTE, D. H. (1967*b*). Separation and measurement of microgram amounts of radioactive polysaccharides in metabolic experiments. *Biochem. J.* **105**, 61.

STREET, H. E. & HENSHAW, G. G. (1963). Cell division and differentiation in suspension cultures of higher plant cells. *Symp. Soc. exp. Biol.* **17**, 234.

TORREY, J. G. (1953). The effect of certain metabolic inhibitors on vascular tissue differentiation in isolated pea roots. *Am. J. Bot.* **40**, 525.

TORREY, J. G. (1959). *Cell, Organism and Milieu*, p. 189. New York: Ronald.

TORREY, J. G. & SHIGEMURA, Y. (1957). Growth and controlled morphogenesis in pea root callus tissue grown in liquid media. *Am. J. Bot.* **44**, 334.

VASIL, V. & HILDEBRANDT, A. C. (1965*a*). Growth and tissue formation from single isolated tobacco cells in microculture. *Science, N.Y.* **147**, 1454.

VASIL, V. & HILDEBRANDT, A. C. (1965*b*). Differentiation of tobacco plants from single isolated cells in microculture. *Science, N.Y.* **150**, 889.

VASIL, V. & HILDEBRANDT, A. C. (1967). Further studies on the growth and differentiation of single isolated cells of tobacco *in vitro*. *Planta*, **75**, 139.

WAREING, P. F. (1958). Interaction between indole-acetic acid and gibberellic acid in cambial activity. *Nature, Lond.* **181**, 1744.

WETMORE, R. H. (1955). Differentiation of xylem in plants. *Science, N.Y.* **121**, 626.

WETMORE, R. H. & RIER, J. P. (1963). Experimental induction of vascular tissue in callus of angiosperms. *Am. J. Bot.* **50**, 418.

WHITE, P. R. (1943). *A Handbook of Plant Tissue Culture.* Lancaster, Pennsylvania: Jacques Cattell.

GROWTH AND DIFFERENTIATION OF ANIMAL CELLS IN CULTURE

JOHN PAUL

The Beatson Institute for Cancer Research, Glasgow

INTRODUCTION

Considering the vast evolutionary gulf between micro-organisms and cultured mammalian cells they exhibit surprisingly many similarities; these have been emphasized during the past 20 years by the direct application of microbiological techniques to the handling of metazoan cells. This 'microbiological phase' developed around 1950 when virologists became seriously interested in the use of cell cultures. Recently there has been a return to the study of those features of animal cells which may be characteristic of them alone.

ORGAN, TISSUE AND CELL CULTURES
ESTABLISHED LINES

Before some details of differentiation and growth of cultured cells are discussed, some terms must be defined. The terminology has constantly been changing, but the confusing situation has at last been clarified by the recommendations of a committee on nomenclature appointed by the American Tissue Culture Association (Federoff, 1967). It will suffice to mention one or two definitions. When cells or tissues are explanted direct from the organism, the culture which results is a primary culture. If this is subcultured, it becomes a cell line at the time of the first subculture. During early passages, a cell line usually exhibits many normal characteristics, including a normal karyotype. It is then referred to as a diploid cell line. If a cell line has been maintained so long that it may be said to demonstrate the potential to be subcultured indefinitely *in vitro*, it is called an established cell line. As a general rule of thumb, a line should be subcultured at least 70 times at intervals of 3 days between subcultures before it can be called established.

A cell strain is a derivative of a primary culture or a cell line which has specific properties or markers.

Primary cultures are by definition derived directly from animal tissues. Because metabolic exchanges in tissue and cell cultures all occur by diffusion, it is necessary to disrupt the tissue to obtain small enough pieces to permit free diffusion. The manner in which tissue is disrupted and subsequently handled distinguishes the three main branches of tissue culture. Organ cultures are either whole embryonic organelles or small pieces of tissue which are kept as intact as possible during removal; the culture conditions are devised in such a way as to prevent tissue structure from being destroyed. Tissue cultures are also initiated from small fragments but no attempt is made to keep the tissue in its original state; hence, particularly owing to cell migration, the structure of the tissue becomes disorganized. Finally, cell cultures are obtained by completely disaggregating the tissue so as to obtain a suspension of single cells. This can be achieved in many ways; the most usual are to disrupt the tissue physically; for example, by sieving through stainless steel gauze; or chemically, usually by the action of proteolytic enzymes, of which the commonest is trypsin.

If primary tissue or cell cultures are maintained for any length of time, the population changes. Some cells do not survive long; for example, the granulocytes of the blood disappear within a day or two. Others survive for a long time, but never increase in number. Neurons, for example, can be maintained in tissue culture for some months; action potentials can be measured during that time, but the cells never exhibit mitosis, and eventually die. Still other cells start to divide soon after explantation and divide continuously. Fibroblasts are typical examples. The dividing cells outgrow other cells in the culture; eventually they are the sole survivors. Cells which have started to grow in this way can be subcultured to give a cell line.

Most lines originating in this way still exhibit many of the constraints of normal tissue cells. Moreover, some of them have a finite life span (Hayflick & Moorhead, 1961). For example, human dermal fibroblasts do not survive in culture for more than about 9 months, although they may double in number every 2 or 3 days during that time. Many primary cell lines, on reaching the end of their lifespan, enter a phase of 'senility' and then die. However, sometimes a dramatic alteration in the properties of senile cells occurs. The diploid cell line is quite suddenly replaced by a rapidly growing cell line which differs from it in many characteristics, especially the number of chromosomes. When this alteration has occurred, the cells which result can usually be subcultured indefinitely and are therefore established cell lines. The nomenclature committee has proposed the use of the term 'culture alteration' to describe this phenomenon, but it is more commonly called 'transformation'. This is an undesirable use of the term, because of confusion with the term 'transformation' in microbial genetics to imply an alteration of properties brought about by transfer of genetic information from one organism to another. Nevertheless, the term 'transformation' is so firmly entrenched in cell-culture literature that it is unlikely that it will disappear very quickly. Microbiologists and geneticists must therefore appreciate that it has a much less precise meaning, in the context of cell culture. Some of the differences between primary diploid cell lines and established cell lines will form the main subjects of subsequent sections in this paper; the main ones are as follows. Primary cell lines are usually diploid, whereas established cell lines are heteroploid. Cell morphology is often retained in primary cell lines, whereas it is considerably modified in established cell lines. The lifespan of primary cell lines is often limited, whereas the lifetime of established cell lines is unlimited. Primary cell lines are easily established, whereas established cell lines are often difficult to initiate.

RETENTION OF DIFFERENTIATED CHARACTERISTICS IN CULTURE

This list intentionally omits mention of 'de-differentiation'. The idea that cells in culture de-differentiate is a very old one. Two classical examples are the disappearance of melanin from cultures of pigmented retinal cells, and the loss of capacity of chondrocytes to synthesize cartilage (Stockdale, Abbott, Holtzer & Holtzer, 1963; Holtzer, Abbott, Lash & Holtzer, 1960). That the situation is not straightforward has been demonstrated recently by the achievements of Coon (1966) and Cahn & Cahn (1966) in growing colonies of cartilaginous and retinal cells from single cells, and demonstrating that the cells in the colonies retained their differentiated characteristics of forming collagen and melanin respectively.

Where differentiated characteristics disappear, they can do so very quickly. For example, Burlington (1959) reported that D-amino acid oxidase and glucose-6-phosphatase disappeared from kidney cells within 4 days in primary culture. Ebner, Hageman & Larsen (1961) found that lactose synthesis ceased within 29 hr. of explanting mammary gland tissue; while UDP-galactose-4-epimerase was completely absent in a few days.

By contrast, there are many examples of established cell lines maintaining differentiated characteristics. Fibroblasts often exhibit the capacity to secrete collagen (Green & Goldberg, 1964; Davidson, 1963). Even L cells (line L929 mouse fibroblasts) which have been growing in continuous culture for a quarter of a century probably retain this property (Merchant & Kahn, 1958). Schindler, Day & Fischer (1959) showed that neoplastic mast cells in continuous culture continue to synthesize 5-hydroxytryptamine and histamine. Moore, Lehner, Kikuchi & Less (1962) established a melanotic cell line from the golden hamster while Sato and his colleagues (1965; Yasamura, Tashjian & Sato, 1966) have established several lines of cells from tumours which produce polypeptide and steroid hormones. Hence there is no clear rule about de-differentiation in culture. In at least some cases, for example melanin production, the failure to synthesize a specific product may merely be due to the absence of a necessary substrate for an enzyme (Doljansky, 1930). Hence the rule may turn out to be that the capacity to perform differentiated functions is not lost, even in continuous culture, provided the necessary nutritional factors are provided.

Species specific antigens clearly persist in all established cell lines, as well as in primary cell lines; indeed, this has become a standard method for identification of the species of origin of cell lines (Franks, Gurner, Coombs & Stevenson, 1963; Brand, 1962; Brand & Syverton, 1962; Franks, Coombs, Beswick & Winter, 1963). However, organ-specific antigens may disappear quite early (Nicol & Beck, 1966). This would seem to apply to blood group antigens also (Högman, 1960).

Cell alteration in culture has often been considered to be associated with the acquisition of malignant characteristics. However, there is no good evidence that this is a necessary association (Paul, 1962).

GENETIC STABILITY OF CELLS IN CULTURE

The development of aneuploidy in established cell lines is dependent on non-disjunction; the striking difference between the behaviour of established cell lines and primary cell lines suggests that there is a constraining mechanism in primary cell lines which prevents deviation from the normal pattern. Some mutations can be recognized by studying chromosomes; for example, certain diseases such as Mongolism and Kleinfelter's syndrome are diagnosed by investigating primary cultures. Other mutations do not involve visible structural alterations, but the metabolic abnormalities involved in acatalasia and galactosaemia, among other conditions, are perpetuated in culture (Krooth, Howell & Hamilton, 1962; Krooth & Weinberg, 1960). These can be used as markers in cells.

New mutations can arise during continuous cultivation in established cell lines. Most of these are recognized by the loss of biochemical functions (Harris & Ruddle, 1961; Szybalski & Smith, 1959; Subak-Sharpe, 1965). Some will be referred to in other contexts.

THE MECHANISM OF CELL ALTERATION
('TRANSFORMATION')

The event which results in the transformation of a primary cell line into an established cell line is still mysterious, although specific factors have now been identified which can initiate it. Before embarking on a discussion of these, it should be noted that many descriptions of transformation in the literature around 1950 are undoubtedly erroneous and due to contamination of cultures of primary lines with cells from established cell lines. The ease with which this could happen was not appreciated until the introduction of immunological methods of recognition of the species of origin of cells. Nowadays workers are well aware of this risk, and in the studies to be discussed it has been excluded.

The most carefully analysed transforming agents of primary cell cultures are viruses. Vogt & Dulbecco (1960) showed that when normal hamster cells were treated with the polyoma virus, the typical morphological changes of transformation occurred. Later studies showed that primary diploid cell lines derived from baby hamster kidneys exhibited marked parallel orientation and contact inhibition, whereas the same cells transformed with polyoma virus deviated towards heteroploidy, were more random in orientation, piled up on each other and grew more vigorously. Whereas the diploid cell lines had a limited lifespan, transformed cells could grow indefinitely. The diploid cell lines proved refractory to growth in suspension, whereas transformed cells grew readily. Moreover, on inoculation into hamsters the diploid cell lines did not readily produce tumours, whereas the transformed ones did. The very large literature on the subject has been reviewed extensively (Sachs, 1967; Stoker, 1962; MacPherson, 1963).

Other DNA viruses produce this transformation. Particular interest has attached to the simian virus 40 (SV 40) because this virus, which can be isolated from monkey tissues, induces transformation in the cells of a wide range of species, including rodents (Black & Rowe, 1963) and humans (Todaro, Wolman & Green, 1963; Jensen, Koprowski & Pontén, 1963). The phenomenon is very similar to the transformation of hamster cells by polyoma virus.

Both the viruses which have been described are DNA viruses, and the presence of virus DNA in the cell after transformation has been demonstrated (Benjamin, 1966; Reich, Black & Weissman, 1966). No infective virus can be recovered from polyoma-transformed hamster cells; the presumption therefore is that the DNA is incorporated into the genome. In SV 40 transformed cells the virus has been recovered by

23-2

promoting fusion between infected cells and susceptible mouse cells. It is not clear whether some of the virus remains in a vegetative form in these cells or whether this represents lysogenic behaviour (Watkins & Dulbecco, 1967).

RNA viruses can also induce transformation. The Rous sarcoma virus (RSV) was found to transform chicken fibroblasts in tissue culture as long ago as 1939 by Halberstaedter & Doljanski. Temin & Rubin (1958) developed an assay for the virus based on this phenomenon. Transformation by RSV has some particularly interesting features. This virus is an avian virus but strains have been developed which will infect mammalian cells; for example, the Schmidt–Ruppin strain can be made to produce typical transformation in hamster cells (MacPherson, 1965, 1966). However, whereas all the progeny of polyoma-transformed cells retain their transformed characteristics, i.e. the transformation is permanently and irreversibly inherited, MacPherson has shown that a proportion of RSV transformed cells always reverts to the normal type. This is the best example of reversibility of transformation.

Some bacteria may also cause transformation; for example, Mac-Pherson & Russell (1966) have described apparent transformation in cells infected by mycoplasmas (see also Russell, 1966).

Viruses and bacteria are not the only transforming agents. There are many clear demonstrations of transformation brought about by treatment of cells with chemical carcinogens. Although these studies go back to Earle's work in 1943, it has required the development of newer techniques to permit a convincing demonstration of transformation, similar to viral transformation, with substances such as 3-4-benzo-pyrene (Berwald & Sachs, 1963), 6-chloro-4-nitroquinoline-N-oxide (Sato & Kuroki, 1966), other quinolines (Kuroki, Goto & Sato, 1967), N-nitrosomethylurea (Sanders & Burford, 1967) and 20-methyl-cholanthrene (Heidelberger & Iype, 1967).

Simple physical factors, too, have been shown to be capable of producing transformation. For example, Borek & Sachs (1966) showed that X-irradiation could produce a typical alteration.

It is little if at all easier to initiate cell lines from tumour tissues than from normal tissues. However, when cell lines have been established from tumour tissues they very often behave as if they are already transformed. In particular they exhibit little contact inhibition, rapid growth which can be sustained in suspension, and heteroploidy.

These factors probably do not explain all cases of transformation. 'Spontaneous' transformation was reported originally by Gey in 1941 and by Earle & Nettleship in 1943. Although these early observations were made by careful and experienced workers, it is difficult to be sure in retrospect that the transformations were not produced either by physical factors or by the inadvertent introduction of a chemical carcinogen or a virus. Where particular care has been taken to exclude these adventitious factors, however, it has still been found that some cell lines tend to undergo transformation spontaneously. The phenomenon seems to vary from species to species, and even from strain to strain. Human skin fibroblasts have never been observed to undergo spontaneous transformation, yet in fibroblasts derived from C_3H mice spontaneous transformation is the rule. For example, Sanford (1967) found that in 22 cell strains which gave negative tests for complement-fixing mouse leukaemia viruses, 16 underwent transformation. In the same laboratory it was observed that cell lines maintained in a chemically defined medium also underwent the same kind of transformation. Moreover, the BHK 21 cell, which is used as a model cell for polyoma transformations, exhibits a typical transformation after maintenance in culture for a long time (Defendi, Lehman & Kraemer, 1963); and no antigens associated with the virus can be found in the transformed cells. Clearly it is much more difficult to exclude the presence of an unidentified agent than it is to recognize the presence of a known one; and therefore there is some scepticism, perhaps justifiable, about the occurrence of spontaneous transformation. However, in view of the fact that such a wide variety of known factors can induce the change, it is certainly not impossible that it might occur as a random event in the absence of such factors. Further evidence for it will be described below.

GROWTH CHARACTERISTICS OF AXENIC CULTURES

The introduction of the dilution method of cloning (Puck & Marcus, 1955) made quantitative growth studies of animal cells possible; most of them have been done with established lines. These all have rather short doubling times, of the order of 12–18 hr. In dilute inoculum they usually show a brief lag phase followed by a prolonged phase of exponential growth. Ultimately a maximum population is reached and the stationary phase is entered. In these respects, therefore, animal cell populations resemble closely bacterial cell populations. It is only when we look at details of the process that interesting differences become apparent.

Cells from established cell lines can be cloned rather easily; plating efficiencies of the order of 100 % are often approached. However, the efficiency of plating of primary cell lines is much lower (Zarof, Sato & Mills, 1961; Pious, Hamburger & Mills, 1964). Moreover, established cell lines commonly exhibit a very long exponential phase and a high final cell density, whereas the growth of primary cell lines frequently shows signs of diverging from the exponential pattern quite early, and the final cell population is much lower (see, for example, Harris, 1957). Moreover, whereas the medium from a stationary culture of established cells will very often not support any further growth, the medium from a stationary culture of primary cells often will. It would seem that the constraints on growth of established cell lines are often simply exhaustion of metabolites or accumulation of waste products; a different explanation must be sought for primary cells. Evidence for contact effects has been obtained and will be discussed in the next section.

Perhaps the most dramatic difference between established and primary lines is the observation that primary lines often have a restricted lifespan, whereas established lines can probably continue to multiply indefinitely in satisfactory conditions. The best descriptions of this phenomenon are by Hayflick & Moorhead (1961), and by Todaro & Green (1963). Hayflick & Moorhead initiated 25 primary fibroblastic lines from human tissues. The cells showed a marked tendency to array themselves in parallel strands throughout the entire course of maintenance and remained diploid. They grew with a doubling time of 2 or 3 days for some months; degenerative changes then gradually appeared and increased over a period of from 1 to 3 months. In no case did the cells survive longer than 11 months (unless they were transformed by SV 40 virus). Since this could have been due to the inadvertent killing of the cells by accidentally introduced factors, experiments were undertaken to exclude this. These consisted of taking samples of cells at intervals during cultivation and storing them at $-70°$C. When the original culture had died out, some of the frozen stock was recovered. These cells began to grow at a rapid rate, but after a time they, too, died out. In all cases irrespective of the passage frozen, the cells died after approximately the same total cumulative passages.

The cells studied by Todaro & Green (1963) were derived from Swiss mouse embryos. Whereas the human cells studied by Hayflick & Moorhead (1961) deteriorated after several months of rapid growth, the multiplication rate of the mouse embryo cells began to decline almost immediately. On the other hand, within about 3 months of culture, they invariably showed evidence of spontaneous transformation. At this time the cells began to multiply faster, although they were morphologically unchanged. The growth rate continued to increase until it was as fast as, or faster than, the original cultures. The chromosomes were at first perfectly normal after the transformation had occurred. Later, however, they became tetraploid before becoming aneuploid, as originally described by Levan & Biesele (1958).

These two examples represent extremes of behaviour.

METABOLISM OF AXENIC CULTURES

Extensive studies of the metabolism of axenic cultures have been made. In general carbohydrate, protein and nucleic acid metabolism (Paul, 1965; Levintow & Eagle, 1961; Seed, 1965) follow the general patterns of metabolism in the whole animal. One or two particular features may be mentioned. Whereas most mammals have an essential requirement for eight amino acids, their cells in culture require thirteen. Also there is no requirement for fat-soluble vitamins in established cell lines, whereas there are requirements for several in the whole animal. Established cell lines have a generally glycolytic carbohydrate metabolism which, however, varies from cell line to cell line. pH, Pasteur and Crabtree effects have been demonstrated.

A question which has exercised investigators for many years is whether normal and malignant cells show different patterns of carbohydrate metabolism, and in particular whether glycolysis is greater in malignant cells than in normal ones. In this connection, Paul, Broadfoot & Walker (1966) found considerable differences in the glycolytic capacities of a line of baby hamster kidney cells, and the same cells which had undergone transformation after treatment with polyoma virus.

During the lag phase of growth there is rapid synthesis of RNA, DNA and protein. Mean values, per cell, remain high during exponential growth and tend to diminish during the stationary phase (Salzman, 1959; Kruse & White, 1961). There seems to be some coupling between DNA synthesis and the synthesis of nucleic acids and proteins in these cells (Salzman & Sebring, 1962; Paul & Hagiwara, 1962).

In eukaryotic cells the cell cycle is divided into four periods. In many animal cells the S period (i.e. the period of DNA synthesis) is about 7 or 8 hr. The G2 period which immediately follows varies from about 2 to 6 hr., but is commonly about 4. Mitosis itself occupies about 1 hr. and the G1 phase between mitosis and S occupies the remainder of the cycle. It is the most variable and may be as short as 1 hr. or several days long (Defendi & Manson, 1963).

The development of reliable methods for obtaining synchronized populations has made it possible to study the different phases of the cell cycle (Terasima & Tolmach, 1963; Johnson & Holland, 1965; Tobey, Petersen, Anderson & Puck, 1966). In mitosis, protein synthesis and RNA synthesis are almost completely suppressed; it can be shown that cell-free protein synthesis and RNA polymerase activity are depressed in extracts of metaphase cells. In the G1 phase most synthetic activities proceed normally. Some of the enzymes involved in DNA synthesis—for example, thymidine kinase and DNA polymerase—probably increase just before DNA synthesis starts. During the S phase DNA synthesis and protein synthesis proceed in the nucleus at the same rate, although Seed (1965) claims that this is true only of normal cells, and that in malignant cells the rates are dissociated. DNA synthesis does not proceed uniformly in all the chromosomes (Taylor, 1960). Radioautographic studies show that DNA synthesis can be almost complete in some chromosomes before it is initiated in others. The work of Tobey et al. (1966) suggests that during the G2 phase some proteins essential for mitosis are synthesized. In the Chinese hamster ovary cells which they studied there was evidence that m-RNA, essential for mitosis, was produced 1·9 hr. before the completion of division, while protein essential for mitosis was apparently synthesized right up to its initiation.

So far as has been determined, this general metabolic behaviour seems to be typical both of primary cell lines and established cell lines. Much of it might have been expected from experience with other systems. However, a uniquely interesting phenomenon emerges when the synthetic patterns in human diploid cells approaching confluence are studied. As soon as cells begin to establish contact with each other the rates of DNA and RNA synthesis diminish. DNA synthesis may virtually disappear whereas in established cell lines DNA synthesis continues even in very crowded cultures. RNA synthesis is also very much reduced, although not to the same extent. Both changes can be reversed simply by diluting the cells or by infecting with a transforming virus.

INTERACTION PHENOMENA

That the cessation of growth when cells came into close apposition was not simply due to crowding was first recognized by Abercrombie & Heaysman (1953), who observed that when certain kinds of normal cells came into contact they were immediately immobilized. Malignant connective tissue cells, by contrast, did not show this phenomenon (Abercrombie, 1961; Abercrombie & Ambrose, 1958). That contact effects involve more than inhibition of movement has gradually come to be recognized (Pace & Aftonomos, 1957; Levine, Becker, Boone & Eagle, 1965; Eagle, 1965; Stoker, 1967). These authors showed that cell division and DNA synthesis diminished or ceased when the cell sheet became confluent. It seems not unlikely that this behaviour reflects a normal cell reflex. The mechanism of contact effects remains obscure but several pieces of evidence combine to suggest that they are mediated by chemical substances over a distance of some hundreds of microns (Rubin, 1962). Bürk (1966) claims to have isolated a substance from BHK 21 cells which inhibits growth and is neutralized by a factor in serum. Cells which have been transformed with polyoma virus do not produce this substance.

Other growth regulatory substances have been described; most are stimulatory rather than inhibitory in action. Virolainen & Defendi (1967) have found that cultures of connective tissue cells produce a substance which can induce repeated mitosis in rodent macrophages; these do not normally show any mitosis in culture. The nerve growth factor (Levi-Montalcini, 1966) has been highly characterized. This substance is probably a polypeptide; it has a highly specific effect on sympathetic nervous tissue. Of a similar nature may be erythropoietin; it is produced by the anoxic kidney and has the function of stimulating maturation of erythroid cells.

Recognition of like cells is an important property of normal animal cells. It was first observed by Holtfreter (1943), who disaggregated embryonic frog cells in calcium- and magnesium-free salt; he then showed that, on restoring calcium and magnesium, they not only aggregated but separated into groups of like cells, and eventually reconstituted the embryo. Similar observations have been made by Moscona (1962). He and his colleagues showed that embryonic chick and mammalian tissues could be disaggregated by tryptic digestion. When the cells were allowed to aggregate they subsequently segregated into groups of like cells and eventually reconstructed the tissue architecture. The nature of the interacting forces has not been established. There is no lack of theories, but it is not easy to reconcile all the facts with any of them. Moscona has adduced experimental evidence to suggest that some extracellular material mediates in the reaction, while Steinberg (1962, 1963; Steinberg & Roth, 1964) has postulated that a general mechanism may depend mainly on the free surface energy of the cells.

Until quite recently it was believed that, with the exception of syncytia such as muscle, animal cells did not establish any direct contact with each other. Recently this has been proven to be wrong. Lowenstein and his colleagues (1965) first showed direct electrical coupling between cells in the whole animal, and subsequently Potter, Furshpan & Lennox (1966) were able to demonstrate that in confluent cells in tissue culture similar electrical coupling became established. Direct transfer between cells has been shown convincingly in experiments by Subak-Sharpe, Bürk & Pitts (1966).

It has been postulated that there might be direct transfer of genetic information from one cell to another, analogous to transformation in bacteria. Something of this sort was suggested by Benitez, Murray & Chargaff in 1959, when they observed heteromorphic changes in fibroblasts cultivated in a medium containing ribonucleoproteins. Subsequently Niu and his colleagues (1962) claimed that they could induce permanent changes in the pattern of protein synthesis by treatment of cells with RNA. Other investigators have not been able to substantiate this claim. Bensch, Gordon & Miller (1964) conducted experiments to determine whether DNA could enter cells intact, and Robins & Taylor (1968) showed that intact exogenous DNA appeared in the nucleus but was probably not integrated into chromosomes. However, other claims have been made to have observed phenomena analogous to transformation. In particular, Amos & Moore (1963) provided some evidence for the promotion of protein synthesis in primary cultures of chicken fibroblasts by the addition of large amounts of RNA to the medium. The most precise claim to have produced DNA-mediated heritable transformation in cell lines was made by Szybalska & Szybalski (1962). These workers used drug-resistant derivatives of a cell line. For example, they use an 8-aza-hypoxanthine resistant derivative which lacked the enzyme inosine-phosphate-pyro-phosphorylase. DNA from wild-type cells or cells with different mutations apparently restored the inosine-phosphate-pyro-phosphorylase activity. The studies were carefully conducted, but other workers have not been able to confirm them, so that the question of transformation remains open.

The occurrence of transduction by viruses has not been proven genetically, but the encapsidation of cellular DNA into polyoma virus particles has been demonstrated (Winocour, 1967; Kaye & Winocour, 1967). This host DNA can be separated from the viral DNA by chemical means; it is not known whether it is a random or non-random piece of DNA nor whether it can be transcribed in the infected cell.

CELL FUSION; HETEROKARYONS AND HYBRIDS

Many of the important observations about the regulation of β-galactosidase were obtained from mating experiments in *Escherichia coli*. Hopes that similar information might be obtained from cultured somatic animal cells have been raised by the fairly recent finding that cells of different types can be caused to fuse. Fusion of cells in tissue culture was observed a long time ago (Lewis, 1927), although this was usually amongst cells of like type, to form giant cells. That this occurred as a regular random event of low frequency in cultures of cell lines was discovered by Barski, Sorieul & Cornefert (1961). Many hybrid lines formed by the fusion of cells from different mouse cell lines have now been characterized (Yoshida & Ephrussi, 1967). The possibility that heterokaryons could be derived from different species was demonstrated by Yaffe & Feldman (1965), who exploited the natural tendency of skeletal myoblasts to form multinucleated giant cells; they obtained hybrids between rat and chick myoblasts, which formed normally functioning muscle fibres. At about the same time, Harris & Watkins (1965) reported a general method for promoting cell fusion by treating a mixed culture with inactivated Sendai virus. Using this trick they prepared many interesting hybrids, including, for example, hybrids between HeLa cells (of human origin) and nucleated avian erythrocytes (see Harris, Watkins, Ford & Schoefl, 1966). Cells which have been caused to fuse by Sendai virus form heterokaryons in which the ratio of one kind of nucleus to another can be manipulated. Generally these do not go on to reproduce, whereas many of the hybrid lines formed by spontaneous fusion can replicate indefinitely.

The results of experiments designed to investigate interaction among genes using heterokaryons and hybrids have been rather confusing and have not yet led to definite conclusions. Generally speaking, the studies on heterokaryons have given evidence of positive controls inasmuch as DNA or RNA synthesis can be evoked in cells which have ceased to perform the functions when they are fused with other cells. Little evidence is as yet available from studies of hybrids. When a mutant is hybridized with a wild-type cell the wild-type characteristic predominates (Littlefield, 1966). Also when transformed cells are hybridized with untransformed cells, the characteristics of the transformed cells predominate. For example, hybrids between wild-type mouse cells and mouse cells which had been transformed by polyoma virus were invariably capable of producing tumours (Defendi, Ephrussi, Koprowski & Yoshida, 1967).

DIFFERENTIATION IN PRIMARY CULTURES

The problems of differentiation in a whole animal are forbidding in their complexity. By isolating parts of the system and maintaining them in controlled conditions it was hoped to obtain more clear-cut answers than could be obtained otherwise.

Development of the organ culture technique permitted the demonstration that many embryonic tissues and organs preserved the capacity to differentiate when cultured *in vitro*. The action of some hormones on these developing rudiments has been investigated. Wolff & Wolff (1952) showed that the development of the syrinx of the embryonic duck could be diverted towards the typical male or female morphology by manipulating the concentration of sex hormones in the medium. The effects of hormones on organ cultures have been particularly studied in mouse mammary tissue (Rivera & Bern, 1961; Stockdale, Juergens & Topper, 1966). The differentiation of mammary gland tissue requires the hormones, insulin, hydrocortisone and prolactin; it culminates in the synthesis of the milk proteins casein, α-lactalbumin and β-lactoglobulin. It has been found that DNA synthesis and cell proliferation are necessary for the completion of the process and that insulin is apparently essential for the augmentation of DNA synthesis and proliferation (Lockwood, Voytovich, Stockdale & Topper, 1967).

Besides hormones, other substances have been found to have an inductive effect. Most striking is the action of vitamin A in promoting the development of ciliated mucus-secreting columnar epithelium in skin explants which would form a squamous keratinizing epithelium in its absence (Fell & Mellanby, 1953).

Although in organ cultures we are dealing with an experimental situation which is very much simpler than a whole animal, it is still complex and it is difficult to resolve questions concerning the interaction of cells. Techniques have been worked out, therefore, for studying the behaviour of these cells in isolation; from the observations made, two kinds of reactions have been recognized. Homotypic induction effects are those which arise in isolated pieces of tissue of uniform cell type. It is possible, for example, to isolate myoblasts and to demonstrate that these will develop into muscle fibres in the absence of other kinds of cells. However, it has been observed that if the number of cells in the piece of tissue is reduced beyond a minimum, cytodifferentiation does not occur (Wilde, 1961). In other words, there has to be a critical mass of cells for this type of induction to proceed.

On the other hand, it has been established that some cells have to be grown in close proximity to a different kind of tissue to develop fully. This is called 'heterotypic induction'. The phenomenon has been investigated, particularly in relation to epithelial and mesenchymal tissues. Grobstein (1953, 1965) has shown that epithelial tissues from a variety of embryonic organs do not differentiate at all if cultured alone. But if they are grown in close proximity to mesenchymal tissue, they develop into epithelium typical of the tissue and form structures such as alveoli. The specificity of interaction is sometimes low; any kind of mesenchyme will produce the induction. In other instances the specificity is rather high; for example, Grobstein (1953) found that, whereas submandibular gland mesenchyme would induce the development of submandibular gland epithelium, and kidney mesenchyme would induce the development of kidney epithelium, neither induction would occur if the epithelia were cultured with mesenchymes from the different organs. The induction can occur across a millipore membrane of porosity too small to permit passage of cell processes. The nature of the material responsible is not yet certain, but it appears to be a large macromolecule and is probably a mucopolysaccharide.

More satisfactory systems than those described have been sought. Ideally, the experimenter would like a pure line of cells which would respond to a clearly defined stimulus by a clearly defined and measurable response. The system which probably comes nearest to this exploits the observation that erythropoietin can stimulate the synthesis of haemoglobin in cultured erythropoietic tissue. In searching for an assay system for erythropoietin, Krantz, Gallien-Lartigue & Goldwasser (1963) found that erythropoietin could promote the maturation of erythropoietic tissue in bone marrow tissue culture. These investigators realized that they might have an interesting tool with which to study the control of gene expression in animal cells. At least two criteria were met; the initiating stimulus was a reasonably well-defined protein and at least one of the results of the stimulation, increase in haemoglobin synthesis, involved an extremely well-defined substance which was easily measured. Moreover, the reaction could be carried out in replicate cultures.

A possibly better experimental system was obtained as the result of studies by Cole & Paul (1966). They found that in the foetal mouse liver there is an accumulation of erythropoietin-responsive cells in the early stages of liver development.

Much is known about the behaviour of this system in the whole animal. The erythropoietin-sensitive cells derive from stem cells which are the common progenitors of granulocytes and erythrocytes. The immediate derivative of the stem cell has not been identified, but it undergoes rapid transition in sequence to cells designated pro-erythroblasts, basophilic erythroblasts, polychromatic erythroblasts, normochromic erythroblasts, reticulocytes and erythrocytes. During this maturation several cell divisions occur. At first much ribonucleic acid is made; later protein synthesis predominates over RNA synthesis. In mammals the nucleus is extruded from the normochromic erythroblasts; most of the haemoglobin of the mature cell is then formed in the reticulocyte. Erythropoietin is thought to act not only on the hypothetical erythropoietin-sensitive cell, but also on the maturing cells at other stages in the pathway. This has been borne out in tissue culture experiments. These have also revealed that erythropoietin must be continuously present to produce an increasing rate of haemoglobin synthesis. If at any point it is removed, haemoglobin synthesis continues at much the same rate, or a slightly diminishing rate, for some hours.

One of the more important questions in relation to differentiation is whether erythropoietin promotes the transcription of more genes, or whether it merely increases the rate of transcription of those which have already been transcribed. We (Paul & Gilmour, 1966) described a technique to test for this by transcribing RNA from chromatin *in vitro* and to test the product by molecular hybridization. Using this test, we have shown that more genes are transcribed in response to erythropoietin stimulation. It has proved possible to investigate steps in the chain of events following treatment of the cells with erythropoietin. Synthesis of haemoglobin is preceded by synthesis of large amounts of RNA, presumably including the globin messengers. This RNA synthesis in turn is preceded by DNA synthesis, which is an absolute prerequisite for the RNA synthesis. When DNA synthesis is blocked, for example by FUdR, no measurable increase in RNA synthesis occurs in response to erythropoietin. On the other hand, merely blocking cell division with, for example, colchicine does not block the increase in haemoglobin synthesis. The wave of DNA synthesis is preceded by synthesis of a very small amount of protein which in turn is dependent on an extremely small amount of RNA synthesis. Hence the tissue culture system has revealed that the steps involved in gene derepression may be complex (Paul & Hunter, 1968).

In the foetus at the time when erythropoiesis increases there is a switch in the kind of haemoglobin formed and the activities of the haem-synthesizing enzymes increase. The tissue culture studies show that erythropoietin does not determine the kind of haemoglobin formed.

It merely increases the rate of synthesis of that kind of haemoglobin for which the cells have been determined. The results of measuring haem-synthesizing enzymes are somewhat paradoxical. The mitochondrial enzymes, D-aminolaevulinic acid synthetase and ferro-chelatase either do not increase in amount or diminish. On the other hand, the enzyme D-aminolaevulinic dehydratase increases in response to erythropoietin.

This system has opened up many possibilities for study but its shortcomings are obvious. The cell population is mixed, it does not grow continuously *in vitro* and therefore it is unlikely to be possible to produce mutants until this technical problem can be overcome.

ENZYMIC INDUCTION IN ESTABLISHED CELL LINES

For many years attempts have been made to discover inducible enzymes in established cell lines to provide a system analogous to the induction of β-galactosidase in *Escherichia coli*. The first clear-cut evidence for regulation of the activity of an enzyme in an established cell line was obtained by De Mars (1958). He showed that glutamyl-transferase could be repressed by glutamine in HeLa cells. The same phenomenon was demonstrated by Paul & Fottrell (1961) in L cells. The latter authors pointed out that stabilization of the enzyme might play a part in leading to enhanced activity. Weissman, Smellie & Paul (1960) also demonstrated substrate induction; they showed that thymidine induced enhanced activities of thymidine kinase and thymidylate kinases in strain L cells and pointed out that the activity of the enzyme fluctuated during a growth cycle. Bojarski & Hiatt (1960) suggested that stabilization of the enzyme played a part in this induction. Schimke (1962) performed a thorough study on repression of the enzymes of the arginine biosynthetic pathway. He found that in three different established cell lines (HeLa, KB, L929) the specific activities of argininosuccinate synthetase and arginino-succinase could be caused to increase greatly by lowering the concentrations of citrulline or arginine in the medium. Arginase activity could be augmented either by increasing the arginine-concentration in the medium or by the addition of manganese. Schimke (1964a, b) adduced evidence that at least two factors were involved; one being the stabilization of the enzyme against breakdown.

In the whole animal glucocorticoid hormones have been implicated in enzyme induction. A related system has now been developed in an established cell line; tyrosine transaminase can be induced in cultures of minimum deviation hepatoma cells by glucocorticoids. This phenomenon was first described by Pitot, Peraino, Morse & Potter (1964) and has been exploited mainly by Thompson, Tomkins & Curran (1966). These workers originated a tissue culture cell line from an ascites tumour of a minimum deviation Morris hepatoma. In this cell line a marked and reproducible increase in tyrosine transaminase activity can be regularly induced with dexamethasone phosphate. Induction of the enzyme is blocked by inhibitors of both protein and RNA synthesis. This indicates that the activity of this enzyme is regulated to a large extent through the rate of transcription of m-RNA. In subsequent studies (Pitot *et al.* 1964; Peterkofsky & Tomkins, 1967) it proved possible to obtain estimates of the half-life of both the messenger RNA which was found to be quite long, of the order of a few hours, and of the protein, which was found to be about seven hours.

These studies of enzyme induction in animal cells have revealed that the regulatory mechanism is rather complex. Besides control of the rate of RNA transcription, the stability of m-RNA, regulation of the rate of translation, and stability of the final protein, all seem to contribute. Moreover, there is always measurable enzyme activity in the cells; the increases which can be induced are of the order of 10- to 20-fold. Hence, the situation is very much less clear-cut than with β-galactosidase in *Escherichia coli*; although in theory the problem of enzyme regulation in animal cells can be tackled by the same means, it seems likely that unravelling the details will be a formidable task.

These observations provide evidence for mechanisms rather similar to those in bacteria. The main question which arises is, however, whether additional mechanisms have evolved in animal cells. Induction of 'true cytodifferentiation', i.e. the diversion of a cell to the synthesis of an entirely new kind of protein, has not yet been demonstrated in an established cell line, and it is by no means certain that the phenomenon is the same as that of enzyme induction. Indeed there is a good deal of evidence to suggest that it is different. Attempts have been made (Cole, Edwards & Paul, 1966) to culture cells from very young embryos in the hope of obtaining undifferentiated cell lines. However, even in these studies the cells already seem to have become established to develop in one direction or another, and the studies of Konigsberg (1963), Coon (1966) and Cahn & Cahn (1966) also suggest that differentiation is quite stable, provided cells have all the factors necessary to carry out the processes performed in the differentiated cells.

CONCLUSION

What parallels can be drawn between the behaviour of animal cells in culture and the behaviour of micro-organisms? Clearly there is much in common; the general kind of growth cycle is much the same, the general form of the genetic code used by the organisms seems to be the same; and the general protein synthetic machinery is the same. Moreover, some evidence has been presented in this paper that some of the control mechanisms may be very much the same. These similarities have evoked the aphorism that 'what is true for *Escherichia coli* must be true for the elephant'. However, while this may contain more than a grain of truth, it is at least equally likely that much must be true of the elephant which is not true of *E. coli*. Watson (1965) has suggested that because there is about a 1000 times as much DNA in animal cells as in *E. coli*, the problems must be approximately 1000 times more difficult. Although it may exaggerate the complexity of organization in animal cells, this statement helps to put the problems in perspective and indicates that we must be prepared to look for entirely new mechanisms in growth and differentiation in eukaryotes which may have been superimposed upon those occurring in micro-organisms.

REFERENCES

ABERCROMBIE, M. (1961). Behavior of normal and malignant connective tissue cells *in vitro*. In *Canadian Cancer Conference*, vol. IV, p. 101. New York: Academic Press.

ABERCROMBIE, M. & AMBROSE, E. J. (1958). Interference microscope studies of cell contacts in tissue culture. *Expl Cell Res.* 15, 332.

ABERCROMBIE, M. & HEAYSMAN, J. (1953). Observations on the social behaviour of cells in tissue culture. I. Speed of movement of chick heart fibroblasts in relation to their mutual contacts. *Expl Cell Res.* 5, 111.

AMOS, H. & MOORE, M. O. (1963). Influence of bacterial ribonucleic acid on animal cells in culture. *Expl Cell Res.* 32, 1.

BARSKI, G., SORIEUL, S. & CORNEFERT, FR. (1961). 'Hybrid' type cells in combined cultures of two different mammalian cell strains. *J. natn. Cancer Inst.* 26, 1269.

BENITEZ, H. H., MURRAY, M. R. & CHARGAFF, E. (1959). Heteromorphic change of adult fibroblasts by ribonucleoproteins. *J. biophys. biochem. Cytol.* 5, 25.

BENJAMIN, T. L. (1966). Virus-specific RNA in cells productively infected or transformed by polyoma virus. *J. molec. Biol.* 16, 359.

BENSCH, K., GORDON, G. & MILLER, L. (1964). The fate of DNA-containing particles phagocytized by mammalian cells. *J. Cell Biol.* 21, 105.

BERWALD, Y. & SACHS, L. (1963). *In vitro* cell transformation with chemical carcinogens. *Nature, Lond.* 200, 1182.

BLACK, P. H. & ROWE, W. P. (1963). SV-40-induced proliferation of tissue culture cells of rabbit, mouse, and porcine origin. *Proc. Soc. exp. Biol. Med.* 114, 721.

BOJARSKI, T. B. & HIATT, H. H. (1960). Stabilization of thymidylate kinase activity by thymidylate and by thymidine. *Nature, Lond.* **188**, 1112.

BOREK, C. & SACHS, L. (1966). *In vitro* cell transformation by X-irradiation. *Nature, Lond.* **210**, 276.

BRAND, K. G. (1962). Persistence and stability of species-specific haemagglutinogens in cultivated mammalian cells. *Nature, Lond.* **194**, 752.

BRAND, K. G. & SYVERTON, J. T. (1962). Results of species-specific haemagglutination tests on 'transformed', non-transformed, and primary cell cultures. *J. natn. Cancer Inst.* **28**, 147.

BÜRK, R. R. (1966). Growth inhibitor of hamster fibroblast cells. *Nature, Lond.* **212**, 1261.

BURLINGTON, H. (1959). Enzyme patterns in cultured cells. *Am. J. Physiol.* **197**, 68.

CAHN, R. D. & CAHN, M. B. (1966). Heritability of cellular differentiation: Clonal growth and expression of differentiation in retinal pigment cells *in vitro*. *Proc. natn. Acad. Sci. U.S.A.* **55**, 106.

COLE, R. J., EDWARDS, R. G. & PAUL, J. (1966). Cytodifferentiation and embryogenesis in cell colonies and tissue cultures derived from ova and blastocysts of the rabbit. *Devl Biol.* **13**, 385.

COLE, R. J. & PAUL, J. (1966). The effects of erythropoietin on haem synthesis in mouse yolk sac and cultured foetal liver cells. *J. Embryol. exp. Morph.* **15**, 245.

COON, H. G. (1966). Clonal stability and phenotypic expression of chick cartilage cells *in vitro*. *Proc. natn. Acad. Sci. U.S.A.* **55**, 66.

DAVIDSON, E. H. (1963). Heritability and control of differentiated function in cultured cells. *J. gen. Physiol.* **46**, 983.

DEFENDI, V., EPHRUSSI, B., KOPROWSKI, H. & YOSHIDA, M. C. (1967). Properties of hybrids between polyoma-transformed and normal mouse cells. *Proc. natn. Acad. Sci. U.S.A.* **57**, 299.

DEFENDI, V., LEHMAN, J. & KRAEMER, P. (1963). 'Morphologically normal' hamster cells with malignant properties. *Virology,* **19**, 592.

DEFENDI, V. & MANSON, L. A. (1963). Analysis of the life-cycle in mammalian cells. *Nature, Lond.* **198**, 359.

DE MARS, R. (1958). The inhibition by glutamine of glutamyl transferase formation in cultures of human cells. *Biochim. biophys. Acta,* **27**, 435.

DOLJANSKY, L. (1930). Observations sur l'évolution des éléments endothéliaux dans les cultures du foie *in vitro*. *C. r. Séanc. Soc. Biol.* **103**, 858.

EAGLE, H. (1965). Metabolic controls in cultured mammalian cells. *Science, N.Y.* **148**, 42.

EARLE, W. R. (1943). Changes induced in a strain of fibroblasts from a C_3H mouse by the action of 20-methylcholanthrene. *J. natn. Cancer Inst.* **3**, 555.

EARLE, W. R. & NETTLESHIP, A. (1943). Production of malignancy *in vitro*. V. Results of injection of cultures into mice. *J. natn. Cancer Inst.* **4**, 213.

EBNER, K. E., HAGEMAN, E. C. & LARSEN, B. L. (1961). Functional biochemical changes in bovine mammary cell cultures. *Expl Cell Res.* **25**, 555.

FEDEROFF, S. (1967). Proposed usage of animal tissue culture terms. *Expl Cell Res.* **46**, 642.

FELL, H. B. & MELLANBY, E. (1953). Metaplasia produced in cultures of chick ectoderm by high vitamin A. *J. Physiol.* **119**, 470.

FRANKS, D., COOMBS, R. R. A., BESWICK, T. S. L. & WINTER, M. M. (1963). Recognition of the species of origin of cells in culture by mixed agglutination. III. Identification of the cells of different primates. *Immunology,* **6**, 64.

FRANKS, D., GURNER, B. W., COOMBS, R. R. A. & STEVENSON, R. (1963). Results of tests for the species of origin of cell lines by means of the mixed agglutination reaction. *Expl Cell Res.* **28**, 608.

GEY, G. O. (1941). Cytological and cultural observations on transplantable rat sarcomata produced by the inoculation of altered normal cells maintained in continuous culture. *Cancer Res.* **1**, 737.

GREEN, H. & GOLDBERG, B. (1964). Collagen and cell protein synthesis by an established mammalian fibroblast line. *Nature, Lond.* **204**, 347.

GROBSTEIN, C. (1953). Epithelio-mesenchymal specificity in the morphogenesis of mouse sub-mandibular rudiments *in vitro*. *J. exp. Zool.* **124**, 383.

GROBSTEIN, C. (1965). Differentiation: Environmental factors, chemical and cellular. In *Cells and Tissues in Culture*, vol. I. Ed. E. N. Willmer. London: Academic Press.

HALBERSTAEDTER, L. & DOLJANSKI, L. (1939). Transformation *in vitro* of cultures of normal cells treated with Rous sarcoma agent into sarcoma cells. *Nature, Lond.* **143**, 288.

HARRIS, H. & WATKINS, J. F. (1965). Hybrid cells derived from mouse and man: artificial heterokaryons of mammalian cells from different species. *Nature, Lond.* **205**, 640.

HARRIS, H., WATKINS, J. F., FORD, C. E. & SCHOEFL, G. I. (1966). Artificial heterokaryons of animal cells from different species. *J. Cell Science*, **1**, 1.

HARRIS, M. (1957). Quantitative growth studies with chick myoblasts in glass substrate cultures. *Growth*, **21**, 149.

HARRIS, M. & RUDDLE, F. H. (1961). Clone strains of pig kidney cells with drug resistance and chromosomal markers. *J. natn. Cancer Inst.* **26**, 1405.

HAYFLICK, L. & MOORHEAD, P. S. (1961). The serial cultivation of human diploid cell strains. *Expl Cell Res.* **25**, 585.

HEIDELBERGER, C. & IYPE, P. T. (1967). Malignant transformation *in vitro* by carcinogenic hydrocarbons. *Science, N.Y.* **155**, 214.

HÖGMAN, C. F. (1960). Blood group antigens on human cells in tissue culture. *Expl Cell Res.* **21**, 137.

HOLTFRETER, J. (1943). The properties and functions of the surface coat in amphibian embryos. *J. exp. Zool.* **93**, 251.

HOLTZER, H., ABBOTT, J., LASH, J. & HOLTZER, S. (1960). The loss of phenotypic traits by differentiated cells *in vitro*. I. Dedifferentiation of cartilage cells. *Proc. natn. Acad. Sci. U.S.A.* **46**, 1533.

JENSEN, F., KOPROWSKI, H. & PONTÉN, J. A. (1963). Rapid transformation of human fibroblast cultures by Simian virus 40. *Proc. natn. Acad. Sci. U.S.A.* **50**, 343.

JOHNSON, T. C. & HOLLAND, J. J. (1965). Ribonucleic acid and protein synthesis in mitotic HeLa cells. *J. Cell Biol.* **27**, 565.

KAYE, A. M. & WINOCOUR, E. (1967). On the 5-methylcytosine found in the DNA extracted from polyoma virus. *J. molec. Biol.* **24**, 475.

KONIGSBERG, I. R. (1963). Clonal analysis of myogenesis. *Science, N.Y.* **140**, 1273.

KRANTZ, S. B., GALLIEN-LARTIGUE, O. & GOLDWASSER, E. (1963). The effect of erythropoietin upon heme synthesis by marrow cells *in vitro*. *J. biol. Chem.* **238**, 4085.

KROOTH, R. S., HOWELL, R. R. & HAMILTON, H. B. (1962). Properties of acatalastic cells growing *in vitro*. *J. exp. Med.* **115**, 313.

KROOTH, R. S. & WEINBERG, A. N. (1960). Properties of galactosemic cells in culture. *Biochem. biophys. Res. Commun.* **3**, 518.

KRUSE, P. F., Jr., & WHITE, P. B. (1961). Changes in protein content during growth cycle of primary mammalian cell cultures. *Expl Cell Res.* **23**, 423.

KUROKI, T., GOTO, M. & SATO, H. (1967). Malignant transformation of hamster embryonic cells by 4-hydroxyaminoquinoline N-oxide in tissue culture. *Tohoku J. exp. Med.* **91**, 109.

LEVAN, A. & BIESELE, J. J. (1958). Role of chromosomes in cancerogenesis as studied in serial tissue culture of mammalian cells. *Ann. N.Y. Acad. Sci.* **71**, 1022.

LEVI-MONTALCINI, R. (1966). The nerve growth factor: Its mode of action on sensory and sympathetic nerve cells. In *The Harvey Lectures*, Series 60, 1964–65. p. 217. New York: Academic Press.

LEVINE, E. M., BECKER, Y., BOONE, C. W. & EAGLE, H. (1965). Contact inhibition, macromolecular synthesis, and polyribosomes in cultured human diploid fibroblasts. *Proc. natn. Acad. Sci. U.S.A.* **53**, 350.

LEVINTOW, L. & EAGLE, H. (1961). The biochemistry of cultured mammalian cells. *A. Rev. Biochem.* **30**, 605.

LEWIS, W. H. (1927). The formation of giant cells in tissue cultures and their similarity to those in tuberculosis lesions. *Am. Rev. Tuberc. pulm. Dis.* **15**, 616.

LITTLEFIELD, J. W. (1966). The use of drug-resistant markers to study the hybridization of mouse fibroblasts. *Expl Cell. Res.* **41**, 190.

LOCKWOOD, D. H., VOYTOVICH, A. E., STOCKDALE, F. E. & TOPPER, Y. J. (1967). Insulin-dependent DNA polymerase and DNA synthesis in mammary epithelial cells *in vitro. Proc. natn. Acad. Sci. U.S.A.* **58**, 658.

LOWENSTEIN, W. R., SOCOLAR, S. J., HUGASHINO, S., KANNO, Y. & DAVIDSON, N. (1965). Intercellular communication: renal, urinary bladder, sensory, and salivary gland cells. *Science, N.Y.* **149**, 295.

MACPHERSON, I. (1963). Characteristics of a hamster cell clone transformed by polyoma virus. *J. natn. Cancer Inst.* **30**, 795.

MACPHERSON, I. (1965). Reversion in hamster cells transformed by Rous sarcoma virus. *Science, N.Y.* **148**, 1731.

MACPHERSON, I. (1966). Malignant transformation and reversion in virus infected cells. In *Recent Results in Cancer Research*, no. 6, *Malignant Transformation by Viruses*, p. 1. Ed. W. H. Kirsten. New York: Springer-Verlag.

MACPHERSON, I. & RUSSELL, W. (1966). Transformations in hamster cells mediated by mycoplasmas. *Nature, Lond.* **210**, 1343.

MERCHANT, D. J. & KAHN, R. H. (1958). Fiber formation in suspension cultures of L strain fibroblasts. *Proc. Soc. exp. Biol. Med.* **97**, 359.

MOORE, G. E., LEHNER, D. F., KIKUCHI, Y. & LESS, L. A. (1962). Continuous culture of a melanotic cell line from the golden hamster. *Science, N.Y.* **137**, 986.

MOSCONA, A. A. (1962). Analysis of cell recombinations in experimental synthesis of tissues *in vitro. J. Cell Comp. Physiol.* **60**, 65.

NICOL, A. G. & BECK, J. S. (1966). Persistence of an organ-specific antigen in organ and tissue cultures of hyper-plastic human thyroid gland. *Nature, Lond.* **210**, 1227.

NIU, M. C., CORDOVA, C. C., NIU, L. C. & RADBILL, C. L. (1962). RNA-induced biosynthesis of specific enzymes. *Proc. natn. Acad. Sci. U.S.A.* **48**, 1964.

PACE, D. M. & AFTONOMOS, L. (1957). Effects of cell density on cell growth in a clone of mouse liver cells. *J. natn. Cancer Inst.* **19**, 1065.

PAUL, J. (1962). The cancer cell *in vitro. Cancer Res.* **22**, 431.

PAUL, J. (1965). Carbohydrate and energy metabolism. In *Cells and Tissues in Culture*, vol. I, p. 239. Ed. E. N. Willmer. London: Academic Press.

PAUL, J., BROADFOOT, M. M. & WALKER, P. (1966). Increased glycolytic capacity and associated enzyme changes in BHK21 cells transformed with polyoma virus. *Int. J. Cancer*, **1**, 207.

PAUL, J. & FOTTRELL, P. F. (1961). Molecular variation in similar enzymes from different species. *Ann. N.Y. Acad. Sci.* **94**, 668.

PAUL, J. & GILMOUR, R. S. (1966). Template activity of DNA is restricted in chromatin. *J. molec. Biol.* **16**, 242.

PAUL, J. & HAGIWARA, A. (1962). A kinetic study of the action of 5-fluoro-2'-deoxyuridine on synthetic process in mammalian cells. *Biochim. biophys. Acta*, **61**, 243.

PAUL, J. & HUNTER, J. A. (1968). DNA synthesis is essential for increased haemoglobin synthesis in response to erythropoietin. *Nature, Lond.* **219**, 1362.

PETERKOFSKY, B. & TOMKINS, G. M. (1967). Effect of inhibitors of nucleic acid synthesis on steroid-mediated induction of tyrosine aminotransferase in hepatoma cell cultures. *J. molec. Biol.* **30**, 49.

PIOUS, D. A., HAMBURGER, R. N. & MILLS, S. E. (1964). Clonal growth of primary human cell cultures. *Expl Cell Res.* **33**, 495.

PITOT, H. C., PERAINO, C., MORSE, P. A. JUN. & POTTER, VAN R. (1964). Hepatomas in tissue culture compared with adapting liver *in vivo*. *Nat. Cancer Inst. Monogr.*, no. 13, p. 229.

POTTER, D. D., FURSHPAN, E. J. & LENNOX, E. S. (1966). Connections between cells of the developing squid as revealed by electro-physiological methods. *Proc. natn. Acad. Sci. U.S.A.* **55**, 328.

PUCK, P. T. & MARCUS, P. I. (7955). A rapid method for viable cell titration and clone production with HeLa cells in tissue culture. The use of X-irradiated cells to supply conditioning factors. *Proc. natn. Acad. Sci. U.S.A.* **41**, 432.

REICH, P. R., BLACK, P. H. & WEISSMAN, S. M. (1966). Nucleic acid homology studies of SV40 virus-transformed and normal hamster cells. *Proc. natn. Acad. Sci. U.S.A.* **56**, 78.

RIVERA, E. M. & BERN, H. A. (1961). Influence of insulin on maintenance and secretory stimulation of mouse mammary tissues by hormones in organ-culture. *Endocrinology*, **69**, 340.

ROBINS, A. B. & TAYLOR, D. M. (1968). Nuclear uptake of exogenous DNA by mammalian cells in culture. *Nature, Lond.* **217**, 1228.

RUBIN, H. (1962). Response of cell and organism to infection with avian tumor viruses. *Bact. Rev.* **26**, 1.

RUSSELL, W. C. (1966). Alterations in the nucleic acid metabolism of tissue culture cells infected with mycoplasmas. *Nature, Lond.* **212**, 1537.

SACHS, L. (1967). An analysis of the mechanism of neoplastic cell transformation by polyoma virus, hydrocarbons and irradiation. In *Current Topics in Development Biology*. Ed. A. Monroy and A. A. Moscona. New York: Academic Press.

SALZMAN, N. P. (1959). Systematic fluctuations in the cellular protein, RNA and DNA during growth of mammalian cell culture. *Biochim. biophys. Acta*, **31**, 158.

SALZMAN, N. P. & SEBRING, E. D. (1962). The coupled formation of deoxyribonucleic acid, nuclear ribonucleic acid and protein in animal-cell cultures. *Biochim. biophys. Acta*, **61**, 406.

SANDERS, F. K. & BURFORD, B. O. (1967). Morphological conversion of cells *in vitro* by N-nitrosomethylurea. *Nature, Lond.* **213**, 1171.

SANFORD, K. K. (1967). 'Spontaneous' neoplastic transformation of cells *in vitro*: some facts and theories. *Nat. Cancer Inst. Monogr.* no. 26, p. 387.

SATO, H. & KUROKI, T. (1966). Malignization *in vitro* of hamster embryonic cells by chemical carcinogens. *Proc. Jap. Acad.* **42**, 1211.

SATO, G. H., ROSSMAN, T., EDELSTEIN, L., HOLMES, S. & BUONASSISI, V. (1965). Phenotypic alterations in adrenal tumor cultures. *Science, N.Y.* **148**, 1733.

SCHIMKE, R. T. (1962). Repression of enzymes of arginine biosynthesis in mammalian tissue culture. *Biochim. biophys. Acta*, **62**, 599.

SCHIMKE, R. T. (1964a). Enzymes of arginine metabolism in cell culture: studies on enzyme induction and repression. *Nat. Cancer Inst. Monogr.* no. 13, p. 197.

SCHIMKE, R. T. (1964b). Enzymes of arginine metabolism in mammalian cell culture. *J. biol. Chem.* **239**, 136.

SCHINDLER, R., DAY, M. & FISCHER, C. (1959). Culture of neo-plastic mast cells and their synthesis of 5-hydroxytryptamine and histamine *in vitro*. *Cancer Res.* **19**, 47.

SEED, J. (1965). Deoxyribonucleic acid and ribonucleic acid synthesis in cell cultures. In *Cells and Tissues in Culture*, vol. I, p. 317. Ed. W. N. Willmer. London: Academic Press.

STEINBERG, M. S. (1962). Mechanism of tissue reconstruction by dissociated cells. II. Time-course of events. *Science, N.Y.* **137**, 762.

STEINBERG, M. S. (1963). Reconstruction of tissues by dissociated cells. *Science, N.Y.* **141**, 401.

STEINBERG, M. S. & ROTH, S. A. (1964). Phases in cell aggregation and tissue reconstruction. An approach to the kinetics of cell aggregation. *J. exp. Zool.* **157**, 327.

STOCKDALE, F. E., ABBOTT, J., HOLTZER, S. & HOLTZER, H. (1963). The loss of phenotypic traits by differentiated cells. II. Behaviour of chondrocytes and their progeny *in vitro*. *Devl Biol.* **7**, 293.

STOCKDALE, F. E., JUERGENS, W. G. & TOPPER, Y. J. (1966). A histological and biochemical study of hormone-dependent differentiation of mammary gland tissue *in vitro*. *Devl. Biol.* **13**, 266.

STOKER, M. (1962). Characteristics of normal and transformed clones arising from BHK 21 cells exposed to polyoma virus. *Virology*, **18**, 649.

STOKER, M. (1967). Contact and short-range interactions affecting growth of animal cells in culture. In *Current Topics in Development Biology*, vol. II, p. 107. Ed. A. Monroy and A. A. Moscona. New York: Academic Press.

SUBAK-SHARPE, J. (1965). Biochemically masked variants of the Syrian hamster fibroblast line BHK 21 and its derivatives. *Expl Cell Res.* **38**, 106.

SUBAK-SHARPE, J., BÜRK, R. R. & PITTS, J. D. (1966). Metabolic co-operation by cell to cell transfer between genetically different mammalian cells in tissue culture. *Heredity*, **21**, 342.

SZYBALSKI, W. & SMITH, M. J. (1959). Genetics of human cell lines. I. 8-Azaguanine resistance, a selective 'single-step' marker. *Proc. Soc. exp. Biol. Med.* **101**, 662.

SZYBALSKA, E. H. & SZYBALSKI, W. (1962). Genetics of human cell lines. IV. DNA-mediated heritable transformation of a biochemical trait. *Proc. natn. Acad. Sci. U.S.A.* **48**, 2026.

TAYLOR, J. H. (1960). Nucleic acid synthesis in relation to the cell division cycle. *Ann. N.Y. Acad. Sci.* **90**, 409.

TEMIN, H. M. & RUBIN, H. (1958). Characteristics of an assay for Rous sarcoma virus and Rous sarcoma cells in tissue culture. *Virology*, **6**, 669.

TERASIMA, T. & TOLMACH, L. J. (1963). Growth and nucleic acid synthesis in synchronously dividing populations of HeLa cells. *Expl Cell Res.* **30**, 344.

THOMPSON, E. B., TOMKINS, G. M. & CURRAN, J. F. (1966). Induction of tyrosine α-ketoglutarate transaminase by steroid hormones in a newly established tissue culture cell line. *Proc. natn. Acad. Sci. U.S.A.* **56**, 296.

TOBEY, R. A., PETERSEN, D. F., ANDERSON, E. C. & PUCK, T. T. (1966). Life cycle analysis of mammalian cells. III. The inhibition of division in Chinese hamster cells by puromycin and actinomycin. *Biophys. J.* **6**, 567.

TODARO, G. J. & GREEN, H. (1963). Quantitative studies of the growth of mouse embryo cells in culture and their development into established lines. *J. Cell Biol.* **17**, 299.

TODARO, G. J., WOLMAN, S. R. & GREEN, H. (1963). Rapid transformation of human fibroblasts with low growth potential into established cell lines by SV 40. *J. Cell. Comp. Physiol.* **62**, 257.

VIROLAINEN, M. & DEFENDI, V. (1967). Dependence of macrophage growth *in vivo* upon interaction with other cell types. In *Growth Regulating Substances for*

Animal Cells in Culture, p. 67. The Wistar Inst. Symp. Mono. no. 7. Philadelphia: Wistar Inst. Press.

VOGT, M. & DULBECCO, R. (1960). Virus cell interaction with a tumor-producing virus. *Proc. natn. Acad. Sci. U.S.A.* **46**, 365.

WATKINS, J. F. & DULBECCO, R. (1967). Production of SV40 virus in heterokaryons of transformed and susceptible cells. *Proc. natn. Acad. Sci. U.S.A.* **58**, 1396.

WATSON, J. D. (1965). *Molecular Biology of the Gene*. New York: W. A. Benjamin.

WEISSMAN, S. M., SMELLIE, R. M. S. & PAUL, J. (1960). Studies on the biosynthesis of deoxyribonucleic acid by extracts of mammalian cells. IV. The phosphorylation of thymidine. *Biochem. biophys. Acta*, **45**, 101.

WILDE, C. E. (1961). Factors concerning the degree of cellular differentiation in organotypic and disaggregated tissue cultures. *Actes du colloque international sur 'La culture organotypique: Associations et dissociations d'organes en culture in vitro'*, p. 183. Ed. M. E. Wolff. Paris: Centre national de la recherche scientifique.

WINOCOUR, E. (1967). On the apparent homology between DNA from polyoma virus and normal mouse synthetic RNA. *Virology*, **31**, 15.

WOLFF, E. T. & WOLFF, E. M. (1952). Le déterminisme de la différentiation sexuelle de la syrinx du canard cultivée *in vitro*. *Bull. biol. Fr. Belg.* **86**, 325.

YAFFE, D. & FELDMAN, M. (1965). The formation of hybrid multinucleated muscle fibres from myoblasts of different genetic origin. *Devl. Biol.* **11**, 300.

YASAMURA, Y., TASHJIAN, A. H. JUN. & SATO, G. H. (1966). Establishment of four functional, clonal strains of animal cells in culture. *Science, N.Y.* **154**, 1186.

YOSHIDA, M. C. & EPHRUSSI, B. (1967). Isolation and karyo-logical characteristics of seven hybrids between somatic mouse cells *in vitro*. *J. Cell Physiol.* **69**, 33.

ZAROF, F. L., SATO, G. & MILLS, E. E. (1961). Single-cell platings from freshly isolated mammalian tissue. *Expl Cell Res.* **23**, 565.

REGULATION OF
BACTERIAL SPORE FORMATION

J. MANDELSTAM

Microbiology Unit, Department of Biochemistry,
University of Oxford

INTRODUCTION

Endospores are commonly formed by some genera of rod-shaped bacteria: *Bacillus* and *Clostridium*. These bacteria, like any others, will remain in a state of vegetative growth as long as the medium will allow it but, as soon as some essential ingredient has been exhausted and the culture comes into a stationary state, the cells are induced to form spores. In suitable conditions, almost every cell in the population will produce a spore but more usually the yield is far from quantitative.

The longevity of spores, and their great resistance to adverse conditions, are coupled with an ability to germinate very rapidly in a suitable environment. These properties, and the fact that some sporulating species are pathogenic while others cause spoilage of food, have long made them of interest medically and in the food industry. There is consequently a great deal of literature stretching well back into the last century which this article is not in any way intended to cover. There are a number of excellent reviews that already do this (Szulmajster, 1964; Halvorson, 1964; Halvorson, Vary & Steinberg, 1966; Murrell, 1967; Schaeffer, 1968). The present discussion will be restricted to a consideration of the physiology and biochemistry of the sporulation process itself.

The information obtained in the last 10 or 15 years makes it clear that sporulation consists of an ordered sequence of structural changes coupled with an ordered sequence of biochemical events. The spore differs from the mother cell both in its morphology and its biochemical constituents. In this sense it is a primitive example of differentiation with the added advantage, from the experimenter's point of view, that it can be studied genetically in some types of Bacilli. In addition, it will be seen that some of the characteristic features of sporulation have their counterpart in the differentiation of more complicated organisms such as yeasts, slime moulds, *Acetabularia* and sea urchins.

In this article the material is presented in the following order: A brief account of the main structural changes that occur during sporulation, a more detailed description of the biochemical events that accompany them and, finally, a discussion of factors initiating sporulation and of such biological concepts as commitment.

MORPHOLOGY OF SPORULATION

With the light microscope some visible changes can be seen in the sporulating cell during the early stages, but the most distinctive indication is the appearance of refractility in the spore structure when the cells are observed in the phase-contrast microscope. The electron microscope has made it possible to study the process, and especially the early parts of it, in much greater detail.

The morphological changes observed by Young & Fitz-James (1959 a–c) in *Bacillus cereus*, by Ryter (1965) in *B. subtilis* and by Ohye & Murrell (1962) in *B. coagulans* all indicate a similar pattern of events.

Ryter, Schaeffer & Ionesco (1966) defined seven stages of sporulation, and Murrell (1967) has proposed that these should be generally adopted as reference points by other workers so that comparisons can be made between different species. Conventionally the whole process is timed from the end of exponential growth (t_0), and hourly periods after this are denoted t_1, t_2, etc. The two common methods of studying sporulation are the 'exhaustion' experiment in which the culture is allowed to grow until some nutrient becomes limited: the other is the 'replacement' experiment in which growing cells are transferred to a medium which will not support normal growth and which will initiate sporulation. In both types of experiment the course of events is similar and mature spores are formed in about 8 hr. under good conditions. The scheme, set out in Fig. 1 and described in this section and the next, is a composite one built up from these studies: the electron micrographs of Pls. 1 and 2 refer specifically to *Bacillus subtilis*.

Stage O. This represents the vegetative state at the end of growth, in which each cell contains two chromosomes.

Stage I (t_0–$t_{1\frac{1}{2}}$): *axial chromatin.* The two chromosomes apparently cohere and form a fairly broad axial filament which occupies the centre of the cell and is probably connected to the membrane by mesosome attachments.

Stage II ($t_{1\frac{1}{2}}$–$t_{2\frac{1}{2}}$): *septation.* The chromosomes separate, one of them moves to the end of the cell and the spore septum (which is a membranous structure) begins to be formed. The developing septum is attached to a mesosome (Pl. 1).

Stage III ($t_{2\frac{1}{2}}$–$t_{4\frac{1}{2}}$): *formation of the spore protoplast.* Further growth of membranous material now occurs, extending back from the point where the septum touches the membrane of the mother cell towards the end of the cell until, finally, there is a complete spore protoplast enclosed in its own membrane and distinctly free in the cytoplasm.

Stage IV $(t_{4\frac{1}{2}}-t_6)$: *cortex formation.* At this stage the spores first appear as refractile objects when viewed by phase-contrast microscopy. The cortex is formed by deposition of material within the double membrane of the spore protoplast, and is probably the layer in which the mucopeptide of the spore is contained.

Stage V (t_6-t_7): *coat formation.* The coat is first discernible as material being laid down, apparently at some distance from the cortex. It is possible (see Murrell, 1967) that some of the cytoplasm of the mother cell becomes trapped within the spore at this stage.

Stage VI (t_7-t_8): *spore maturation.* The coat material becomes denser and the inner portion of the spore appears to be featureless in the electron microscope, presumably because the fixatives used in preparing the specimen fail to penetrate. Distinct laminations appear and also ridges.

Stage VII. The sporangium disintegrates because of the action of lytic enzymes and the spore is liberated. These events are not relevant to the discussion and this stage is not illustrated.

BIOCHEMISTRY OF SPORULATION

The object of biochemical studies of differentiation is to give a biochemical explanation of a morphological change, and the first step is obviously to correlate the biochemical and structural events. At first sight the scheme set out in Fig. 1 could be regarded as a promising beginning, but it could give a very misleading impression because it includes not only the 'proper' sequence of events needed for making the spore but, interspersed among them, a number of others which are really expressions of vegetative functions of the cell and which only become manifest inadvertently and perhaps for trivial reasons. On *a priori* grounds biochemical substances which appear in the course of sporulation can be subdivided into several classes:

(*a*) Spore-specific components such as dipicolinic acid and also a number of enzymes that are found in spores but not in vegetative cells (see below).

(*b*) Those which are concerned with the synthesis of spore-specific components. An example would be dipicolinic acid synthetase, an enzyme which may not itself be incorporated into the spore but which catalyses the formation of a major spore component.

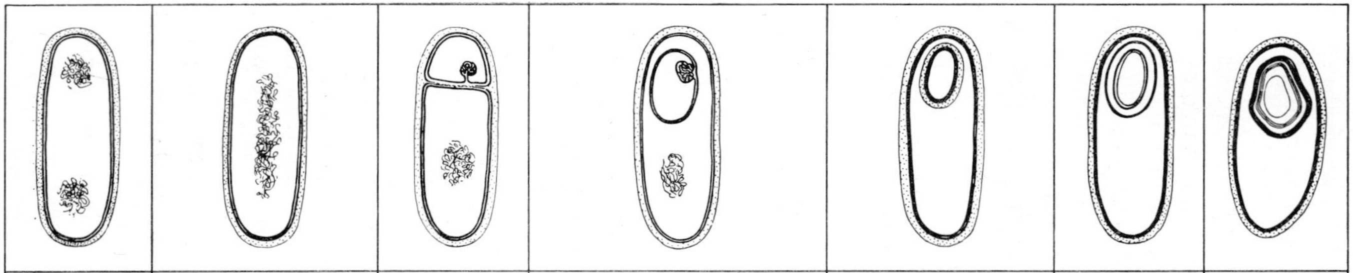

STAGE 0	STAGE I	STAGE II	STAGE III	STAGE IV	STAGE V	STAGE VI
Vegetative cell	Chromatin filament	Spore septum	Spore protoplast	Cortex formation (refractility)	Coat formation	Maturation
	Antibiotic	Alanine dehydrogenase	Alkaline phosphatase	Ribosidase	Cysteine incorporation	Alanine racemase
	Exo-protease		Glucose dehydrogenase	Adenosine deaminase	Octanol resistance	Heat resistance
	Protein turnover		Aconitase	Dipicolinic acid		
	Ribonuclease		Heat-resistant catalase	Uptake of Ca^{2+}		
	Amylase					

Fig. 1. Morphological and biochemical events associated with sporulation in *Bacillus* spp. This is a composite diagram incorporating information obtained by various authors with different types of aerobic spore-forming Bacilli. The time scale is about 8 hr. It is important to note that the correlation is, at best, very approximate and that some of the biochemical events may occur at different times in different types of organisms and may also be affected by the medium in which sporulation occurs (see text).

(*c*) Vegetative enzymes which were repressed by the substrates of a full growth medium but which have become derepressed in the nutritionally defective medium which initiates sporulation. Derepression may account for the general finding that enzymes of the glyoxylate and tricarboxylic acid cycles usually increase during sporulation. This category will also include other vegetative enzymes or intermediary metabolites which are found in exponentially growing cells but which are required to a greater extent during sporulation. For example, it has been shown by Hanson, Srinivasan & Halvorson (1963) that there is a high oxygen demand at certain stages of sporulation and the enzymes concerned might be expected to change in amount with concomitant changes in the levels of other types of molecules, e.g. ATP .

(*d*) Finally, there are biochemical substances that are probably 'by-products'. An example is the brown melanin pigment that characterizes spore-forming colonies of *Bacillus subtilis*. This is very useful for the experimenter who is trying to recognize asporogenous mutants, which remain white, but the pigment appears so long after the majority of the cells have sporulated that it probably plays no part in the process.

It would be useful if we could assign each event to its proper category but, in the following section, it will be seen that in many instances this cannot be done and, until it is, the picture is likely to remain confusing.

The biochemical sequence of events has been investigated in *Bacillus cereus* (Young & Fitz-James, 1959*a–c*) and in *B. subtilis* by Schaeffer, Ionesco, Ryter & Balassa (1965), Szulmajster (1964) and Warren (1968). These and other studies have been discussed by Halvorson *et al.* (1966), Murrell (1967) and Schaeffer (1968).

Changes associated with Stage I

The most notable characteristics at this stage are the formation and release into the medium of antibiotics and of exo-enzymes such as proteases, ribonuclease and amylase.

Antibiotics. Some of the antibiotics have now been isolated and characterized. Examples are bacitracin, which is produced in *Bacillus licheniformis* throughout sporulation (Bernlohr & Novelli, 1960, *a*, *b*; 1964) and bacilysin, which is produced in Stage I by *B. subtilis* and the constitution of which has been determined by Rogers, Lomakina & Abraham (1965). These compounds are peptides yielding only a few amino acids after hydrolysis. Various ideas have been advanced to account for their production by sporulating cells. One of these is teleological: that the antibiotic is produced and discharged into the surrounding medium to destroy other bacteria and thus to ensure a less competitive environment for the spore when it germinates. This hypothesis is not easy to test. Another suggestion, that the antibiotic is a component of the spore coat, is supported by the observation of Bernlohr & Novelli (1960b; 1964) that prelabelled bacitracin is taken up by the cells and incorporated, apparently without degradation into its component amino acids, into the spore coat. It is contradicted by the finding (Bhattacharya & Bose, 1967) that the spore coat of *B. subtilis* does not contain proline and some other of the 13 amino acid residues of mycobacillin. Similarly, Snoke (1964) showed that the composition of the coat of *B. licheniformis* is incompatible with the assumption that it contains significant amounts of the antibiotic which is produced by these bacteria.

At present we have no way of knowing whether the antibiotics serve any function in sporulation or whether they should be regarded as by-products.

Protease. Any one type of organism is likely to produce more than one type of protease (Drenth & Hol, 1967). In *Bacillus subtilis*, however, the enzyme mainly responsible for the extracellular proteolytic activity is probably that characterized by McConn, Tsuru & Yasonobu (1964). This is a zinc-dependent enzyme which can attack most peptide bonds.

It has been known for some time that both the antibiotic and the protease were intimately related to sporulation and that mutants unable to produce these substances (designated ab^- and $prot^-$) are almost invariably also asporogenous (sp^-), and incidentally poorly competent (co^-) in genetic transformation experiments (see Balassa, Ionesco & Schaeffer, 1963; Spizizen, 1965; Schaeffer, 1967).

The functioning of the protease has been examined in wild type *Bacillus subtilis* and in two mutants that were blocked at or before Stage I, and were ab^-, $prot^-$, sp^- (Mandelstam & Waites, 1968). When sporulation was induced by suspending wild-type cells in a deficient medium (glutamate and inorganic ions) protein turnover proceeded at the high rate of 8–10 % per hr. so that, by the time refractile spores had begun to appear in Stage IV ($t_{4\frac{1}{2}}$), a large part of the vegetative protein had been degraded. In the proteaseless mutants there was no breakdown of vegetative protein in resuspension experiments.

One of the mutants reverted easily to the wild-type, apparently as a result of a single mutation. When this occurred, all the lost properties, i.e. exo-protease production, protein turnover, formation of antibiotic, and ability to form spores, were restored. The same enzyme thus appears to be responsible for the extracellular proteolytic activity and for intracellular degradation of protein. The existence of this enzyme accounts for the common finding of a high rate of protein turnover in sporulating bacilli (Foster & Perry, 1954; Young & Fitz-James, 1959*b*; Canfield & Szulmajster, 1964). It accounts also for the fact, noted by several workers independently, that if cells that are prelabelled with an amino acid are allowed to sporulate in a medium with excess of unlabelled amino acid as a 'trap', the proteins of the spore are virtually unlabelled, i.e. they are newly synthesized. In *Bacillus subtilis* the proportion of new protein in the spore is about 95 % (Mandelstam & Waites, 1968) and similarly high values are indicated by the earlier results of Monro (1961), Canfield & Szulmajster (1964), Young & Fitz-James (1959*b*).

All this still leaves unanswered the question why the protease is necessary for sporulation. The possibilities are as follows:

(*a*) The sporulating cell is often in an environment where net synthesis of protein is impossible. When this happens the only way in which the cell can change the pattern of its proteins from that which characterizes the vegetative cells to that which characterizes the spore is by protein turnover (see Mandelstam, 1960). For this, of course, proteolytic activity is required. Although this explanation may hold for cells sporulating in buffer solution or in water, it is not true in the case of cells which are in a resuspension medium containing glutamate and organic ions. *Bacillus subtilis* in such a medium can carry out net synthesis and presumably this could provide the new types of protein needed (Mandelstam, Waites & Warren, 1967; Mandelstam & Waites, 1968). Nevertheless, in such a medium the characters *prot*⁻ and *sp*⁻ are still associated and some other explanation must be sought.

(*b*) During vegetative growth the cells make inhibitors or repressors of sporulation which are proteins. Unless these are assumed to be spontaneously unstable they will have to be inactivated by an effector ('inducer') or a protease will be needed to degrade them. In the latter case the *prot*⁻ mutants would be asporogenous. There is, however, no evidence to support such a hypothesis.

(*c*) Proteolytic activity is not needed for sporulation at all, and the protease is a secondary and incidental consequence of something which occurs in an ordered succession of events (see below). This would mean that the mutants had been damaged in such a way that the primary initiating event failed to occur and that, in consequence, no protease was formed.

Other exo-enzymes. Ribonuclease is produced at the end of the growth phase of *Bacillus subtilis* (Lampen, 1965). It is possible that the enzyme has some role in sporulation because mutants which have lost it may also be asporogenous (Schaeffer, 1968). Since these mutants have also usually lost the exo-protease and the ability to form antibiotics it is difficult to assess the function of the ribonuclease.

Amylase is also produced at this stage (Lampen, 1965), but since amylase-negative mutants retain the ability to sporulate (Schaeffer, 1968) the enzyme can be classified as a by-product.

Changes associated with Stage II

At this stage alanine dehydrogenase is found in *Bacillus subtilis* (Warren 1968). This is a spore component which may be important in germination.

Changes associated with Stage III

In *Bacillus cereus* glucose dehydrogenase is produced at this stage while in *B. subtilis* it is produced in Stage I (Warren, 1968). The difference may reflect a species difference or it may be due to the experimental conditions. But, whatever the explanation, it shows that this enzyme is not concerned with the formation of the axial filament or with the septation that follows it, since both these stages occur quite well in *B. subtilis* without its intervention.

Aconitase. Warren (1968) has found that aconitase is associated with Stage III in experiments with *Bacillus subtilis* grown in a medium where glucose exhaustion is the initiating factor in sporulation. This enzyme is a classic example of one that is likely to appear at some time in the process, can indeed be shown to be necessary for it, and which is nevertheless synthesized at this stage for quite fortuitous reasons.

Under the usual conditions of vegetative growth, i.e. aerobically and in the presence of glucose, the enzyme is not formed because of the combined repressor effects of glucose and of glutamate which are usually present (Hanson & Cox, 1967). Exhaustion of glucose in the medium—frequently the initiating factor in sporulation experiments—relieves the repression, and aconitase begins to be formed as if it were part of the sporulation process. Furthermore, it has been shown that aconitaseless mutants are also asporogenous (Hanson, Blicharska & Szulmajster, 1964). If this were all we knew about aconitase, we would conclude that it is a specific sporulation enzyme. However, if cells are grown in the absence of glucose, e.g. in a hydrolysed casein medium, aconitase is formed while the culture is still in the exponential state and is manifestly a normal vegetative enzyme. The fact that the aconitaseless mutant is also asporogenous might be attributable to a high requirement for ATP and for a functioning citric acid cycle during sporulation. Although this enzyme does not properly belong to the set of events we are interested in, the need for it emphasizes the point that sporulation is an exacting process that will only occur successfully in the cytoplasm of a healthy cell. If the cell is enzymically damaged spore formation may be much more adversely affected than vegetative growth.

Similar effects might be expected from defects in any of a number of other enzymes concerned in the metabolism of carbohydrate and the production of energy. A very large number of mutants of this type have been isolated by Freese & Fortnagel (1967) and they are all characterized by an inability to incorporate uracil into RNA at the normal rate. Many of them respond to the addition of ribose or of glutamate, i.e. these substances restore the incorporation of labelled uracil. An examination of some of them indicates that they are damaged in the enzymes of the TCA cycle (Fortnagel & Freese, 1968). As one might expect, these mutants are oligosporogenous rather than asporogenous; that is, they form spores at a low incidence.

Heat-resistant catalase. This enzyme exemplifies an activity found both in the vegetative cell and in the spore and having different properties in each condition. Other instances will be noted later. The catalase found in spores is markedly more resistant to heat than that of the vegetative cell (Lawrence & Halvorson, 1954). Furthermore, in *Bacillus cereus* it has different immuno-electrophoretic properties (Norris & Baillie, 1964). However the spore catalase is bound to particles while that of the vegetative cell is not. This fact raises the possibility that the spore enzyme is not really a different protein and that the observed differences in heat stability and in immunological properties are attributable to its physical state.

Alkaline phosphatase. This is another enzyme whose appearance is likely to be displaced according to the experimental conditions. When *Bacillus subtilis* is allowed to sporulate because of the exhaustion of glucose in a defined medium the enzyme appears at Stage I (Warren, 1968). But, if the same strain is induced to sporulate in a resuspension experiment, the enzyme appears only at about Stage III (Mandelstam & Sterlini, 1969). Furthermore, in *B. megaterium* the enzyme is produced even later—at about Stage V (Millet, 1963). Its function in sporulation is obscure. The enzyme in *B. subtilis* is repressed by inorganic phosphate and its synthesis is induced by phosphate starvation. This is analogous to the regulation of alkaline phosphatase in other micro-organisms. However, in the sporulating cell it begins to be formed in considerable quantities in spite of the fact that phosphate is present in the medium. Increasing the phosphate concentration several-fold still fails to repress the enzyme formed under these conditions (J. Mandelstam & J. M. Sterlini, unpublished). The interpretation of these findings is that either there are two different alkaline phosphatases, or there is only one enzyme, the production of which is regulated by inorganic phosphate in the vegetative cell and in some undetermined way after sporulation has been initiated.

Alkaline phosphatase, like catalase, poses the question whether the same enzymic activity in spores and in cells is due to two completely different proteins. However, in some instances enzymes that are found in the spore—though they may be newly synthesized—are almost certainly the same proteins as those in the vegetative cell. Elaborate comparisons of properties of enzymes isolated from vegetative cells and spores have been made by Kornberg and his associates, who determined amino acid composition, molecular weight, electrophoretic mobility, substrate specificity, etc., for the following enzymes: inorganic pyro-phosphatase of *Bacillus megaterium* (Tono & Kornberg, 1967); DNA polymerase in *B. subtilis* (Falaschi & Kornberg, 1966) and purine nucleoside phosphorylase of *B. cereus* (Gardner & Kornberg, 1967). All tests failed to reveal any differences between the spore enzyme and its counterpart in the vegetative cell.

Changes associated with Stage IV

Powell & Strange (1956) showed that in sporulating *Bacillus cereus* there is an increase both in ribosidase and in adenosine deaminase. The first of these enzymes is an example of an activity not detectable in vegetative cells.

Apart from the enzymic changes that occur here, further events of an undetermined nature take place which give the spore a refractile appearance in the phase-contrast microscope. Refractility is probably the end-result of a complex of changes of which we know little except that they can be prevented by such inhibitors as chloramphenicol. It was frequently suggested that refractility might be due to dipicolinic acid (DPA) (see, for example, Fitz-James, 1965). This view was questionable because refractility often precedes the appearance of DPA and the matter has now been definitely settled by the isolation of mutants which are unable to make DPA but which are, nevertheless, able to generate refractile spores (see below). The existence of dipicolinate in spores was first demonstrated by Powell (1953). Though it was known for a long time to be derived from aspartic acid, the synthetic pathway has only recently been described (Bach & Gilvarg, 1966; Chasin & Szulmajster, 1967). It is not detectable at all in the vegetative cell and its synthesis only begins in Stage IV. Its appearance coincides with the uptake of large amounts of Ca^{2+} and it may be laid down as the calcium salt. It can constitute as much as 10–20 % of the weight of the spore and it is probably responsible for the heat-resistant properties to which we have already referred. This is thought to be the case because mutants which lack the ability to make DPA give rise to spores which are refractile but heat-sensitive (Wise, Swanson & Halvorson, 1967). Similarly, if cells are allowed to sporulate in a calcium-deficient medium, refractile bodies are formed in the normal way but these too are heat-sensitive (Grelet, 1957).

Changes associated with Stage V: coat formation

This stage of development is marked by the incorporation of a large amount of cysteine into the spore coat (Vinter, 1959; Fitz-James, 1965). The latter author has shown that if chloramphenicol is added to *Bacillus cereus* it inhibits in a comparable manner the incorporation of cysteine and the completion of the coat.

Changes associated with Stage VI: spore maturation

Alanine racemase, an enzyme possibly important in spore germination, is formed here and the enzyme obtained from the spores of *Bacillus cereus* differs from that of the vegetative cell. The enzyme is particulate in the spore and hardly affected by a temperature of 80° for 2 hr., while the enzyme of the vegetative cells is rapidly destroyed at this temperature. However, continued sonic oscillation solubilizes some of the racemase which is then found to be heat-sensitive (Stewart & Halvorson, 1954). It is therefore quite possible that the enzyme is one and the same in both forms of the cell and that the heat stability is due to the fact that it is held on some particulate surface.

Other changes occur at this stage, again of an unidentified nature, which make the spores resistant to octanol and other organic solvents. Finally the spores become virtually anhydrous and also heat-resistant. It has been suggested by a number of authors that insensitivity to heat is, in part at any rate, due to dehydration.

Summary

In addition to the philosophical point that a correlation does not establish a causal relationship, it will be apparent to the reader that, for the most part, we cannot even be sure of the correlation. The fact that some of the bio-chemical events are 'movable' raises the possibility that some of the others might also be. Before any correlation can be regarded as established it would be necessary to show that an event is invariably correlated with a particular morphological stage—irrespective of the medium in which the cells are sporulating and irrespective of the medium in which they were previously grown. Furthermore, the same correlation should hold in other species of bacteria. If it does not, it obviously means that the event is not necessary for the occurrence of the earlier stage; moreover there is no guarantee that it is required for the later stage—or indeed at all.

At the moment we are left with a situation where, even when a correlation seems to be reasonably well established, there is still no mechanistic link between the biochemical and the morphological events. This is a point which will be discussed later.

Spore antigens

Apart from catalase the spores contain a number of substances that differ antigenically from those of the vegetative cell. This was known from the early immunological studies of Doak & Lamanna (1948) on several types of Bacilli and from several later studies (see Norris, 1962; Baillie & Norris, 1964). The latter authors, using immuno-electrophoresis, found that antiserum prepared against spores and run against spore extracts showed seven precipitin lines that were not produced when extracts of vegetative cells were tested. Furthermore, some of these new antigens were heat-stable.

Waites (1968) followed the production of spore antigens, probably protein, in sporulating cultures of *Bacillus subtilis* and found that they began to be formed an hour or so after transferring the cells to a re-suspension medium. In addition, he showed that if the antiserum was first repeatedly adsorbed with an excess of vegetative protein and then tested against spore extracts all the apparently specific lines with one exception were removed. The implication is that, except for this one remaining antigen, the others are present in the vegetative cell but in relatively small quantities. This is reminiscent of the basal and induced levels of inducible enzyme systems and suggests that commitment to sporulation merely accelerates the synthesis of proteins that were being made in small amounts in the vegetative cell.

GENETICS OF SPORULATION

As described above, a number of biochemical and physiological characters are frequently associated with one another, viz. the production of antibiotic (ab^+), exo-protease ($prot^+$), competence in transformation (co^+) and ability to form spores (sp^+). The possibility that the corresponding genes were closely linked and might be on an episome was suggested (Jacob, Schaeffer & Wollman, 1960; Spizizen, 1965) and was consistent with the fact that a single mutation led to simultaneous loss of all these characters. The episome hypothesis was supported by the observation that acridine dyes, which characteristically eliminate episomes, produced a coordinate loss of the characters associated with sporulation (Rogolsky & Slepecky, 1964).

In contradiction to this, a number of sp^- mutations in *Bacillus subtilis* were mapped by transduction and by Sueoka's method by Takahashi, (1965), who found that the markers are not episomal but widely scattered along the chromosome. Even with mutants blocked at the same stage the genetic loci often appear to be unlinked (Rouyard, Ionesco & Schaeffer, 1967). These authors showed that there were at least four loci concerned in Stage II which were unlinked in transformation experiments, and at least five, similarly unlinked, concerned in Stage III. The production of protease in *B. subtilis* is also controlled by more than one gene, and mapping by transduction shows that the genes are widely separated (J. Spizizen, personal communication).

The apparent linkage of the characters *prot*, *ab* and *sp* is therefore probably physiological and not genetic, i.e. there is a dependent sequence of events in which the occurrence of any one event depends on the successful completion of the earlier ones. This assumption, which will be discussed later, is supported by the study of large numbers of asporogenous mutants that have been isolated which are damaged in an early function. In these, the later functions seem not to be expressed (Schaeffer, 1968).

INITIATION OF SPORULATION

It is well established that sporulation is a response to nutritional deprivation and that, provided cells are kept in an adequate growth medium, the vegetative state can be prolonged indefinitely. This can be readily achieved in a continuous culture apparatus. From this fact alone it seems reasonable to believe that regulation of spore formation, like most forms of regulation in the bacterial cell, is negative, i.e. that the cells make an inhibitor or a repressor from some ingredient in the medium which prevents the process from occurring. It can then be assumed that, in a flask culture, when this ingredient is expended the inhibition is released and the cells sporulate. With many species of bacteria glucose, either directly or indirectly, contributes to the inhibition (see, for example, Foster, 1956) and the control mechanism for sporulation is in many ways analogous to the catabolite repression of inducible enzymes. In both types of repression the actual repressor has remained unidentified though attempts to characterize it in sporulating cells by imposing various types of starvation upon cells have been made. For some recent attempts see Jičínská (1964) and Schaeffer, Millet & Aubert (1965).

Concept of a 'threshold' for the inhibitor

If the medium contains some inhibitor of sporulation the ensuing events can be viewed as follows: as growth proceeds the inhibitor will gradually fall in concentration until a threshold value is reached at which sporulation will be initiated in the population. A graph relating spore incidence to the degree of starvation would be expected to show an abrupt transition as the threshold value is reached. Experimentally, however, no such threshold can be found (see below).

A different concept was put forward by Schaeffer, Millet & Aubert (1965). This is based on the finding that cultures are likely to produce small numbers of spores even when they are in a state of active vegetative growth. Schaeffer, Millet & Aubert (1965) grew *Bacillus subtilis* for at least fifteen generations in media containing a variety of carbon and nitrogen sources. In a medium that was 'good', that is one supporting a high growth rate, the spore incidence was very low, while with poorer media the spore incidence increased. The authors proposed that, for a culture growing in a particular medium, there is a defined probability that a cell will be initiated to form a spore and this will be reflected in the spore incidence (p). When both the carbon source and the nitrogen source are good (e.g. glucose and NH_4^+), p is low (3 in 10^4); with lactate and NH_4^+, p rises to 1 in 10^2. With a very poor source of nitrogen, such as histidine, p rises to 8 in 100. These observations prompted the conclusion that the repressor contained both carbon and nitrogen, and that the effect was more complicated than a simple glucose effect.

The relationship between growth rate and sporulation has been examined more quantitatively in continuous flow experiments using *Bacillus subtilis* and a defined medium in a chemostat (Dawes & Mandelstam, 1969). In agreement with the conclusion of Schaeffer, Millet & Aubert (1965) it was shown that apparently both carbon and nitrogen are needed to inhibit sporulation and that a deficiency of either will relieve the inhibition. Of more interest is the fact that there is a continuous relationship between the growth rate and the incidence of spores instead of the discontinuous curve predicted by the threshold hypothesis. It should be noted, in addition, that at the slowest growth rate, when the generation time was 5 hr., the observed incidence of refractile spores was 33 %. It has to be remembered that the sporulation time is several hours and that, during this time, a certain proportion of cells in which sporulation was initiated will have been washed out.

Making allowance for this, it turns out that the real spore incidence was 55 % and, under some conditions, it is possible for the value to approach 100 %. This means that sporulation and cell division can probably go on in the same cell at the same time, a finding that has a bearing on the concept of commitment (see below).

Sporulation and the division cycle

In *Bacillus cereus* about half of the DNA that is present in the cells when growth stops is incorporated into the spore (Young & Fitz-James, 1959 a). It was therefore suggested that the cell has two chromosomes, one of which is enclosed in the spore, while the other remains in the cytoplasm of the mother cell.

Presumably any cell that was in the middle of DNA replication would be unable to make spores containing the complete chromosome complement until replication had been completed. Presumably, too, if the cell were not in this state, sporulation would not be initiated at all. This may account for the net synthesis of 20–50 % of DNA that is observed when cells are abruptly transferred from a growth medium to a medium which will initiate sporulation. The belief that the cell has to have a full complement of completed chromosomes would be a reason for supposing that sporulation can only be initiated at a certain point in the division cycle. This was tested as follows (I. W. Dawes & J. Mandelstam, unpublished): cells growing in a chemostat in steady state at a generation time of 2 hr. in a nitrogen-limited medium were subjected to further nutritional deprivation by reducing the flow rate to impose a generation time of 5 hr. As previously described, this increased the probability of spore initiation. With the restricted flow rate maintained for 30 min. ($\frac{1}{4}$ of a generation time) one would expect that about 25 % of the cells would have been in, or could have got to, the correct point in the division cycle and that some hours later those cells that had been initiated in this period would give rise to a burst of spores. When the experiment was done with alternating 30 min. periods of slow and fast growth, spore formation occurred in a step-wise fashion in which each burst produced about a quarter of the number of spores that would have been obtained by a prolonged period of starvation. The result is consistent with the assumption that initiation is only possible at a specific stage of the division cycle but it gives no indication what that stage might be. To obtain this information it would be necessary to induce sporulation in samples taken from a synchronized culture.

BIOCHEMISTRY OF COMMITMENT

The term commitment can be rather loosely defined as the time at which the metabolism of a cell is channelled in the direction of sporulation to such an extent that addition of growth medium will not reverse it. Hardwick & Foster (1952) determined this as follows: *Bacillus cereus* was grown in a defined medium and transferred to phosphate buffer to induce sporulation. At hourly intervals samples were removed, glucose added to them and incubation continued. The spore incidence was measured after 14 hr. Glucose, added at any time up to 5 hr., suppressed sporulation completely. By 6 hr. about 10 % of the cells had 'escaped', i.e. they gave rise to spores, and by the end of 8 hr. virtually all the cells had become insensitive to inhibition by glucose. Experiments of the same sort have been done with *B. subtilis* resuspended in a medium containing glutamate and inorganic ions to induce sporulation (Mandelstam & Sterlini, 1969). Hydrolysed casein was added at intervals as in Hardwick & Foster's (1952) experiments. Using refractility as an indicator, the point of commitment was found to occur at about $2\frac{1}{2}$ hr.

Now, Aronson & del Valle (1964) had shown that sporulation in *Bacillus cereus* became insensitive to actinomycin D at an early stage. This appeared not to be the case in *B. subtilis*, where similar experiments were carried out by Szulmajster, Canfield & Blicharska (1963). Aronson (1965) carried the work further with *B. cereus* using hybridization and pulse-labelling techniques to analyse the RNA formed during sporulation of these cells. It appeared that both vegetative and spore m-RNA were formed and that after the initiation of sporulation there was a persistently stable fraction of m-RNA which was presumed to be the block of messages required in sporogenesis.

Mandelstam & Sterlini (1969) considered that commitment might be due to the formation of a stable m-RNA. Using *Bacillus subtilis*, they compared the time of commitment, defined as the time at which hydrolysed casein no longer prevented the formation of refractile spores, with the time at which the sporulation process became insensitive to actinomycin D. Two parallel curves were obtained, the first of which—the commitment time measured by hydrolysed casein—always preceded the other by about 20 min. Characters other than refractility were examined in the same way, viz. alkaline phosphatase which occurs earlier than refractility, and formation of DPA and incidence of heat-resistance which occur later. It emerged that there is no single point of commitment for the sporulation process as a whole. Instead there is a separate point of commitment for each of the characters. Each point of commitment is marked by the fact that the process concerned becomes refractory first to hydrolysed casein and then about 20 min. later to actinomycin D. A further point of interest is that in each case the commitment time for a character precedes the expression of the character by 1–2 hr. This has implications which will be discussed in the next section.

MODELS FOR SPORULATION

The facts to be accounted for include the following:

(1) During vegetative growth the spore function is not expressed in the majority of cells, but is initiated by starvation of one kind or another.

(2) Even in a state of active vegetative growth some cells will sporulate, and their incidence is related to the growth rate of the culture.

(3) Commitment is not an 'either-or' phenomenon. In suitable conditions almost all cells can grow and sporulate at the same time (see above).

(4) When sporulation is initiated, there is an ordered sequence in which the occurrence of any event is dependent upon the successful occurrence of those that come earlier.

(5) There is no commitment to sporulation as such. Instead, the cells become committed to carrying out in turn the specific parts of the process. Each commitment point apparently corresponds to the formation of a stable message or block of messages.

(6) The messages are transcribed in the same order as the biochemical events that are their expression and there is a considerable time interval between transcription and translation.

It is convenient to refer to the genes responsible for vegetative growth and for sporulation as the vegetative and spore genomes without making any assumptions about the map positions of the genes. We could then assume that the spore genome was controlled either positively or negatively. The former implies that there is a sporulation initiating factor. The presence of a substance with the requisite properties has been reported by Srinivasan (1965) in a variety of spore-forming organisms, but not in Gram-negative bacteria. The properties attributed to it were those of an inducer which acted by combining with some macromolecular repressor which it thereby converted to an inactive form, thus initiating sporulation (Srinivasan, 1965). If this idea were accepted one would have to explain why it is produced by cells that are sporulating but not by vegetative cells. This poses the problem of the regulation of its biosynthesis and so, basically, it is the initial problem removed one step. This does not, of course, mean that the model is incorrect; it merely means that the alternative assumption of negative control is simpler for purposes of discussion.

On the latter basis we assume that one of the metabolites within the cells acts as an inhibitor of sporulation and that when it is present the whole of the spore genome is switched off. We can also assume that there is a definable relationship between the external concentration of growth substrate and the level of metabolites in the cell cytoplasm. Now the inhibitor must act at some acceptor surface—we can assume one per cell for simplicity—and the reaction could be expected to follow the equation for an adsorption isotherm. In a large population of cells there will be many such surfaces and there will be a finite probability that, even at high levels of substrate and therefore of inhibitor, a few of them will be uncombined at the time when the cells containing them enter the sensitive phase of the division cycle. These cells will begin to sporulate and the lower the level of external substrate and of internal inhibitor the more of them there will be. This mechanism would account for the continuous relationship between growth rate and spore incidence that we have described. There is thus no threshold for the population nor can one be assumed to exist for the individual cell. Instead we can only say that, for a particular concentration of inhibitor, there is a definable probability that the acceptor surface in any one cell will become un-combined at some time during the sensitive phase.

In addition to this, it seems necessary to assume that the vegetative genome is not inhibited at any time. This model explains the retention of the ability of cells to resume growth at any time and it explains the initial event of sporulation itself. It would be an adequate model if activation of the whole spore genome occurred as the result of this single event. Since it does not we have to account for a number of points of commitment and for the fact that we have a dependent sequence.

Now, there are two obvious ways of accounting for a sequence of biochemical events. The first is by the mechanism of sequential induction. In its simplest form (see Stanier, 1951) this entails an enzyme sequence in which the first substrate induces the first enzyme, the reaction product induces the second enzyme, etc. A model for sporulation involving sequential induction was proposed by Halvorson (1965) and it has the merit of explaining the dependence of each event upon all those preceding it. A prediction which can be tested is that there should be cross-feeding between different types of mutant. Thus, a mutant blocked at an early stage should sporulate when grown together with a mutant blocked at some later stage. Ryter et al. (1966) tried to demonstrate cross-feeding by growing together sporulation mutants blocked at six different stages. There are fifteen binary combinations and in none of them was there any evidence of cross-feeding. A similar attempt was made by Mandelstam (unpublished), who used agar plates on each of which eight mutants were streaked from the centre outwards. The inner zone where the streaks converged to within a few millimetres of one another might have been expected to show cross-feeding. A much larger number of binary combinations could be examined since each plate tested twenty-eight $(7 + 6 + \ldots + 1 = 28)$. Several hundred combinations from forty different mutants were plated and no signs were found of cross-feeding between the mutants or between any mutant and the wild-type. Negative results do not, of course, rule out the sequential induction model. It is always possible that synthesis of the inducers is closely regulated so that one strain never produces enough to cross-feed another. Alternatively the cell membranes may be impermeable to these substances.

Another model considered by Halvorson (1965) is that of sequential transcription. The model is based on the assumption that the order of the genes on the chromosome corresponds with that of the biochemical sequence, and that this is how the order of events is determined. There is supporting evidence for such a hypothesis in other systems. For example, in synchronized cultures of yeast there is step-wise enzyme production which occurs for a particular enzyme at a particular stage in the division cycle (Halvorson, 1964). At present there seems no good reason for preferring the model of sequential transcription to that of sequential induction. It is apparent that a proper genetic analysis of asporogenous mutants would be of great help in resolving this problem. If the order of the genes followed that of the biochemical sequence it would strongly favour the idea of sequential transcription: if it did not, the argument for sequential induction would be stronger.

There is no good reason for producing yet another model at this stage, but it will be useful to recapitulate the salient features of sporulation and to compare sporulation with other differentiating systems.

(*a*) The initial precipitating factor is starvation of the cells which, in some unknown way, initiates the whole train of events.

(*b*) An early event is the production or activation of a proteolytic enzyme. This is not specifically something connected with sporulation. It occurs at a more primitive level when a culture of *Eschericha coli* is starved and stops growing, and is probably part of a primitive adaptation mechanism that allows an organism to change its protein pattern under conditions where net synthesis is impossible (see above and Mandelstam, 1960). Proteolysis is important in differentiation further up the evolutionary scale, e.g. during sporulation of yeast (Miller & Hoffman-

Ostenhof, 1964) and at an early stage in the development of the slime mould *Dictyostelium discoideum* (Sussman, 1966 and this Symposium; Wright, 1966). In higher organisms we encounter the same phenomenon in the pupa which is changing into a moth, and in amphibian metamorphosis where the tail of the tadpole is proteolysed and the amino acids that are liberated are used for the manufacture of body tissue of the frog (Bennett & Frieden, 1962). In all these instances the rearrangement of the protein pattern occurs during starvation of the organism concerned and can thus be accomplished only through protein turnover.

(*c*) An ordered series of biochemical events occurs concomitantly with a series of morphological changes. For any particular species of *Bacillus* the order is fixed, given the conditions of sporulation. However, with different media and with different types of Bacilli, the times of appearance of some of the biochemical events can be altered both in relation to other biochemical events and in relation to the morphological time scale. There is, in consequence, no clear correlation between morphology and biochemistry.

It has already been argued that a knowledge of the genetic map should enable us to distinguish between the models of sequential induction or sequential transcription. But it is obvious that an essential prerequisite is to know which events are essentially concerned in sporulation and to know what their real order is. For a start one should perhaps exclude from consideration those biochemical events which do not invariably occur at a particular morphological stage, irrespective of the medium and irrespective of the species of *Bacillus*.

(*d*) Whatever the mechanism which produces the order it results in the synthesis of a number of stable species of m-RNA. Although these are apparently produced in the same sequence as the proteins they code for, they are not immediately translated but remain, presumably inactive, in the cytoplasm for periods of an hour or longer. The fact that the RNA begins to be translated only after a fairly long period means that there must be regulation at this level in addition to that which is exerted at the level of the DNA and which determines whether the gene will be expressed in the first place.

The occurrence of stable m-RNA is probably not common in bacteria. Most experiments on inducible enzymes or done with actinomycin D are compatible with, and have been quoted as evidence for, the instability of m-RNA. However, the m-RNA molecules involved in sporulation appear, as we have seen, to be stable for long periods. Again there are parallels in higher organisms, particularly in the developing slime mould. Addition of actinomycin D at a fairly early stage (8–12 hr.) allows aggregation of cells to continue in the normal way, while later stages are inhibited by the drug. If the actinomycin is added at 14 hr., development proceeds for a further 4 hr. to the next stage and then stops. Similarly, later additions each allow a few hours of development to take place (Sussman, 1966 and this Symposium). The results indicate that here, as in *Bacillus subtilis*, there may be a series of points of commitment, each point corresponding to the formation of a stable message or an array of messages. In sporulating slime moulds there is also a time gap between the formation of the m-RNA and its translation into protein. Here, as in sporulating Bacilli, it seems that regulation exists both at the level of transcription and of translation. The same is true of the alga *Acetabularia* in which extensive developmental changes occur some weeks after removal of the nucleus of the cell (see Harris, 1968). In developing sea-urchins' eggs and in amphibian eggs there is again evidence that the m-RNA is stable and that the early stages of growth take place through regulation exerted at this level (see Gross & Cousineau, 1964; Duprat, Zalta & Beetschen, 1966).

This very brief summary suggests that, notwithstanding the enormous diversity in the organisms concerned, the bacterial cell, the yeast cell, the slime mould and even amphibia exhibit, as part of the process of differentiation, certain basic biochemical patterns which are common.

LIMITATIONS OF THE BIOCHEMICAL APPROACH

Despite the difficulties we have stressed it seems only a matter of time before we have a proper correlation of the biochemistry and morphology of sporulation; that is, we will know which biochemical events are relevant and we will know which morphological stages they are associated with. Once this point is reached—and it might be quite soon if enough biochemists became interested—we will still be left with the problem of knowing how a fluctuation in one particular group of chemical substances produces, or is produced by, say, the spore septum while another causes the chromosomes to form an axial filament.

This is a problem that will ultimately confront everyone who is trying to explain a morphological change—be it differentiation or cell division or pinocytosis—in biochemical terms. What seems to be needed is a great deal of information about the physical properties of large molecules in solution and in the presence of high concentrations of the solutes of the sort which are found in the cytoplasm. This is an aspect of physics that experimenters and theoreticians alike seem to have avoided.

As long as they continue to do so, our understanding of phenomena of this type will be limited to correlations which have no real explanatory value.

ACKNOWLEDGEMENTS

Plates 1 and 2 are reproduced by kind permission of the authors and the editors of the *Biochemical Journal*. I am indebted to Dr P. Schaeffer for a prepublication copy of his review and to many of my colleagues for advice and criticism.

REFERENCES

ARONSON, A. I. (1965). Characterization of messenger RNA in sporulating *Bacillus cereus*. *J. molec. Biol.* **11**, 576.

ARONSON, A. I. & DEL VALLE, M. R. (1964). RNA and protein synthesis required for bacterial spore formation. *Biochim. biophys. Acta*, **87**, 267.

BACH, M. L. & GILVARG, C. (1966). Biosynthesis of dipicolinic acid in sporulating *Bacillus megaterium*. *J. biol. Chem.* **241**, 4563.

BAILLIE, A. & NORRIS, J. R. (1964). Antigen changes during spore formation in *Bacillus cereus*. *J. Bact.* **87**, 1221.

BALASSA, G., IONESCO, H. & SCHAEFFER, P. (1963). Preuve génétique d'une relation entre la production d'un antibiotique par *Bacillus subtilis* et sa sporulation. *C. r. hebd. Séanc. Acad. Sci., Paris*, **257**, 986.

BENNETT, T. P. & FRIEDEN, E. (1962). Metamorphosis and biochemical adaptation in amphibia. In *Comparative Biochemistry*, **4**, 483. Ed. by M. Florkin and H. S. Mason. New York: Academic Press.

BERNLOHR, R. W. & NOVELLI, G. D. (1960*a*). Some characteristics of bacitracin production by *Bacillus licheniformis*. *Archs Biochem. Biophys.* **87**, 232.

BERNLOHR, R. W. & NOVELLI, G. D. (1960*b*). Uptake of bacitracin by sporangia and its incorporation into the spores of *Bacillus licheniformis*. *Biochim. biophys. Acta*, **41**, 541.

BERNLOHR, R. W. & NOVELLI, G. D. (1964). Bacitracin biosynthesis and spore formation: the physiological role of an antibiotic. *Archs Biochem. Biophys.* **103**, 94.

BHATTACHARYYA, P. & BOSE, S. K. (1967). Amino acid composition of cell wall and spore coat of *Bacillus subtilis*. *J. Bact.* **94**, 2079.

CANFIELD, R. E. & SZULMAJSTER, J. (1964). Time of synthesis of deoxyribonucleic acid and protein in spores of *B. subtilis*. *Nature, Lond.* **203**, 596.

CHASIN, L. A. & SZULMAJSTER, J. (1967). Biosynthesis of dipicolinic acid in *Bacillus subtilis*. *Biochem. biophys. Res. Commun.* **29**, 648.

DAWES, I. W. & MANDELSTAM, J. (1969). Biochemistry of sporulation of *Bacillus subtilis* 168: continuous culture studies. *Proceedings 4th Int. Symp. Continuous Cultivation of Micro-organisms* (*Prague*) (in the Press).

DOAK, B. W. & LAMANNA, C. (1948). On the antigenic structure of the bacterial spore. *J. Bact.* **55**, 373.

DRENTH, J. & HOL, W. G. J. (1967). A comparison of crystallographic data of the subtilopeptidase B and C. *J. molec. Biol.* **28**, 543.

DUPRAT, A.-M., ZALTA, J.-P. & BEETSCHEN, J.-C. (1966). Action de l'actinomycine D sur la différenciation de divers types de cellules embryonnaires de l'amphibien *Pleurodeles waltlii* en culture *in vitro*. *Expl Cell Res.* **43**, 358.

FALASCHI, A. & KORNBERG, A. (1966). Biochemical studies of bacterial sporulation. II. Deoxyribonucleic acid polymerase in spores of *Bacillus subtilis*. *J. biol. Chem.* **241**, 1478.

FITZ-JAMES, P. C. (1965). Spore formation in wild and mutant strains of *B. cereus* and some effects of inhibitors. *Colloq. int. Centre natl Rech. Sci., Marseille*, p. 529.

FORTNAGEL, P. & FREESE, E. (1968). Analysis of sporulation mutants. II. Mutants blocked in the citric acid cycle. *J. Bact.* **95**, 1431.

FOSTER, J. W. (1956). Morphogenesis in bacteria: some aspects of spore formation. *Q. Rev. Biol.* **31**, 102.

FOSTER, J. W. & PERRY, J. J. (1954). Intracellular events occurring during endotrophic sporulation in *Bacillus myscoides*. *J. Bact.* **67**, 295.

FREESE, E. & FORTNAGEL, P. (1967). Analysis of sporulation mutants. I. Response of uracil incorporation to carbon sources, and other mutant properties. *J. Bact.* **94**, 1957.

GARDNER, H. & KORNBERG, A. (1967). Biochemical studies of bacterial sporulation and germination. V. Purine nucleoside phosphorylase of vegetative cells and spores of *Bacillus cereus*. *J. biol. Chem.* **242**, 2383.

GRELET, N. (1957). Growth limitation and sporulation. *J. appl. Bact.* **20**, 315.

GROSS, P. R. & COUSINEAU, G. H. (1964). Macromolecule synthesis and the influence of actinomycin on early development. *Expl Cell Res.* **33**, 368.

HALVORSON, H. O. (1964). Genetic control of enzyme synthesis. *J. exp. Zool.* **157**, 63.

HALVORSON, H. O. (1965). Sequential expression of biochemical events during intracellular differentiation. *Symp. Soc. gen. Microbiol.* **15**, 343.

HALVORSON, H. O., VARY, J. C. & STEINBERG, W. (1966). Developmental changes during the formation and breaking of the dormant state in bacteria. *A. Rev. Microbiol.* **20**, 169.

HANSON, R. S., BLICHARSKA, J. & SZULMAJSTER, J. (1964). Relationship between the tricarboxylic acid cycle enzymes and sporulation in *Bacillus subtilis*. *Biochem. biophys. Res. Commun.* **17**, 1.

HANSON, R. S. & COX, D. P. (1967). Effect of different nutritional conditions on the synthesis of tricarboxylic acid cycle enzymes. *J. Bact.* **93**, 1777.

HANSON, R. S., SRINIVASAN, V. R. & HALVORSON, H. O. (1963). Biochemistry of sporulation. I. Metabolism of acetate by vegetative and sporulating cells. *J. Bact.* **85**, 451.

HARDWICK, W. A. & FOSTER, J. W. (1952). On the nature of sporogenesis in some aerobic bacteria. *J. gen. Physiol.* **35**, 907.

HARRIS, H. (1968). *Nucleus and Cytoplasm*. Oxford: Clarendon Press.

JACOB, F., SCHAEFFER, P. & WOLLMAN, E. L. (1960). Episomic elements in bacteria. *Symp. Soc. gen. Microbiol.* **10**, 67.

JIČÍNSKÁ, E. (1964). Sporulation of auxotrophic mutants of *Bacillus subtilis* in suboptimal concentrations of essential amino acids. *Folia Microbiol.* **9**, 73.

KAY, D. & WARREN, S. C. (1968). Sporulation in *Bacillus subtilis*. Morphological changes. *Biochem. J.* **109**, 819.

LAMPEN, J. O. (1965). Secretion of enzymes by micro-organisms. *Symp. Soc. gen. Microbiol.* **15**, 115.

LAWRENCE, N. & HALVORSON, H. O. (1954). Studies on the spores of aerobic bacteria. IV. A heat-resistant catalase from spores of *Bacillus terminalis*. *J. Bact.* **68**, 334.

MANDELSTAM, J. (1960). The intracellular turnover of protein and nucleic acids and its role in biochemical differentiation. *Bact. Rev.* **24**, 289.

MANDELSTAM, J. & STERLINI, J. M. (1969). Commitment to sporulation in *Bacillus subtilis*. In *Spores*, IV, (in the Press).

MANDELSTAM, J. & WAITES, W. M. (1968). Sporulation in *Bacillus subtilis*. The role of exo-protease. *Biochem. J.* **109**, 793.

MANDELSTAM, J., WAITES, W. M. & WARREN, S. C. (1967). Exo-protease activity and its relationship to some other biochemical events involved in the sporulation of *Bacillus subtilis. Abstr. 7th int. Congr. Biochem., Tokyo*, Symp V–2, 1, 253.

McCONN, J. D., TSURU, D. & YASUNOBU, K. T. (1964). *Bacillus subtilis* neutral proteinase. I. A zinc enzyme of high specific activity. *J. biol. Chem.* **239**, 3706.

MILLER, J. J. & HOFFMANN-OSTENHOF, O. (1964). Spore formation and germination in *Saccharomyces. Z. allgem. Mikrobiol.* **4**, 273.

MILLET, J. (1963). La phosphatase alcaline de *Bacillus megaterium. C. r. hebd. Séanc. Acad. Sci., Paris*, **257**, 784.

MONRO, R. E. (1961). Protein turnover and the formation of protein inclusions during sporulation of *Bacillus thuringiensis. Biochem. J.* **81**, 225.

MURRELL, W. G. (1967). The biochemistry of the bacterial endospore. *Advanc. Microb. Physiol.* **1**, 133.

NORRIS, J. R. (1962). Bacterial spore antigens: a review. *J. gen. Microbiol.* **28**, 393.

NORRIS, J. R. & BAILLIE, A. (1964). Immunological specificities of spore and vegetative cell catalase of *Bacillus cereus. J. Bact.* **88**, 264.

OHYE, D. F. & MURRELL, W. G. (1962). Formation and structure of the spore of *Bacillus coagulans. J. cell Biol.* **14**, 111.

POWELL, J. F. (1953). Isolation of dipicolinic acid (pyridine-2:6-dicarboxylic acid) from spores of *Bacillus megaterium. Biochem. J.* **54**, 210.

POWELL, J. F. & STRANGE, R. E. (1956). Biochemical changes occurring during sporulation in *Bacillus* species. *Biochem. J.* **63**, 661.

ROGERS, H. J., LOMAKINA, N. & ABRAHAM, E. P. (1965). Observations on the structure of bacilysin. *Biochem. J.* **97**, 579.

ROGOLSKY, M. & SLEPECKY, R. A. (1964). Elimination of a genetic determinant for sporulation of *Bacillus subtilis* with acriflavin. *Biochem. biophys. Res. Commun.* **16**, 204.

ROUYARD, J.-F., IONESCO, H. & SCHAEFFER, P. (1967). Classification génétique de certains mutants de sporulation de *Bacillus subtilis*, Marburg. *Annls Inst. Pasteur, Paris*, **113**, 675.

RYTER, A. (1965). Étude morphologique de la sporulation de *Bacillus subtilis. Annls Inst. Pasteur, Paris*, **108**, 40.

RYTER, A., SCHAEFFER, P. & IONESCO, H. (1966). Classification cytologique, par leur stade de blocage, des mutants de sporulation de *Bacillus subtilis* Marburg. *Annls Inst. Pasteur, Paris*, **110**, 305.

SCHAEFFER, P. (1967). Asporogenous mutants of *Bacillus subtilis* Marburg. *Folia Microbiol*, **12**, 291.

SCHAEFFER, P. (1968). Sporulation and the production of antibiotics, exoenzymes and exotoxins. *Bact. Rev.* (in the Press).

SCHAEFFER, P., IONESCO, H., RYTER, A. & BALASSA, G. (1965). La sporulation de *Bacillus subtilis*: étude génétique et physiologique. *Colloq. int. Centre natl Rech. Sci., Marseille*, p. 553.

SCHAEFFER, P., MILLET, J. & AUBERT, J-P. (1965). Catabolic repression of bacterial sporulation. *Proc. natn. Acad. Sci. U.S.A.* **54**, 704.

SNOKE, J. E. (1964). Amino acid composition of *Bacillus licheniformis* spore coat. *Biochem. biophys. Res. Commun.* **14**, 571.

SPIZIZEN, J. (1965). Analysis of asporogenic mutants in *Bacillus subtilis* by genetic transformation. In *Spores*, III, p. 125. Ed. L. L. Campbell and H. O. Halvorson. Ann Arbor, Michigan: American Society for Microbiology.

SRINIVASAN, V. R. (1965). Intracellular regulation of sporulation of bacteria. In *Spores*, III, p. 64. Ed. L. L. Campbell and H. O. Halvorson. Ann Arbor, Michigan: American Society for Microbiology.

STANIER, R. Y. (1951). Enzymatic adaptation in bacteria. *A. Rev. Microbiol.* **5**, 35.

STEWART, B. T. & HALVORSON, H. O. (1954). Studies on the spores of aerobic bacteria. II. The properties of an extracted heat-stable enzyme. *Archs Biochem. Biophys.* **49**, 168.

SUSSMAN, M. (1966). Some genetic and biochemical aspect of the regulatory program for slime mold development. *Current Topics in Developmental Biology*, **1**, 61.

SZULMAJSTER, J. (1964). Biochimie de la sporogénèse chez *B. subtilis*. *Bull. Soc. Chim. biol.* **46**, 443.

SZULMAJSTER, J., CANFIELD, R. E. & BLICHARSKA, J. (1963). Action de l'actinomycine *D* sur la sporulation de *Bacillus subtilis*. *C. r. hebd. Séanc. Acad. Sci., Paris*, **256**, 2057.

TAKAHASHI, I. (1965). Localization of spore markers on the chromosome of *Bacillus subtilis*. *J. Bact.* **89**, 1065.

TONO, H. & KORNBERG, A. (1967). Biochemical studies of bacterial sporulation. III. Inorganic pyrophosphatase of vegetative cells and spores of *Bacillus subtilis*. *J. biol. Chem.* **242**, 2375.

VINTER, V. (1959). Sporulation of bacilli. VII. The participation of cysteine and cystine in spore formation by *Bacillus megaterium*. *Folia Microbiol.* **4**, 216.

WAITES, W. M. (1968). Sporulation in *Bacillus subtilis*. Antigenic changes during spore formation. *Biochem. J.* **109**, 803.

WARREN, S. C. (1968). *Sporulation in Bacillus subtilis*. Biochemical changes. *Biochem. J.* **109**, 811.

WISE, J., SWANSON, A. & HALVORSON, H. O. (1967). Dipicolinic acid-less mutants of *Bacillus cereus*. *J. Bact.* **94**, 2075.

WRIGHT, B. E. (1966). Multiple causes and controls in differentiation. *Science, N.Y.* **153**, 803.

YOUNG, E. I. & FITZ-JAMES, P. C. (1959*a*). Chemical and morphological studies of bacterial spore formation. I. Formation of spores in *Bacillus cereus*. *Biophys. biochem. Cytol.* **6**, 467.

YOUNG, E. I. & FITZ-JAMES, P. C. (1959*b*). Chemical and morphological studies of bacterial spore formation. II. Spore and parasporal protein formation in *Bacillus cereus* var. *Alesti*. *Biophys. biochem. Cytol.* **6**, 483.

YOUNG, E. I. & FITZ-JAMES, P. C. (1959*c*). Chemical and morphological studies of bacterial spore formation. III. The effect of S-azaguanine on spore and parasporal protein formation in *Bacillus cereus* var. *Alesti*. *Biophys. biochem. Cytol.* **6**, 499.

EXPLANATION OF PLATES

Stages in sporulation of *Bacillus subtilis* (Kay & Warren, 1968). Bacteria were fixed, sections were cut, and stained with lead citrate.

PLATE 1

(1) Stage I. The cell has two nuclei which have formed an axial filament (*A*).

(2). Stage II. A spore septum (*B*) has formed at one end of the cell. It is thinner than the division septum (*C*).

(3, 4). Stage III. The septum (*B*) bulges inwards and synthesis of the membrane extends towards the pole of the cell until a completed membrane has been formed which is detached from that of the mother cell. Nuclear material, appearing white, is enclosed in the developing spore which also contains a mesosome (*D*).

PLATE 2

(1). Stage IV. The cortex (*G*) is beginning to be formed and vesicles (*J*) can be seen on the membrane. The spore coat (*H*) is becoming visible.

(2). Stage V. The coat (*H*) develops as a layered structure containing granules and a less dense coat (*K*) also begins to form.

(3). Stage V (*cont.*). The layered spore coat is much denser and, at the same time, less detail is discernible inside the spore. The spore membrane (*E*) can, however, still be seen.

(4). Stage VI. The spore shows much more plainly the layered structures (*H* and *K*) of the outer and inner spore coats, and it has the typical corrugated appearance of the mature spore.

PLATE 1

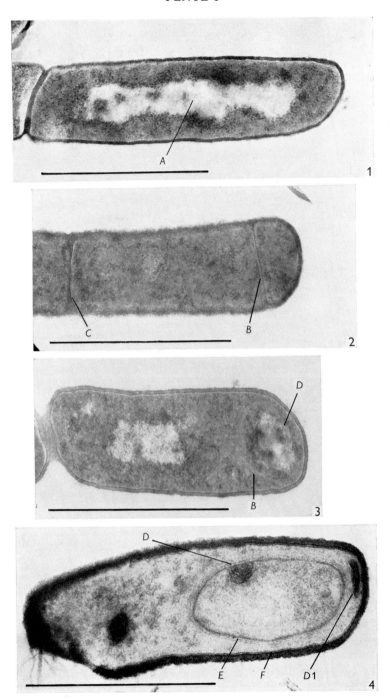

PLATE 2

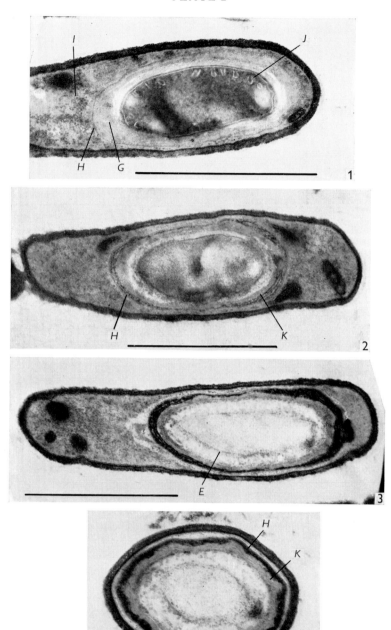

PATTERNS OF RNA SYNTHESIS AND OF ENZYME ACCUMULATION AND DISAPPEARANCE DURING CELLULAR SLIME MOULD CYTODIFFERENTIATION

MAURICE SUSSMAN AND RAQUEL SUSSMAN

Department of Biology, Brandeis University, Waltham, Mass. 02154, U.S.A.

I used to smile at biologists who spend their entire lives studying the differentiation of slime moulds or the behaviour of flatworms, but I have learnt better now.

MAX FERDINAND PERUTZ[1]

INTRODUCTION

Modern biology has provided us with a formal model to account for the expression of a phenotypic character by a cell or a multicellular assembly. According to this model, a specific region of the DNA is transcribed into RNA and this in turn is translated into a particular protein or proteins by the appropriate synthetic machinery. The protein(s) so fabricated acts and is acted upon in a specific manner within a matrix of activities of all the other cell proteins and thereby creates the character we observe.

It is also imagined that a set of controls operates upon this succession of events and stipulates when (in the life cycle of the cell or cell assembly) the phenotypic character will in fact be expressed; where (topographically) it will emerge; to what extent (quantitatively) it will occur. In principle these controls might be exerted at all levels of the succession, i.e. to regulate the transcription of DNA into RNA, and/or the translation of RNA into protein, and/or the activity of the protein as an enzyme, transport agent, hormone, antibody, structural subunit, etc.

How much do we now understand about these controls? A fair amount in the case of transient phenotypic expressions in bacteria (Gorini, 1963; Jacob & Monod, 1964; Gilbert & Müller-Hill, 1966; Ames & Hartman, 1963). Thus, the genes for groups of functionally related activities are themselves physically clustered and single, multicistronic RNA messages are transcribed from these clusters. Whether a given message will or will not be fabricated is determined by

---

[1] 'The living organism', *Encycl. of the Life Sciences*, I, Doubleday and Co., 1965.

one or more regulatory genes whose protein products can interfere with the initiation of the transcriptive act. Small molecular weight effectors can alter the activity of the regulatory proteins thereby permitting parts of the metabolic machinery to exert feed-back controls upon the genes which govern them and these effectors can also intervene directly in the activity of the metabolic machinery by allosteric effects upon the component proteins (Umbarger, 1956; Monod, Changeaux & Jacob, 1963; see Clarke & Lilly, this Symposium). The studies of bacteria also show that the m-RNA once fabricated is very unstable and is destroyed by first-order kinetics with a half life of a few minutes at most (Levinthal, Fan, Higa & Zimmerman, 1963). All of this confers upon the system enormous flexibility in phenotypic expression, a quality obviously necessary if bacterial cells are to be able to adjust rapidly to the environmental fluctuations which they ordinarily face during their natural existence.

In contrast to the above, we have only begun to acquire a detailed understanding of the control mechanisms involved in more complex, programmed sequences of phenomic expression, especially those exhibited by multicellular assemblies of eucaryotic cells (i.e. fruiting-body construction and other colonial organizations in Protista and Lower Metazoa and Metaphytes, embryogenesis in plants and animals, metamorphosis, regeneration, etc.) but also corresponding processes in solitary pro- and eucaryotes (i.e. the cell-division cycle, sporulation and encystation, viral development, etc.). We do not know if the controls that bacteria employ for transient, aperiodic alterations of phenotype are applicable to the coordinated sequences of phenotypic expression mentioned above in which the individual steps must occur in a fixed chronology and independent of external cues (beyond the initial signal which triggers the entire proceedings). We are not even sure at what levels the controls operate. If there is any sort of physical relationship among groups of genes involved in specific developmental programmes (which might for example help to explain the temporal relationships among the individual events), the pattern has not even begun to be elucidated except in the case of phage development (Epstein et al. 1963).

In recent years the cellular slime moulds have emerged as prime material for studies of this kind. They are eucaryotic soil protists, easily cultivated either axenically or in association with bacteria. They cooperate to construct organized multicellular assemblies with functionally and structurally differentiated components, complicated enough to be interesting but simple enough to be presently capable of genetic, biochemical, and morphological definition. The morphogenetic sequence can be carried out under rigorously defined experimental conditions by synchronized cell populations in sufficient numbers to permit the application of conventional biochemical procedures. Because the morphogenesis begins only after growth ceases, the metabolic machinery of the one can be studied unencumbered by the operation of the other and mutant strains have been isolated that grow normally but exhibit any of a wide variety of developmental deficiencies and aberrations (Sussman, 1966a, b). The growing cells are capable of genetic recombination (Sussman & Sussman, 1963), probably parasexual in nature, and thus provide the potential for genetic definition of the developmental programme.

At present the questions being asked of this developmental cycle are simple, naïve ones. What proteins are developmentally regulated and can serve as useful markers in studying the molecular bases of the regulatory programme? To what extent are the intra- and extracellular concentrations of small molecular weight metabolites involved in regulating the activities of these proteins by allosteric effects, protection against degradation, mass action, etc.? To what extent are RNA and protein syntheses involved in the observed patterns of enzyme accumulation and disappearance and can one delineate relevant periods of genetic transcription and translation?

The present communication will attempt to review some of the answers to some of these questions and will summarize in detail recent investigations carried out in our own laboratory.[1]

THE DEVELOPMENTAL CYCLE

Figure 1 schematically summarizes the gross morphogenetic aspect of cellular slime mould development carried out under the experimental conditions employed in our laboratory. *Dictyostelium discoideum* has thus far been the species of choice, but *Polysphondylium pallidum* has also been employed. They are both grown either in association with *Aerobacter aerogenes* or axenically (Sussman, 1966a; Sussman & Sussman, 1967) in complex liquid media containing components such as proteose peptone, yeast extract, etc. When the amoebae attain the stationary growth phase, they are harvested, washed, and resuspended and aliquots are dispensed on a solid substratum so that they may begin the morphogenetic sequence. The substratum is a 47 mm. Milli-

[1] The data reported here represent the combined efforts of a number of individuals notably including J. M. Ashworth, J. Inselburg, W. F. Loomis Jr., M. J. Osborn, E. Rayner and R. Roth and, more recently, S. M. Cocucci, J. Ellingson, J. Franke, P. C. Newell and M. Schwalb.

pore filter resting on an absorbent pad saturated with a buffer–salts–streptomycin solution inside a 60 mm. Petri dish. The dishes are then incubated at 22°C. in a water-saturated atmosphere. Each filter supports 10^8 cells (*ca.* 5 mg. dry weight at the start) which then proceed through the morphogenetic sequence as shown with a very high degree of synchrony to form *ca.* 1000 fruiting bodies of approximately equal size (Sussman, 1966 *a*). The Millipore filters have recently been replaced by Whatman no. 50 filters, which are less expensive and more convenient.

As shown, the cells initially appear as a homogeneous lawn but then collect into aggregates, at first amorphous blobs, but subsequently conical, organized structures. These extend apically into fingers, then

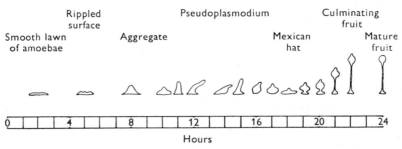

Fig. 1. The developmental cycle of *Dictyostelium discoideum* (Sussman & Sussman, 1967). The scale shows the hours of development.

settle down laterally into migrating slugs or pseudoplasmodia which are sensate, responding in a purposeful manner to light, heat (Bonner, Clark, Neeley & Slifkin, 1950), and pH (P. C. Newell & M. Sussman, unpublished data) and display evidence of both histological and functional organization (Raper, 1940; Bonner, Chiquoine & Kolderie, 1955). After rising apically once again, the slugs proceed through a complex sequence of morphogenetic movements to construct fruiting bodies consisting of a mass of spores at the top and a cellulose ensheathed, cellular, stalk below, all resting upon a basal disc. The spores, stalk and basal disc cells are differentiated biochemically, morphologically and functionally.

As a substitute for the filter-paper system, washed cells can be spread on non-nutrient agar surfaces in order to construct fruiting bodies (Wright & Anderson, 1960; White & Sussman, 1961). However, the sequence takes *ca.* 8 hr. longer under these conditions and the development is not as synchronous.

Alternatively, one can examine single aggregates, migrating slugs and fruits by sensitive immunochemical and histochemical assays (Gregg, 1965).

SOME BIOCHEMICAL ACTIVITIES ACCOMPANYING MORPHOGENESIS AND SOME QUESTIONS CONCERNING THEM

The morphogenetic sequence can begin only after cessation of growth and is accompanied by at most a negligible level of nuclear and cell division (Bonner & Frascella, 1952; Sussman & Sussman, 1960). At any time, if the developing multicellular aggregates are dispersed and nutrient materials (such as the bacterial prey) are reintroduced, the cells revert to the vegetative state (Raper, 1956). The nature of the signal that shifts the cells from growth-associated activities to morphogenetic ones is not clear. Speculation might centre about the levels of various catabolites as being the determining factors. One might equally well look to the DNA synthesizing apparatus for the location of the big switch, particularly to the availability of deoxythymidine or its phosphorylated derivatives (Hotta & Stern, 1961). Thus, the prolonged absence of the latter might trigger the start of the morphogenetic programme. Still other possibilities exist. As long as the amoebae had to be grown in association with bacteria a rigorous approach to this problem was impractical. The availability of axenic media for at least two species now makes it possible.

The normal process of fruiting body construction can be carried out in the absence of any exogenous materials. Hence, the energy and raw materials required for cell movements, morphogenesis, and biosyntheses can be supplied exclusively by catabolic turnover of pre-existing endogenous constituents. As a result, the cells lose a considerable fraction of their dry weight (approaching 50%, and sometimes even more) during the course of morphogenetic sequence (Gregg & Bronsweig, 1956; Wright & Anderson, 1960; White & Sussman, 1961). This includes proportionate losses in protein and RNA content.

A detailed catabolic map is not yet available. For example, the pathways of nucleotide catabolism are completely unknown. However, some parts of the map have been investigated. A high level of glutamate dehydrogenase activity has been detected and the carbon skeleton apparently feeds through an operative Krebs cycle (Brühmüller & Wright, 1963). Presumably other amino acid dehydrogenases or deaminases exist, since ammonia is evolved in considerable amounts

(Gregg & Bronsweig, 1965). Certainly the observed net loss in protein content makes it clear that amino acid oxidation must be a major catabolic pathway. Furthermore, isotopic studies suggest an increase in the efficiency with which carbon dioxide is evolved by amino acid degradation during the morphogenetic sequence and a concomitant decrease in its evolution from glucose (Wright, 1964). A complete glycolytic pathway via lactate dehydrogenase has been demonstrated (Cleland & Coe, 1968) but it is not clear whether it operates in the direction of glycolysis or gluconeogenesis or both. At least one shunt enzyme, glucose-6-phosphate dehydrogenase, has been detected. The details of protein and RNA breakdown are also lacking. Vegetative amoebae are meat-eaters and contain proteolytic and nucleolytic enzymes and phosphatases probably in vesicular compartments (Gezelius, 1968; D. R. Patterson, unpublished data). The question is whether during morphogenesis these enzymes assist in the turnover of endogenous cell components as well? Finally, very little is known about the electron transport system. This is particularly vital information, since, while vegetative amoebae grow as facultative anaerobes, the morphogenetic sequence is known to be obligately aerobic (Bradley, Sussman & Ennis, 1956) and there is no current understanding of how profound a metabolic shift this fact may reflect.

The process of cell aggregation comprises two different sets of activities: adhesion and chemotaxis. The adhesion of the cells is specific and involves cell recognition and sorting out phenomena reminiscent of those observed in embryonic cell aggregates (Moscona, 1961). Concomitant changes in the cell surface have been observed. These include the acquisition of specific areas of 'stickiness' (Mercer & Shaffer, 1960) and the appearance of a new lipoprotein antigen in the cell membrane fraction (Sonneborn, Sussman & Levine, 1964). Only species which coaggregate with *Dictyostelium discoideum* contain cross-reacting antigens and the presence of the antibody in the agar substratum specifically inhibited aggregation of these species without disturbing cell viability. The second set of activities involves the chemotactic attraction of the cells when they are widely separated and must first collect together before they can come in contact with and stick to one another. The chemotactic agent had previously been detected (Shaffer, 1956), and shown to involve a complex of at least two substances, one of which was converted into the other by an extracellular enzyme (Sussman, Lee & Kerr, 1956). Recent studies have demonstrated that the amoebae are strongly attracted by very low concentrations of 3',5'-cyclic AMP and contain significant quantities of it at the time of aggregation (Konijn, Van de Meene, Bonner & Barkley, 1967). They also appear to release an extracellular phosphodiesterase (Chang, 1968) during this period. The possibility therefore exists that the chemotactic complex may consist of this enzyme plus its substrate 3′,5′-cyclic AMP and its product, 5′-AMP. An adenylcyclase has yet to be demonstrated. It is possible that the latter might be the cell surface antigen described above, since adenylcyclases have invariably

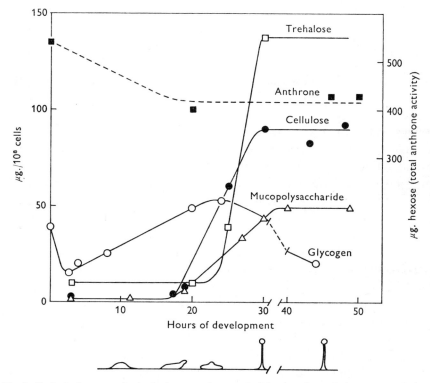

Fig 2. Carbohydrate synthesis during morphogenesis. The data for trehalose accumulation were published in tabular form by Ceccarini & Filosa (1965). The remainder are drawn from White & Sussman (1961, 1963 a, b). These experiments were performed with cells incubated on non-nutrient agar.

been found to be cell membrane bound (Sutherland, Øye & Butcher, 1965). Clearly this entire subject represents a most important area of investigation, not only from the point of view of slime mould chemotaxis but because of the widespread implication of cyclic AMP in both microbial (Perlman & Pasten, 1968) and metazoan (Sutherland *et al.* 1965) control systems.

Once the cells have entered into multicellular aggregates, carbohydrate metabolism represents one of their major-known activities. They synthesize at least three polysaccharides. The first is cellulose (Raper & Fennell, 1952; Gezelius & Ranby, 1957), which, in the mature fruiting bodies, constitutes the integument of the stalk and is also present in the spore coats. The second is a glucose polymer resembling glycogen (White & Sussman, 1963a). Though present early in the sequence as a water soluble, TCA soluble component, it is also found later in a TCA insoluble, alkali insoluble fraction complexed with cellulose with which it serves as part of the stalk sheath and spore coat material (Wright, 1966). The third is an acid mucopolysaccharide, composed of galactose, N-acetylgalactosamine, and galacturonic acid, found associated exclusively with the spores in the mature fruit (White & Sussman, 1963b). In addition, a considerable amount of the disaccharide trehalose (glucose $1,1,\alpha$-D-glucoside) accumulates in the spores and apparently serves as a carbon and energy source during germination (Clegg & Filosa, 1961; Ceccarini & Filosa, 1965).

Figure 2 summarizes the kinetics and logistics of these synthetic events. It will be seen that 10^8 cells (ca. 5 mg. dry weight) start the morphogenetic sequence with ca. 500 μg. of anthrone reactive material and end up with ca. 400 μg. in the mature fruit. Of this total more than 300 μg. can be accounted for as trehalose (ca. 6% of the dry weight), cellulose (ca. 4–5%), and glycogen and mucopolysaccharide (ca. 2% each). As noted elsewhere (Cleland & Coe, 1968), interconversion of pre-existing sugars would seem to be sufficient for these needs and gluconeogenesis may be an activity of minor significance.

One of the last synthetic activities in *Dictyostelium discoideum* is the elaboration of a lemon yellow pigment associated with the spore mass and identified as a zeta carotene derivative (Staples & Gregg, 1967). Presumably then a biosynthetic pathway which encompasses such intermediates as mevalonic acid, isopentenyl pyrophosphate, geranyl pyrophosphate, etc., is operative at this time.

Syntheses of RNA and protein are maintained at significant rates throughout the morphogenetic sequence and do not subside until the appearance of the mature fruiting body. These activities will be described in detail in succeeding sections but some rough estimates of magnitude are given below.

In *Polysphondylium pallidum* 10^8 vegetative amoebae contain about 750 μg. total RNA (ca. 15% of the dry weight). During the entire fruiting process ca. 500 μg. of this is degraded. At least 250 μg. of new RNA is synthesized over the same period so that, in the mature fruit, the cells contain ca. two-thirds of the total amount of RNA they used to have, of which about 50 % is newly fabricated. Since the latter is not completely stable either, it is clear that significantly more than 250 μg. RNA is in fact synthesized (R. R. Sussman, 1967). The same level of synthesis probably operates in *Dictyostelium discoideum* as well (Inselburg & Sussman, 1967).

An estimate of the extent of protein turnover in *Dictyostelium discoideum* has been made by Wright & Anderson (1960) by isotope dilution experiments involving ^{35}S-methionine. According to their data, the turnover rate remained approximately constant at *ca.* 7 % per hour throughout the sequence. Assuming a first-order decay, as was observed for total protein by White & Sussman (1961), this would mean that slightly less than 12 % of the protein originally present in the vegetative cells could have remained intact by the end of fruiting body construction. Thus 10^8 cells having started out with 3000 μg. protein would have retained only *ca.* 350 μg. of it intact and would have replaced the degraded fraction with approximately 1650 μg. of newly synthesized protein. There is good reason to believe that the latter also turns over.

At a minimum, then, the RNA synthesized during morphogenesis comprises *ca.* 8 % of the dry weight of the mature fruit and the newly synthesized protein may exceed 55 %.

THE DEVELOPMENTAL KINETICS OF SEVERAL SLIME MOULD ENZYMES

Figure 3 shows the developmental kinetics of three enzymes which have been characterized in detail in our laboratory. The specific enzyme activities were determined in crude extracts of cells harvested after different periods of incubation on the filters as described above and are expressed as percentages of their respective peak activities.

(*a*) *Trehalose-6-phosphate synthetase*. As in yeast (Cabib & Leloir, 1958), trehalose is synthesized by the following reactions:

$$\text{G-6-P} + \text{UDPG} \longrightarrow \text{trehalose-6-P} + \text{UDP}, \qquad \text{(I)}$$

$$\text{Trehalose-6-P} + \text{H}_2\text{O} \longrightarrow \text{trehalose} + \text{P}_i. \qquad \text{(II)}$$

Reaction (I) is catalysed by the synthetase; reaction (II) by a phosphatase, present in excess in slime mould extracts. The synthetase has been assayed by measuring the rate of UDP generation and also by following ^{14}C-glucose transfer from either UDPG or G-6-P to trehalose (Roth, Ashworth & Sussman, 1968; Roth & Sussman, 1968), both assays yielding identical results. As shown, the synthetase activity is undetectable in vegetative amoebae and appears initially after 5 hr. incubation, i.e. at the first stages of cell aggregation. The specific activity rises linearly to a peak at 16–17 hr. (17 mμmoles/min./mg. total cell protein at 37° C.) and then declines to a negligible level.

(*b*) *UDPG pyrophosphorylase.* In view of the volume of polysaccharide and disaccharide synthesis that attends the fruiting process,

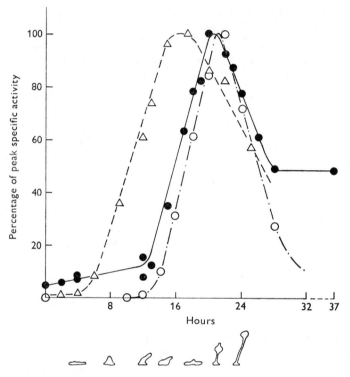

Fig. 3. The developmental kinetics of three enzymes in *Dictyostelium discoideum*: trehalose-6-phosphate synthetase ($\triangle$); UDP glucose pyrophosphorylase ($\bullet$); UDP galactose: polysaccharide transferase ($\circ$). (Roth, Ashworth & Sussman, 1968.)

it might be expected that UDPG would be a major metabolite. As in other organisms, its synthesis is catalysed by a pyrophosphorylase according to the following reaction:

$$G\text{-}1\text{-}P + UTP \rightleftharpoons UDPG + PP_i.$$

The pyrophosphorylase has been assayed in slime mould extracts in the forward direction (Ashworth & Sussman, 1967) by following the conversion of [^{14}C]glucose-1-P to UDPG-[^{14}C] and in the reverse direction by measuring PP_i-dependent glucose-1-phosphate generation (Newell,

Ellingson & Sussman, 1968; Newell & Sussman, 1968). Both assays yielded the results shown. The vegetative amoebae display a low level of activity which for a time remains constant or perhaps rises slightly. After 11–12 hr. incubation, i.e. at the end of cell aggregation, the specific activity begins to increase linearly over an 8 hr. period, reaching a peak at 20 hr. (*ca.* 250 mμmoles/min./mg. total cell protein in the forward direction at 37° C.; *ca.* 500 mμmoles/min./mg. in the reverse direction), at which time the cells are in the terminal stages of fruit construction. Part of this activity, the fraction associated with the stalk cells, disappears. The remainder, associated exclusively with the spores, is retained (Ashworth & Sussman, 1967).

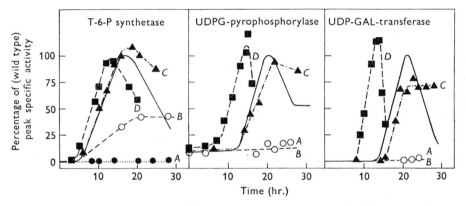

Fig. 4. Developmental kinetics of the three enzymes (see Fig. 3) in morphogenetically aberrant mutants. Solid lines represent the patterns observed in the wild type and are taken from Fig. 3. Dotted lines are fitted to data obtained with mutants representative of the four classes (*A, B, C, D*) described in the text. (Sussman & Osborn, 1964; Yanagisawa, Loomis & Sussman, 1967; Ashworth & Sussman, 1967; Roth & Sussman, 1968.)

(*c*) *UDP galactose:polysaccharide transferase.* This enzyme is responsible for the incorporation of galactose into the mucopolysaccharide described above. It is assayed by following the transfer of galactose-^{14}C from UDP Gal to the mucopolysaccharide acceptor which is present in standard concentration (Sussman & Osborn, 1964; Loomis & Sussman, 1966). The activity is undetectable in vegetative amoebae, first appears at 12·5 hr. and then accumulates over an 8 hr. period to an activity level of 5 mμmoles galactose transferred/hr./mg. total cell protein at 30° C. (in the presence of excess mucopolysaccharide acceptor). Within the next hour or two virtually all the enzyme formerly associated with cells is released by them into the extracellular space, i.e. the activity is now found in the supernatant fluid when cells at this stage are harvested in buffer and centrifuged (Sussman & Lovgren, 1965). The release is preferential, i.e. the cells jettison all of the activity within a quantity of protein comprising *ca.* 3 % of their total. The released enzyme is then rapidly destroyed as shown in Fig. 3. In contrast, the mucopolysaccharide product remains associated with the cells long after the enzyme has been extruded. The general significance of this release phenomenon for processes of cellular differentiation is discussed elsewhere (Sussman, 1965*a*).

The alterations in enzyme activities and locations described above appear to be controlled by the over-all developmental programme. As shown in Fig. 4, mutants which are developmentally aberrant display changes in the normal patterns of enzyme accumulation and disappearance consistent with the stages at which the particular morphogenetic aberration is manifested. The performances of these mutants can be grouped as follows.

A (example: AGG-206). This mutant strain grows normally but, after entrance into the stationary phase, the cells make only an abortive attempt to aggregate and then stop. They accumulate none of the enzymes in question.

B (examples: Agg-204; Fr-2; KY-11). Agg-204 proceeds through the initial stages of cell aggregation, a few hours further along than Agg-206, and then stops. Fr-2 and KY-11 construct conical aggregates but stop development short of the pseudoplasmodial stage. As seen in Fig. 4, such mutants can accumulate trehalose-6-phosphate synthetase activity but do not lose it; they do not accumulate the other two enzymes.

C (examples: KY-3; KY-19). KY-3, when incubated under the same conditions as the wild-type, produces normal migrating pseudoplasmodia which never construct fruiting bodies. In this strain, the synthetase activity accumulates and disappears in normal fashion; both UDPG pyrophosphorylase and UDP Gal transferase appear at approximately the normal time and accumulate at approximately the normal rate but neither one disappears. Strain KY-19 proceeds through the precursor stages in normal fashion but then produces a fruiting body with a spore mass but no stalk. This strain accumulates UDPG pyrophosphorylase activity to the usual level but having few stalk cells loses none of it. Recently Loomis (1968*a*) has examined a mutant that constructs a normal fruiting body within which the final transformation into spores is not accomplished. In this strain the UDP Gal transferase accumulates, is preferentially released, and disappears just as in the wild-type.

D (example: Fr-17). This is a temporally deranged mutant in which morphogenetic events generally occur and specific products accumulate too soon and too fast in comparison with the wild-type. (It is also topographically deranged, ultimately producing flat, papillated masses in which spores and stalk cells are mixed in chaotic disarray.) As shown in Fig. 4, all three enzymes accumulate and disappear too soon and too fast.

A consistent pattern emerges from the survey described above. It is seen that the regulatory programmes for the three enzymes are affected in a manner consistent with the temporal relationships among them. For example, some mutants accumulate an early enzyme but not a later one but no mutant which accumulates a later enzyme fails to accumulate the earlier ones. Secondly, the steps in the regulatory programme for any given enzyme are affected in a manner consistent with the timing of the morphogenetic block in the mutant strain, i.e. a mutant blocked at a particular morphogenetic stage can exert those regulatory controls which are normally exerted by the wild-type prior to that block but not those normally exerted after it. All of this implies that the developmental performance of a cell, though it may be triggered by an initial external signal, is in fact guided by a continuing sequence of internal cues without which a variety of subsequent steps cannot be taken.

A brief digression into methodology

How significant are the observed changes in specific enzyme activity in the wild-type and mutant strains? In terms of reproducibility under the experimental and assay conditions employed, $\pm$ *ca.* 10–15 % with respect to the absolute levels of specific activity and $\pm$ 0·5 hr. with respect to the times at which activities begin to increase or the peaks are reached. However, the question can be asked, to what extent do the measurements of enzyme activity in crude cell extracts accurately reflect *in vivo* enzyme concentrations? Specifically, it might and has been argued (Wright & Dahlberg, 1968) that dramatic increases in enzyme activity observed in this manner during development could conceivably be due to either (*a*) differential instability of the enzyme in crude extracts of cells harvested early in development as contrasted with those harvested late; and/or (*b*) the presence of inhibitors in early cell extracts which mask the activity. A careful examination of these arguments has recently been made by Newell & Sussman (1968) in connection with UDPG pyrophosphorylase and they do not seem to be significant. This conclusion is reached on the following grounds.

(*a*) Regardless of developmental stage, the pyrophosphorylase activity of crude cell extracts remains constant in the reaction mixture[1] for at least 30 min. at 37°C. (The assay itself requires *ca.* 5 min.). When cells at early or late developmental stages were harvested and broken in the complete reaction mixture thereby insuring stability from the moment of breakage, the specific enzyme activities were the same as by the routine procedure employed in the studies reported above, i.e. harvesting the cells in cold 0·1 M tricine, pH 7·4 buffer, breaking in the cold, assaying immediately, and bringing the extract to temperature only after addition of the reaction mixture. Therefore it seems highly unlikely that in the reported experiments significant enzyme activity could have been lost between the time of cell breakage and the time of assay. It should be noted that the reports of the instability of UDPG pyrophosphorylase (Wright & Dahlberg, 1968) were based on experiments in which extracts were incubated at 37°C. for periods up to 30 min. in the absence of all reaction components including substrates and in the presence of a low concentration (0·01 M) of Tris buffer at pH 8·5 in which the enzyme is known to be far less stable than in 0·1 M Tricine, pH 7·4 (Newell & Sussman, 1968). It should be noted that three methods of breakage have been used: French pressure cell; Branson sonifier; a detergent, Cemulsol NPT-12. All have yielded similar results unaccompanied by significant activity losses even when each treatment was made more rigorous (i.e. increased times or intensity of treatment) than is normally employed.

(*b*) Mixtures in varying proportions have been made of relatively inactive crude extracts from cells harvested at early developmental stages and highly active extracts from cells harvested at later stages. The mixtures were assayed after varying periods of incubation in 0·1 M Tricine at 37°C. In all cases the activities of the mixtures were the precise sums of the activities of the separate extracts incubated under the same conditions, i.e. the presence of material in excess in the early extracts that might reduce the activity of the late ones, or, conversely, of material in the late extracts that might stimulate the early ones could not be demonstrated. Wright & Dahlberg (1968) have reported corresponding experiments under somewhat different experimental conditions (of buffer, pH, etc.) which indicated that inactivating agents were present in early cell extracts, but a previous report by Wright & Anderson (1959) in which mixed extracts were assayed under the same

[1] The mixture contained in 1 ml.: 1·0 μmole UDPG, 1·4 μmole, TPN, 4·0 μmoles MgCl₂, 2·0 μmoles Na pyrophosphate, 0·15 units glucose-6-phosphate dehydrogenase (Boehringer), 0·05 unit phosphoglucomutase (Boehringer), 100 μmoles Tricine (Calbiochem) buffer pH 7·5. To this was added 10–50 μl. of crude extract.

conditions yielded data which conclusively proved the reverse. We eagerly await the resolution of this controversy.

The conclusions derived above concerning the absence of inhibitors in early extracts confirm previous findings using the radioisotopic assay for UDPG pyrophosphorylase (Ashworth & Sussman, 1967), as well as from studies of UDP Gal:polysaccharide transferase (Sussman & Osborn, 1964) and trehalose-6-phosphate synthetase activities (Roth & Sussman, 1968). However, it should be noted that these experiments can only test for the presence of inhibitors (or activators) which are present in excess. They cannot test the existence of limited amounts of specific masking agents bound tightly in stoichiometric proportions to the proteins which render them inactive catalytically. Such arguments, however, must be handled with care, for, stated in extreme forms, they are untestable.

The last and most difficult question to answer is one which is germane to all physiological investigations, namely to what extent is the activity of an enzyme as measured in cell extracts (where, for example, the spatial restrictions of the intact cell are eliminated) relevant to its actual performance when acting *in vivo* as part of the functional metabolic machinery? The compartmentalization of enzymes in mitochondria, lysosomes, glyoxosomes, and other vesicular and granular arrangements makes this question particularly important if one is attempting to deal with anything more than simply accumulation and disappearance of the enzymes in question. In slime moulds, structural studies of this kind have only just begun (Ashworth, private communication).

Some properties of the system regulating enzyme accumulation and disappearance

A requirement for protein synthesis

Cycloheximide is a reversible inhibitor of protein synthesis in *Dictyostelium discoideum* (Sussman 1965*b*), as well as other organisms (Ennis & Lubin, 1964). For each of the three enzymes studied thus far (trehalose-6-P synthetase, UDPG pyrophosphorylase, UDP-Gal transferase), addition of cycloheximide immediately prior to the usual period of accumulation prevented it. Addition at any time during the period immediately stopped further accumulation and prevented subsequent disappearance of the activity already accumulated (Sussman, 1965*b*; Ashworth & Sussman, 1967; Roth, Ashworth & Sussman, 1968). It also inhibited further morphogenesis. Where tested, the removal of cycloheximide was followed by resumption of the normal programme.

27

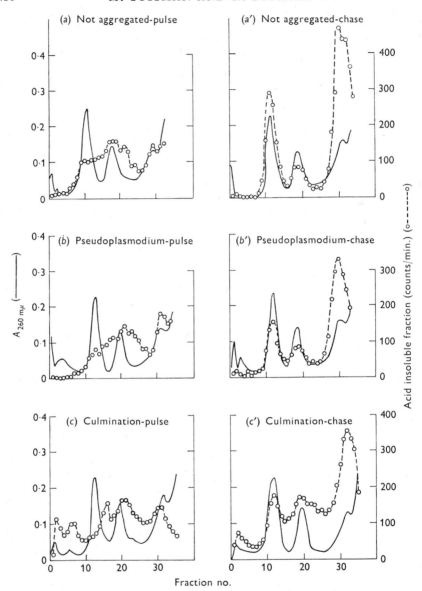

Fig. 5. Sucrose density gradient analysis of pulse labelled and stable RNA at different stages of morphogenesis of *Polysphondylium pallidum*. Cells at specified stages on Millipore were pulsed with [³H]uracil for 20 min. Some filters were harvested and others (chase) were switched to support pads containing excess cold uracil and incubated further for several hours before harvesting. The harvested cells were lysed with sodium dodecyl-sulphate and centrifuged in a sucrose gradient (15–30%). Tube contents were emptied from the bottom with a finger pump, passed through a Gilford recording spectrophotometer and collected in 1 ml. fractions to determine acid precipitable, RNase sensitive counts. Solid line: $A_{260m\mu}$. Dotted line: radioactivity (R. R. Sussman (1967). *Biochim. Biophys. Acta*, **149**, 407–421, fig. 2.)

A requirement for RNA synthesis

In both *Polysphondylium pallidum* (R. R. Sussman, 1967) and *Dictyostelium discoideum* (Inselburg & Sussman, 1967) RNA synthesis as reflected by incorporation of labelled uracil or uridine proceeds at an appreciable rate throughout morphogenesis. Figure 5 shows that this RNA is heterogeneous with respect to sedimentation in a sucrose

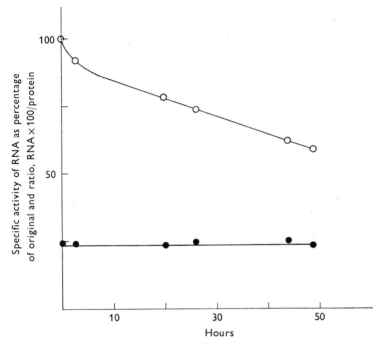

Fig. 6. Decay of stable [^{14}C] RNA during morphogenesis in *Polysphondylium pallidum*. Cells were grown in the presence of [^{14}C]uracil, harvested, washed and grown further in fresh medium and then dispensed on Millipore filters (zero time). At intervals, cells were harvested to determine acid precipitable, RNase sensitive radioactivity; total RNA content by the orcinol reaction; and total protein by the Lowry procedure. (R. R. Sussman (1967). *Biochim. Biophys. Acta*, **149**, 407–21, fig. 3.) ○, Specific activity of RNA (percentage of original); ●, ratio, RNA × 100/protein.

gradient and with respect to stability. The part that remains stable when the exogenous label is removed and the cells are incubated further in the presence of excess cold precursor sediments like the bulk transfer and ribosomal RNA of the cells and appears to be identical in the matter of the base composition as well (R. R. Sussman, 1967). The less stable component contains fractions heavier than and lighter than r-RNA and presumably includes m-RNA and ribosomal precursors.

It should be recalled that such synthesis occurs in the face of a net loss of total RNA per cell. A measure of the turnover was made in axenically grown *Polysphondylium pallidum*. The cells were labelled during growth with [^{14}C]uracil and grown a few generations further in cold medium to chase out the unstable components. Figure 6 shows the fate of the remainder when the cells were dispensed on Millipore filters and allowed to construct fruiting bodies in the presence of excess cold uracil. Had the preformed RNA simply been destroyed and

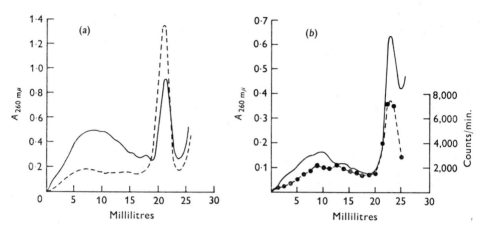

Fig. 7. Cells were harvested during growth or morphogenesis, broken with detergent Cemulsol NPT 12, and separated into nuclear and cytoplasmic fractions. The latter were layered over a sucrose density gradient and centrifuged. Tube contents were emptied from below with a finger pump through a Gilford recording spectrophotometer and in one instance fractions were collected in order to measure acid precipitable radioactivity. (*a*) Absorbance profiles of cells harvested at densities of 1.7×10^6/ml. (solid curve) and 1.6×10^7/ml. (dashed curve). (*b*) Profiles of absorbance (solid curve) and acid precipitable radioactivity (dashed curve) of cells incubated on filters 8 hr. in the presence of [^{3}H] uridine and 1.5 hr. in the presence of excess cold uridine (Cocucci & Sussman, 1968).

no new RNA synthesized during the morphogenetic sequence, the specific radioactivity of the RNA would have remained a its original level. In fact, the specific activity fell appreciably as the old, labelled RNA was diluted by new, cold RNA. The extent of the decrease indicates that at least two thirds of the RNA present in the vegetative cells was lost during fruiting and an amount equivalent to at least the remaining third was synthesized *de novo*. It should be noted that the new ribosomes precisely resemble the old ones with respect to the base composition of the r-RNA and its specific hybridizability with the DNA (R. R. Sussman, 1967) and to the over-all composition of ribosomal proteins (J. M. Ashworth, unpublished data).

In recent experiments (Cocucci & Sussman, 1968) with *Dictyostelium discoideum*, measurements have been made of the levels of polyribosomes in cells during growth and morphogenesis. Figure 7 summarizes some of the results. The cells were broken with detergent Cemulsol NPT-12 under conditions which yielded intact nuclei free of ribosomes

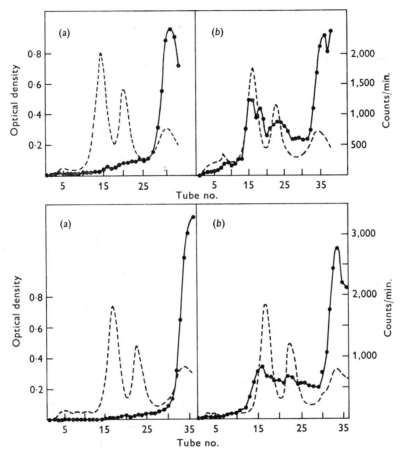

Fig. 8. Effect of actinomycin D on RNA synthesis in *Dictyostelium discoideum*. Cells were incubated on filters for 2 hr. (top) and 16 hr. (bottom) and then exposed to ^{32}P-inorganic phosphate for 2 hr. in the presence of actinomycin (left) and in its absence (right). The cells were then harvested and treated as described in Fig. 5. Dashed curves: $A_{260m\mu}$. Solid curves: acid precipitable radioactivity.

and incapable of amino acid incorporation and a cytoplasmic fraction containing all the ribosomes. The latter was centrifuged in a sucrose density gradient and the trace of optical density yielded profiles of polyribosomes and free monomers as shown in Fig. 7. The polyribosome regions could be completely eliminated by a short exposure to ribonuclease in the cold and could be shown to be preferentially associated with nascent protein by amino acid incorporation studies. In cells growing exponentially at a density of 1.7×10^6 cells/ml. in axenic medium, at least 74 % of the ribosomes were found in the polysomal region. In cells harvested from the culture at a density of 1.6×10^7 cells/ml., i.e. entering the stationary phase, the proportion of poly-somes had sunk to 44 % and was negligible in cells that had reached a

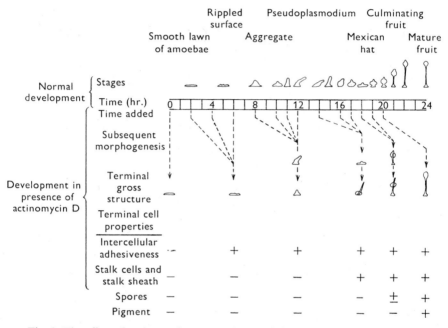

Fig. 9. The effect of actinomycin on morphogenesis in *Dictyostelium discoideum.*

density of 2.8×10^7/ml. and were approaching the death phase. In contrast, cells harvested at a density of 1.6×10^7/ml. and incubated on filter papers for 9.5 hr. showed a very substantial level of polysomes (*ca.* 50 %). The level of polysomes did not fall in such cells until about 18–20 hr., near the end of the morphogenetic sequence. Comparison between the optical density profile and the distribution of [³H]uridine that had been incorporated into RNA over an 8 hr. period during morphogenesis and then chased for 1.5 hr. with cold uridine showed that the newly made RNA is preferentially located in the light poly-ribosomal region.

The foregoing makes it clear that RNA synthesis and mobilization represent major cell activities during morphogenesis. Experiments with

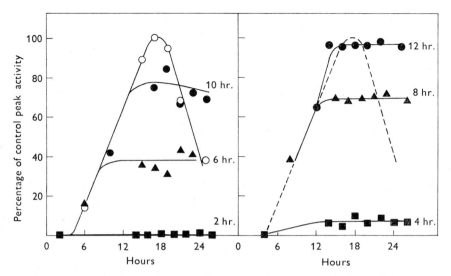

Fig. 10. Effect of actinomycin on trehalose-6-P synthetase accumulation and disappearance. Washed cells on filters were exposed to actinomycin starting at the times shown. Cell samples were collected at intervals thereafter for enzyme assays. A control curve (open circles) obtained with untreated cells is shown on the left and reproduced as a dashed line on the right.

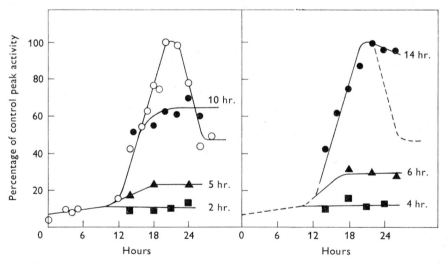

Fig. 11. Effect of actinomycin on UDPG pyrophosphorylase accumulation and disappearance. See legend to Fig. 10 for experimental details.

actinomycin D indicate that these activities are required for normal fruit construction (Sussman, Loomis, Ashworth & Sussman, 1967). As Fig. 8 shows, the exposure of *Dictyostelium discoideum* to actinomycin stopped the synthesis of high molecular weight RNA, residual incorporation being confined to the 4S region presumably reflecting exchange at the CCA terminus of t-RNA (Franklin, 1963). In contrast, protein synthesis continued unimpeded for at least 4 hr. after addition of the agent. No evidence of non-specific cell damage or of the interference of actinomycin with the extraneous metabolic activities was apparent. However, Fig. 9 shows that the agent did interfere with morphogenesis in a specific manner depending on the time of administration. The pattern of inhibition suggests the existence of morphogenetic plateaus such that, once past any one plateau, the development of the cells did not stop after addition of actinomycin until they reached the next plateau. Most interesting is the inhibitory pattern of cells exposed to the drug after 14–17 hr. of development. In these the prestalk cells developed normally and produced a typical parenchymatous-like stalk ensheathed in cellulose but the prespore cells remained at the base and did not transform into spores. Finally, when added after *ca.* 20 hr. the agent could no longer interfere with subsequent development.

The accumulation of each of the three enzymes was shown to be sensitive to prior inhibition of RNA synthesis by actinomycin (Sussman & Sussman, 1965; Ashworth & Sussman, 1967; Roth *et al.* 1968). Each transcriptive period was delineated by adding the drug at later and later times during the developmental sequence in order to determine how much, if any, of the corresponding enzyme accumulated thereafter. Figures 10 to 14 summarize the results.

In the case of trehalose-6-P synthetase which accumulates during the period between 5 and 16 hr., addition of actinomycin prior to 4 hr. entirely prevented its appearance and addition after 13 hr. permitted accumulation to the normal peak of specific activity. When the drug was added between times, the enzyme accumulated at the normal rate but to an intermediate level such that the later the addition, the higher was the level attained (Fig. 10). For UDPG pyrophosphorylase which accumulates during the period between 11–12 and 20 hr., addition of actinomycin prior to 5 hr. prevented any increase, addition after 12 hr. permitted the accumulation to occur at the normal rate to a level slightly higher than the untreated control. As with the synthetase, addition between times permitted accumulation of intermediate levels (Fig. 11). For UDP Gal transferase, which accumulates between 12·5 and 21 hr., addition prior to 7·5 hr. entirely prevented its appearance, whereas addition after 14–15 hr. was without effect. Again, intermediate levels were attained by addition between these times (Fig. 12). The precise relationships between the time at which actinomycin was

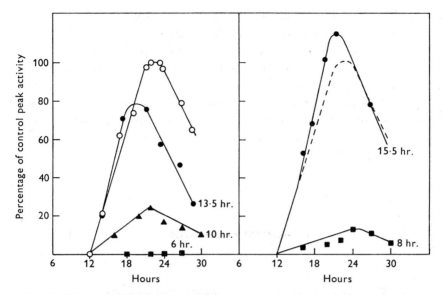

Fig. 12. Effect of actinomycin on UDP Gal:polysaccharide transferase accumulation and disappearance. See legend to Fig. 10 for experimental details.

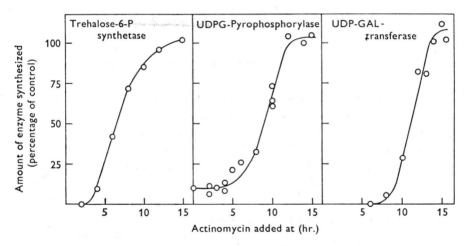

Fig. 13. Relation between times of actinomycin addition and peak levels of enzyme that subsequently accumulated expressed as percentages of the peak level in untreated cells. The points are calculated from the data shown in Figs. 10–12 and from experiments not shown.

added and how much of each enzyme later appeared are shown in Fig. 13 and are seen to be linear or perhaps sigmoid.

Figure 14 schematically summarizes temporal relations among the three transcriptive periods. In discussing them, we tentatively assume that each reflects the synthesis of an m-RNA species and that the accumulation of corresponding enzyme activity is the result of the translation of the former. These transcriptive periods are seen to occupy restricted portions of the total time over which morphogenesis occurs and over which RNA is synthesized and they clearly do not coincide. This suggests that both the initiation of a transcriptive event and its duration are under developmental control and this conclusion is supported by the finding (Sussman & Sussman, 1965) that in the

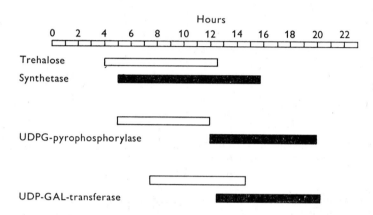

Fig. 14. A schematic summary of transcriptive (shown in white) and translative (shown in black) periods for accumulation of the three enzymes derived from Figs. 10–13.

temporally deranged mutant Fr-17, the transcriptive period required for the accumulation of the transferase begins sooner than in the wild-type and is of much shorter duration (4 versus 7–8 hr.) in correspondence with the difference in the periods over which the enzyme itself accumulates. The data also indicate that time lags of (respectively) 1, 7, and 5 hr. exist between the initiation of transcriptive periods and the appearance of the corresponding enzymes. The possibility that these lags are required for preformed, enzymically inactive subunits to polymerize is made unlikely by the fact that the addition of cycloheximide during the periods of enzyme accumulation froze each of the activity levels immediately, indicating the absence of significant pools of enzymically inactive subunits. The extents of the lags (up to 7 hr.) and the differences among them would therefore seem to imply the existence of specific translational controls or perhaps controls over the times in which m-RNA species migrate through the nuclear envelope to the site of translation. The lags also imply functional life times for m-RNA considerably in excess of the estimates for bacteria (Levinthal *et al.*

1963) and more in keeping with those postulated for eucaryotic differentiating cells (Hotta & Stern, 1961; Marks, Burka, Rifleind & Danon, 1963).

It is also noteworthy that the disappearance of trehalose-6-P synthetase and UDPG pyrophosphorylase both required a prior period of RNA synthesis. For the transferase this requirement was demonstrable in mutant Fr-17 (Sussman & Sussman, 1965) if not in the wild-type (where the periods for accumulation and disappearance appear to coincide).

Do the cells accumulate a particular enzyme simultaneously or sequentially?

The periods during morphogenesis over which the three enzyme activities accumulate have been seen to range between 8 and 11 hr. Such extended time spans could arise in one of two ways. It might be that, though all the cells accumulate a given enzyme simultaneously, they do so very slowly. Alternatively it might be that, though any one cell takes only a few minutes to accumulate its complement of an enzyme, they do so at different times and it takes many hours for all of them to finish. In principle, a crucial distinction between these alternatives might be made by examining the cell population at a developmental stage when only a relatively small fraction of the enzyme has as yet accumulated. According to the first model, all of the cells should contain the same, relatively low specific enzyme activity. According to the second model, the cells should contain significantly different levels, i.e. those that had already accumulated their complement of enzyme should display peak specific enzyme activity; those yet to accumulate their complement should display only basal levels. Few, if any (those in the midst of accumulation), would display intermediate values.

These alternatives were tested in the following manner (Newell, Ellingson & Sussman, 1968). Single pseudoplasmodia were dissected and separated into segments as described below. The segments were each collected, frozen, thawed, and lysed under a dissecting microscope by addition of detergent. Such extracts were then assayed for UDPG pyrophosphylase activity by the spectrophotometric method. Protein content was determined by measuring the amount of TCA-insoluble [^{14}C]-leucine present in the extracts (the cells had been grown in the presence of the labelled amino acid) and this in turn was calibrated in terms of Folin activity. Table 1 is a summary of the results. The over-all enzyme content of the pseudoplasmodia increased at the usual rate (consistent with the data shown in Fig. 3) and the levels were quite in

Table 1. *Comparison between enzyme contents of different parts of the fruiting body*

Time harvested (hr.)	Number of slugs assayed	Segmentation pattern		Relative proportions of segments	Specific enzyme activity* (mμ mole G-I-P-/ min./mg. protein)
14	15		top	0·20	193±20%
			bottom	0·80	191± 7
15	15		top	0·50	233±10%
			bottom	0·50	219±11
16	18		top	0·50	294±12%
			bottom	0·50	289±12
16	20		inside	0·10	259±12%
			outside	0·90	258±15
19	15		whole	—	402± 8%

* UDPG pyrophosphorylase.

keeping with the times at which the slugs were harvested. The assay of specific enzyme activity is seen to have been precise within a standard error of ± *ca.* 10 %. The data reveal that at a time when the slug as a whole had accumulated approximately half of the peak enzyme activity it eventually would have done, all parts of the slug showed the same intermediate levels of enzyme content. This held true when the central core of the cells was compared with the periphery, although if a distribution of activity did exist it might be expected to have proceeded radially. It also held true when the top half of the slug or even the apical 20 % was compared with the bottom. This is particularly meaningful, since the cells have been shown to take up and retain a position with respect to the longitudinal axis in accord with the order in which they originally entered the cell aggregate and their developmental fates are so determined, i.e. the apical 20 % of the cells form the stalk of the mature fruit and the cells posterior to them become the spores (Raper, 1940). It should be borne in mind that since peak and basal levels of specific activity are, respectively, *ca.* 500 and *ca.* 35–50 mμmoles G-1-P/min./mg. protein, if a temporal distribution had existed along either axis it surely would have been detected. The only remaining distribution not yet tested is a random one. We intend to examine this possibility by obtaining fluorescent-labelled antibody preparations against the purified enzymes and assaying sectioned pseudoplasmodia.

In any case we tentatively conclude that the component cells of the developing multi-cellular assembly act in concert at least as far as the accumulation of specific enzymes is concerned and presumably as well with respect to the transcriptive events that precede them.

The relevance of morphogenetic topography

When cells, permitted to develop normally on the filter papers, are harvested, disaggregated, and redeposited, they quickly reaggregate and accomplish in a few hours what may have taken them many hours to do the first time. If redeposited in the presence of EDTA, reaggregation is drastically impeded. At best, they remain a smooth lawn of cells. At worst, they enter into loose amorphous cell masses but in either case do not transform into spores or stalk cells. This confirms previous demonstrations of the requirement of normal cell aggregation for di-valent cations (De Haan, 1959; Gerisch, 1961). Loomis & Sussman (1966) demonstrated that when cells were allowed to develop un-disturbed on the filters for *ca.* 8 hr. and then disaggregated and re-deposited in the presence of EDTA, UDP Gal transferase accumulated and disappeared in normal fashion. Thus neither of these activities depends upon the gross or fine morphogenetic topography which is unique to normal fruiting body construction or spore and stalk cell development. However, accumulation of the enzyme did require at least intimate juxtaposition of the cells if not topographical precision since redeposition at low cell densities in the presence of EDTA did not permit the expected increase. Also necessary was the initial period of development in the absence of EDTA prior to disaggregation, since cells incubated from zero time in its presence accumulated only a very low level of enzyme (1–10 % of the peak activity).

The preservation of the transcriptive product; the commutation ticket theory of m-RNA stability

Cycloheximide, while inhibiting protein synthesis, does not interfere significantly with the rate of RNA synthesis (Ennis & Lubin, 1964). However, sucrose density centrifugation of this material shows it to lack high molecular weight components normally present and have a surplus of low molecular weight components (Sussman, 1966*b*). Pre-sumably this reflects degradation rather than differential synthesis, as already demonstrated in the case of chloramphenicol (Levinthal *et al.* 1963).

In mutant Fr-17, cycloheximide was added for 2 hr. periods at the beginning of, during, and immediately after the transcriptive period for the accumulation of UDPG Gal transferase. The results, reviewed else-where in detail (M. Sussman, 1967), indicated that any of the trans-criptive product already fabricated prior to the addition of cyclo-heximide could, after removal of the drug, be utilized to provide the same level of enzyme that it normally would. But any RNA synthesized in the presence of cycloheximide did not lead to accumulation of the enzyme after removal of the drug. The possibility has been raised that cycloheximide acts in this manner by virtue of its ability to prevent the appearance of a new protein whose fabrication may be mandatory if nascent m-RNA is to be preserved and transported through the nuclear envelope (Sussman, 1966 b).

In Fr-17, as in the wild-type, there is a 4–5 hr. lag between the start of the transcriptive period and the appearance of UDP Gal transferase activity. Hence, as mentioned previously, the transcriptive product must be at least that stable. When cycloheximide was added at the end of the transcriptive period and then removed after two hours, the transferase appeared after an additional one hour lag and accumulated at the normal rate and to the normal level (even though additional RNA synthesis was prevented during this time). Thus the 4–5 hr. lag, normally operative, could be extended to 7–8 hr. without significant detriment to the amount of enzyme that accumulated.

This result may imply that an m-RNA molecule, once engaged with ribosomes, does not decay through random nucleolytic attack as previously envisaged (Levinthal et al. 1963; Marks et al. 1963) but instead may be systematically destroyed as a consequence of its use as a template for protein synthesis. In other words each m-RNA molecule may carry a commutation ticket which is punched every time that a ribosome traverses it, a ticket good only for a given number of ribo-somal rides after which destruction of the message automatically ensues. The ticket might consist of a redundant nucleotide sequence at the beginning of the m-RNA molecule; for example, a repeating set of chain initiating triplets. Each time a ribosome passed by this region, one such triplet might be excised by an endonuclease bound to the ribosome. Only as many ribosomal passages would be allowed as there were redundant triplets originally present and the excision of the last one would divest the message of ribosomes and result in its rapid destruc-tion. It is noteworthy that Robertson, Webster & Zinder (1968) have provided evidence that one of the ribosomal proteins is in fact an endonuclease.

The idea of the commutation ticket has much to recommend it, particularly for differentiating cells, since it would provide an automatic quantitative control over phenotypic expression. It would also provide a neat device by which m-RNA molecules of differing stabilities could exist simultaneously in the cell, i.e. the base sequence of a DNA cistron would not only dictate the amino acid sequence of the corresponding protein but also the amount of that protein to be synthesized and different cistrons might begin with different numbers of the redundant sequence. Mutation and selection would therefore be able to guide not only the kinds of proteins to be made but also the amounts thereof.

The developmental patterns of some other enzymes

The specific activities of two other enzymes have been shown to change markedly during the morphogenetic sequence and these changes are sensitive to cycloheximide and to actinomycin D. Ceccarini (1967) has found that trehalase activity accumulates in the spores either just prior to or during germination. A low level of enzyme persists in the vegetative amoebae but this is preferentially released by them during aggregation. When the aggregated cells were dispersed and incubated with bacteria, they reaccumulated the enzyme. Loomis (1968b) recently demonstrated the presence of an enzyme that hydrolyses p-nitrophenyl N-acetyl glucosamine. It is undetectable in the vegetative amoebae and its accumulation is one of the first observed events of the morphogenetic sequence.

Rosness (1968) has studied two enzymes that catalyse the breakdown of cellulose and are presumably involved in spore germination. One of them, β-glucosidase, is present in the vegetative amoebae and drops precipitously during aggregation. (The possibility of preferential release was not examined.) It then accumulates to a high level at the later stages of fruit construction. The second enzyme is a cellulase which also accumulates at that time starting from an undetectable level. However, the activity appears to be reversibly masked by a protein present in earlier extracts and this may obscure the true starting point.

Solomon, Johnson & Gregg (1964) examined crude extracts by starch gel electrophoresis and detected at least two varieties of alkaline phosphatase, five of acid phosphatase, and ten of esterase using sodium naphthyl-phosphate and naphthyl-acetate as chromogenic substrates. The levels of some of these multiple forms change markedly during the morphogenetic sequence. In an earlier study Bonner et al. (1955) examined migrating slug histochemically and demonstrated significant differences in alkaline phosphatase activity between the prestalk and prespore regions thereby raising the possibility of differential accumulation. Gezelius (1968) has presented evidence implying the existence of at least two forms of pyrophosphatase. Hence the presence of isozymes may be widespread and the maintenance of relatively constant levels of a particular catalytic function during morphogenesis may in fact mask dramatic changes in protein constitution.

Four enzymes probably associated with digestion of the bacterial prey have been examined. One, an acid phosphatase, appears to increase ca. fourfold in specific activity during development (Gezelius, 1966). The second is a protease with a pH optimum of 2·3 (D. R. Patterson, unpublished data). It is the only proteolytic enzyme detectable at any stage of development using the chromogenic substrate, azoalbumin, and remains at a constant level of activity throughout the morphogenetic sequence. The last two are DNases, one with a pH optimum of 4·5 and the other 8·5. Both are preferentially destroyed during the first 10 hr. of morphogenesis and are undetectable in the cell aggregates (D. R. Patterson, unpublished data).

REFERENCES

AMES, B. & HARTMAN, P. (1963). The histidine operon. *Cold Spring Harb. Sym. quant. Biol.* **28**, 349.

ASHWORTH, J. M. & SUSSMAN, M. (1967). Appearance and disappearance of UDPG pyrophosphorylase activity during the differentiation of *D. discoideum. J. biol. Chem.* **242**, 1696.

BONNER, J. T., CLARK, W., NEELEY, C. & SLIFKIN, M. (1950). Orientation to light and temperature gradients in the slime mold *D. discoideum. J. cell. comp. Physiol.* **36**, 149.

BONNER, J., CHIQUOINE, A. D. & KOLDERIE, M. (1955). A histochemical study of differentiation in the cellular slime molds. *J. exp. Zool.* **130**, 133.

BONNER, J. & FRASCELLA, E. (1952). Mitotic activity in relation to differentiation in *D. discoideum. J. exp. Zool.* **121**, 61.

BRADLEY, S. G., SUSSMAN, M. & ENNIS, H. L. (1956). Environmental factors affecting slime mold aggregation. *J. Protozool.* **3**, 33.

BRÜHMÜLLER, M. & WRIGHT, B. E. (1963). Glutamate oxidation in the differentiation slime mold. II. *Biochim. biophys. Acta,* **71**, 50.

CABIB, E. & LELOIR, L. F. (1958). The biosynthesis of trehalose phosphate. *J. biol. Chem.* **231**, 259.

CECCARINI, C. (1967). The biochemical relationship between trehalase and trehalose during growth and differentiation in *D. discoideum. Biochim. biophys. Acta,* **148**, 114.

CECCARINI, C. & FILOSA, M. (1965). Carbohydrate content during development of the slime mold *D. discoideum. J. cell. comp. Physiol.* **66**, 135.

CHANG, Y. (1968). An extracellular cyclic AMP phosphodiesterase produced by *D. discoideum. Science, N.Y.* (in the Press).

CLEGG, J. & FILOSA, M. (1961). Trehalose in the cellular slime mold *D. mucoroides. Nature, Lond.* **192**, 1077.

CLELAND, S. V. & COE, E. L. (1968). Activities of glycolytic enzymes during early stages of differentiation of *D. discoideum. Biochim. biophys. Acta,* **156**, 44.

Cocucci, S. M. & Sussman, M. (1968). (In preparation.)

DeHaan, R. L. (1959). Effects of EDTA on cell adhesion in the slime mould, *D. discoideum. J. Embryol. exp. Morph.* **7**, 335.

Ennis, H. L. & Lubin, M. (1964). Cycloheximide: aspects of inhibition of protein synthesis in animal cells. *Science, N.Y.* **146**, 1475.

Epstein, R. H., Bolle, A., Steinberg, C., Kellenberger, E., Boy de la Tour, Chevalley, R., Edgar, R. S., Sussman, M., Denhardt, G. H. & Lielausis, A. (1963). Physiological studies of conditional lethal mutants of bacteriophage T4D. *Cold Spring Harb. Sym. quant. Biol.* **28**, 375.

Franklin, R. M. (1963). Inhibition of RNA synthesis in mammalian cells by Actinomycin D. *Biochim. biophys. Acta,* **72**, 555.

Gerisch, G. (1961). Zell funktionen in der Entwicklung von *D. discoideum. Expl Cell Res.* **25**, 535.

Gezelius, K. (1966). Acid phosphatase in *D. discoideum. Physiologia Pl.* **21**, 946.

Gezelius, K. (1968). Hydrolysis of inorganic pyrophosphate in *D. discoideum. Physiologia Pl.* **21**, 35.

Gezelius, K. & Ranby, B. (1957). Morphology and fine structure of *D. discoideum. Expl Cell Res.* **12**, 265.

Gilbert, W. & Müller-Hill, B. (1966). Isolation of the Lac repressor. *Proc. natn. Acad. Sci. U.S.A.* **56**, 1891.

Gorini, L. (1963). Control by repression of a biochemical pathway. *Bact. Rev.* **27**, 182.

Gregg, J. H. (1965). Regulation in the cellular slime molds. *Devl Biol.* **12**, 377.

Gregg, J. H. & Bronsweig, R. (1956). Biochemical events accompanying stalk formation in the slime mold, *D. discoideum. J. cell. comp. Physiol.* **48**, 293.

Hotta, Y. & Stern, H. (1961). Transient phosphorylation of deoxyribosides and regulation of DNA synthesis. *J. biol. biochem. Cytol.* **11**, 311.

Inselburg, J. & Sussman, M. (1967). Incorporation of ³H-uridine into RNA during cellular slime mold development. *J. gen. Microbiol.* **46**, 59.

Jacob, F. & Monod, J. (1964). Mécanismes biochimiques et génétiques de la régulation dans la cellule bactérienne. *Bull. Soc. Chim. biol., Paris,* **46**, 1499.

Konijn, T. M., Van de Meene, Bonner, J. T. & Barkley, D. S. (1967). Acrasin activity of 3′, 5′ cyclic AMP. *Proc. natn. Acad. Sci. U.S.A.* **58**, 1152.

Levinthal, C., Fan, D. P., Higa, A. & Zimmerman, R. A. (1963). Decay and protection of m-RNA in bacteria. *Cold Spring Harb. Sym. quant. Biol.* **28**, 183.

Loomis, W. F., Jun. (1968a). Relationship between cytodifferentiation and inactivation of a developmentally controlled enzyme. *Expl Cell Res.* (In the Press).

Loomis, W. F., Jun. (1968b). (In preparation.)

Loomis, W. F., Jun. & Sussman, M. (1966). Commitment to the synthesis of a specific enzyme during cellular slime mold development. *J. molec. Biol.* **22**, 401.

Marks, P. A., Burka, E. R., Rifleind, R. & Danon, D. (1963). Polyribosomes active in reticulocyte protein synthesis. *Cold Spring Harb. Symp. quant. Biol.* **28**, 223.

Mercer, E. H. & Shaffer, M. B. (1960). Electronmicroscopy of solitary and aggregated slime mold cells. *J. biochem. biophys. Cytol.* **7**, 353.

Monod, J., Changeaux, J. P. & Jacob, F. (1963). Allosteric proteins and cellular control systems. *J. molec. Biol.* **6**, 306.

Moscona, A. A. (1961). Rotation mediated histogenetic aggregation of dissociated cells. *Expl Cell Res.* **22**, 455.

Newell, P. C., Ellingson, J. & Sussman, M. (1968). (In preparation.)

Newell, P. C. & Sussman, M. (1968). (In preparation.)

Perlman, R. & Pasten, I. (1968). Cyclic AMP: stimulation of β-galactosidase and tryptophorase induction. *Biochem. biophys. Res. Commun.* **30**, 656.

RAPER, K. B. (1940). Pseudoplasmodium formation and organization in *D. discoideum*. *J. Elisha Mitchell sci. Soc.* **56**, 241.

RAPER, K. B. (1956). Factors affecting growth and differentiation in simple slime molds. *Mycologia*, **48**, 169.

RAPER, K. B. & FENNELL, D. I. (1952). Stalk formation in *Dictyostelium*. *Bull. Torrey Bot. Club*, **79**, 25.

ROBERTSON, H. D., WEBSTER, R. E. & ZINDER, N. D. (1968). Purification and properties of ribonuclease III from *E. coli*. *J. biol. Chem.* **243**, 82.

ROSNESS, P. (1968). *J. Bact.* (in the Press).

ROTH, R. M., ASHWORTH, J. M. & SUSSMAN, M. (1968). Periods of genetic transcription required for the synthesis of 3 enzymes during cellular slime mold development. *Proc. natn. Acad. Sci. U.S.A.* **59**, 1235.

ROTH, R. M. & SUSSMAN, M. (1968). Trehalose-6-phosphate synthetase and its regulation during slime mold development. *J. biol. Chem.* **243**, 5081.

SHAFFER, B. M. (1956). Properties of acrasin. *Science, N.Y.* **123**, 1172.

SOLOMON, E. P., JOHNSON, E. M. & GREGG, J. H. (1964). Multiple forms of enzymes in a cellular slime mold during morphogenesis. *Devl Biol.* **9**, 314.

SONNEBORN, D. R., SUSSMAN, M. & LEVINE, L. (1964). Serological analyses of cellular slime mold development. *J. Bact.* **87**, 1321.

STAPLES, S. & GREGG, J. H. (1967). Carotenoid pigments in *D. discoideum*. *Biol. Bull. mar. biol. Lab. Woods Hole*, **132**, 413.

SUSSMAN, M. (1965*a*). Temporal, spacial and quantitative control of enzyme activity during slime mold cytodifferentiation. *Brookhaven Symp.* **18**, 66.

SUSSMAN, M. (1965*b*). Inhibition of actidione of protein synthesis and UDP Gal: polysaccharide transferase accumulation in *D. discoideum*. *Biochem. biophys. Res. Commun.* **18**, 763.

SUSSMAN, M. (1966*a*). Biochemical and genetic methods in the study of cellular slime. mold development. *Methods in Cell Physiology*, II, 397. New York: Academic Press.

SUSSMAN, M. (1966*b*). Protein synthesis and the temporal control of genetic transcription during slime mold development. *Proc. natn. Acad. Sci. U.S.A.* **55**, 813.

SUSSMAN, M. (1967). Evidence for temporal and quantitative control of genetic transcription and translation during slime mold development. *Fedn. Proc.* **26**, 77.

SUSSMAN, M., LEE, F. & KERR, N. S. (1956). Fractionation of acrasin. *Science*, **123**, 1171.

SUSSMAN, M., LOOMIS, W. F. JUN., ASHWORTH, J. M. & SUSSMAN, R. R. (1967). The effect of actinomycin d on cellular slime mold morphogenesis. *Biochem. biophys. Res. Commun.* **26**, 353.

SUSSMAN, M. & LOVGREN, N. (1965). Preferential release of UDP Galactose: polysaccharide transferase. *Expl Cell Res.* **38**, 97.

SUSSMAN, M. & OSBORN, M. J. (1964). UDP Galactose: polysaccharide transferase in the cellular slime mold, *D. discoideum*. *Proc. natn. Acad. Sci. U.S.A.* **52**, 81.

SUSSMAN, M. & SUSSMAN, R. R. (1965). Regulatory program for UDP Gal: polysaccharide transferase activity during slime mold differentiation. *Biochim. biophys. Acta*, **108**, 463.

SUSSMAN, R. R. (1967). RNA metabolism during cytodifferentiation in *P. pallidum*. *Biochim. biophys. Acta*, **149**, 407.

SUSSMAN, R. R. & SUSSMAN, M. (1960). Dissociation of morphogenesis from cell division in *D. discoideum*. *J. gen. Microbiol.* **23**, 287.

SUSSMAN, R. R. & SUSSMAN, M. (1963). Ploidal inheritance in *D. discoideum*: Haploidization and genetic segregation of diploid strains. *J. gen. Microbiol.* **30**, 349.

SUSSMAN, R. R. & SUSSMAN, M. (1967). Cultivation of *D. discoideum* in axenic medium. *Biochem. biophys. Res. Commun.* **29**, 53.

SUTHERLAND, E. W., ØYE, I. & BUTCHER, R. W. (1965). Action of epinephrine and the role of adenyl cyclase system in hormone action. *Recent Prog. Horm. Res.* **21**, 623.

UMBARGER, H. E. (1956). Evidence for a negative feedback mechanism in the biosynthesis of isoleucine. *Science, N.Y.* **123**, 848.

WHITE, G. J. & SUSSMAN, M. (1961). Metabolism of major cell components during slime mold morphogenesis. *Biochim. biophys. Acta,* **53**, 285.

WHITE, G. J. & SUSSMAN, M. (1963 *a*, *b*). Polysaccharides involved in slime mold development I and II. *Biochim. biophys. Acta,* **74**, 173, 179.

WRIGHT, B. E. (1964). Biochemistry of the Acrasiales. *Biochemistry and Physiology of Protozoa,* **2**, 341. New York: Academic Press.

WRIGHT, B. E. (1966). Multiple causes and controls in differentiation. *Science, N.Y.* **153**, 830.

WRIGHT, B. E. & ANDERSON, M. L. (1959). Biochemical differentiation in the slime mold. *Biochim. biophys. Acta,* **31**, 316.

WRIGHT, B. E. & ANDERSON, M. L. (1960). Protein and amino acid turnover during differentiation in the slime mold I. *Biochim. biophys. Acta,* **43**, 62.

WRIGHT, B. E. & DAHLBERG, D. (1968). Stability *in vitro* of UDPG pyrophosphorylase in *D. discoideum. J. Bact.* **95**, 983.

YANIGASAWA, K., LOOMIS, W. F. JUN. & SUSSMAN, M. (1967). Developmental regulation of the enzyme UDP Gal: polysaccharide transferase. *Expl Cell Res.* **46**, 328.

INDEX